1st Multiflow Summer Workshop 2013

Journal of Physics: Conference Series Volume 506

Madrid, Spain
10 June – 12 July 2013

Editors:

J. Jimenez

ISBN: 978-1-63439-286-0
ISSN: 1742-6588

1st Multiflow Summer Workshop 2013

Journal of Physics: Conference Series Volume 506

Madrid, Spain
10 June – 12 July 2013

ISBN: 978-1-63439-286-0
ISSN: 1742-6588

IOP Institute of Physics

Printed from e-media with permission by:

Curran Associates, Inc.
57 Morehouse Lane
Red Hook, NY 12571

Printed by Curran Associates, Inc. (2014)

For permission requests, please contact the Institute of Physics
at the address below.

Institute of Physics
Dirac House, Temple Back
Bristol BS1 6BE UK

Phone: 44 1 17 929 7481
Fax: 44 1 17 920 0979

techtracking@iop.org

Additional copies of this publication are available from:

Curran Associates, Inc.
57 Morehouse Lane
Red Hook, NY 12571 USA
Phone: 845-758-0400
Fax: 845-758-2634
Email: curran@proceedings.com
Web: www.proceedings.com

TABLE OF CONTENTS

LES / NUMERICS

STOCHASTIC SELF-ENERGY SUBGRID MODEL FOR THE LARGE EDDY SIMULATION OF TURBULENT CHANNEL FLOWS 1
V Kitsios, J A Sillero, J Soria, J S Frederiksen

SCALING OF VELOCITY FLUCTUATIONS IN OFF-WALL BOUNDARY CONDITIONS FOR TURBULENT FLOWS 16
M P Encinar, R García-Mayoral, J Jiménez

NUMERICAL ISSUES IN LAGRANGIAN TRACKING AND TOPOLOGICAL EVOLUTION OF FLUID PARTICLES IN WALL-BOUNDED TURBULENT FLOWS 29
C Atkinson, J Hackl, P Stegeman, G Borrell, J Soria

THE ENERGY CASCADE AND REDUCED-ORDER MODELS

TURBULENCE IN THE HIGHLY RESTRICTED DYNAMICS OF A CLOSURE AT SECOND ORDER: COMPARISON WITH DNS 43
N C Constantinou, A Lozano-Durán, M-A Nikolaidis, B F Farrell, P J Ioannou, J Jiménez

THE ATTACHED REVERSE AND DETACHED FORWARD CASCADES IN WALL-TURBULENT FLOWS 61
A Cimarelli, E De Angelis, A Talamelli, C M Casciola, J Jiménez

GRANGER CAUSALITY IN WALL-BOUNDED TURBULENCE 71
G Tissot, A Lozano-Durán, L Cordier, J Jiménez, B R Noack

WALL-BOUNDED FLOWS

ANALYSIS OF A TURBULENT BOUNDARY LAYER SUBJECTED TO A STRONG ADVERSE PRESSURE GRADIENT 86
A G Gungor, Y Maciel, M P Simens, J Soria

HAIRPIN VORTICES IN TURBULENT BOUNDARY LAYERS 100
G Eitel-Amor, O Flores, P Schlatter

SKIN-FRICTION CRITICAL POINTS IN WALL-BOUNDED FLOWS 115
J I Cardesa, J P Monty, J Soria, M S Chong

TURBULENT PIPE FLOW: STATISTICS, RE-DEPENDENCE, STRUCTURES AND SIMILARITIES WITH CHANNEL AND BOUNDARY LAYER FLOWS 130
G K El Khoury, P Schlatter, G Brethouwer, A V Johansson

SCALING OF PRESSURE SPECTRUM IN TURBULENT BOUNDARY LAYERS 144
S S Patwardhan, O N Ramesh

POSSIBLE MODIFICATION OF THE LARGE-SCALE FLOW STRUCTURES BY VORTICAL STRUCTURAL INTERACTIONS 154
M V Goudar, G E Elsinga

OTHER SHEAR FLOWS

CHARACTERIZATION OF THE SHEAR LAYER IN A MACH 3 SHOCK/TURBULENT BOUNDARY LAYER INTERACTION 164
C Helm, M P Martin, P Dupont

INFLUENCE OF SOLID BOUNDARY CONDITIONS ON THE EVOLUTION OF FREE AND WALL-BOUNDED TURBULENT FLOWS 178
P D'Addio, D Sassun, O Flores, P Orlandi

CHARACTERISTICS OF THE TURBULENT/NONTURBULENT INTERFACE IN BOUNDARY LAYERS, JETS AND SHEAR-FREE TURBULENCE .... N/A
C B Da Silva, R R Taveira, G Borrell

Author Index

Journal of Physics: Conference Series **506** (2014) 011001 doi:10.1088/1742-6596/506/1/011001

Preface

First Multiflow Summer School on Turbulence

Editor: Javier Jiménez

Multiflow is a research program, funded by the European Research Council, whose goal is to improve our understanding of the multiscale dynamics of turbulence in fluids. Its first Summer School on Turbulence took place at the School of Aeronautics of the Universidad Politécnica de Madrid over the months of June and July of 2013, with the goal of providing a meeting place for theoreticians, experimentalists and simulators, in which to develop and test new ideas on turbulence physics and structure. Around forty, mostly young, participants from twenty international groups met for five weeks of collaborative work, primarily using the computational data archived in the receiving institution but, in many cases, also contributing their own. Although the format included a few invited formal seminars and periodic plenary meetings, most of the work took place in small groups that, in many cases, changed their composition during the workshop. The papers in these proceedings reflect the results of the work of those groups which, in many cases, later continued in the form of new collaborations.

Journal of Physics: Conference Series **506** (2014) 011001 doi:10.1088/1742-6596/506/1/011001

Workshop organising committee:

Rafael Gómez Blanco
Javier Jiménez
Vassilios Theofilis
School of Aeronautics, Universidad Politécnica de Madrid
28040 Madrid, Spain

Acknowledgements

The organizers are grateful to the European Research Council for their funding of this meeting, and to the School of Aeronautics of the Universidad Politécnica for their help in hosting it. The meeting would not have been possible without the dedication of those participants who acted as hosts to the visiting groups, facilitating everything from everyday problems to computer access and data usage: Guillem Borrell, José I Cardesa, Miguel P Encinar, Oscar Flores, Ricardo García-Mayoral, Ayse G Gungor, Jason F Hackl, Adrián Lozano-Durán and Juan A Sillero. Very especially, we are indebted to Aurora García Sarrión, without whose administrative help nothing would have worked.

Picture credits

The cover figure displays a high-velocity structure from a simulation of a zero-pressure- gradient turbulent boundary layer at $\delta^+ \approx 1800$, as well as the smaller surrounding intense vortices. The height of the structure is roughly equal to the thickness δ of the layer. The figure is due to Juan A Sillero. The participants picture was graciously provided by the press department of the School of Aeronautics–UPM.

Peer review statement

All papers published in this volume of *Journal of Physics: Conference Series* have been peer reviewed through processes administered by the proceedings Editors. Reviews were conducted by expert referees to the professional and scientific standards expected of a proceedings journal published by IOP Publishing.

Stochastic self-energy subgrid model for the large eddy simulation of turbulent channel flows

V Kitsios[1,2], JA Sillero[3], J Soria[1,4] and JS Frederiksen[2]

[1] Laboratory For Turbulence Research in Aerospace and Combustion, Department of Mechanical and Aerospace Engineering, Monash University, Clayton 3800, AUSTRALIA
[2] Centre for Australian Weather and Climate Research, CSIRO Marine and Atmospheric Research, 107-121 Station St, Aspendale 3195, AUSTRALIA
[3] School of Aeronautics, Universidad Politécnica de Madrid, Pza. Cardenal Cisneros 3, E-28040 Madrid, SPAIN
[4] Department of Aeronautical Engineering, King Abdulaziz University, Jeddah, KINGDOM OF SAUDI ARABIA.

E-mail: vassili.kitsios@monash.edu, sillero@torroja.dmt.upm.es, julio.soria@monash.edu, jorgen.frederiksen@csiro.au

Abstract. This paper presents the large eddy simulation (LES) of turbulent channel flow using a self-energy (SE) subgrid model with coefficients determined from reference direct numerical simulations (DNSs). In contrast to standard approaches that develop subgrid models based upon physical hypotheses, in the present SE approach the model coefficients are determined from the subgrid statistics of a DNS, with physical interpretations made apostiori. This technique is applied here for the first time to wall-bounded flows, specifically channel flow. The stochastic SE subgrid model consists of a meanfield shift, deterministic drain dissipation acting on the resolved field and a stochastic backscatter force. The deterministic SE subgrid model comprises of a net dissipation representing the net effect of the drain and backscatter. We present LESs that reproduce the time-averaged kinetic energy spectra of the DNS within the resolved scales. The direction and magnitude of the energy transfers in scale space can then be determined from the coefficients of the SE subgrid model. Results are presented for friction velocity based Reynolds numbers up to $Re_\tau = 950$.

1. Introduction

With the current level of algorithmic and computational hardware technology, it is not possible to simulate industrial nor geophysical flows at typical Reynolds numbers of interest by explicitly resolving all scales of motion. Instead one resorts to large eddy simulation (LES), where the large eddies are explicitly resolved on a computational grid and the unresolved subgrid-scale interactions are parameterised (or modelled). There have been many approaches taken to parameterise the subgrid interactions. The empirical Smagorinsky subgrid model [1] is one of the most widely adopted and celebrated. In this model the subgrid interactions are represented by an eddy viscosity determined from the product of the resolved strain rate, the local grid spacing and a specified tuning parameter. The next major development in this area was the Dynamic Smagorinsky model [2, 3], where the eddy viscosity was calculated from the scales resolved in a test filter at each step, and again assuming a similar relationship between the subgrid stresses and the resolved strain rate. In other physically justified approaches subgrid models have been developed that are representative of idealised vortex structures observed in isotropic homogeneous turbulence and wall-bounded flows [4, 5]. Regardless of the approach, if the subgrid interactions are not self-consistently represented then an increase in resolution will not necessarily increase the accuracy of the explicitly resolved scales.

As in general it is only possible to parameterise the statistical effects of the subgrid eddies [6], statistical dynamical closure theory is the natural formulation for developing self-consistent subgrid models. In closure calculations, equations are solved for the time varying statistical moments of the flow, as opposed to the instantaneous fields. The pioneering study in this area was the parameter-free Eulerian direct interaction approximation (DIA), developed for isotropic homogeneous turbulence [7]. In this closure the tendency of the transient component consists of a nonlinear damping and a nonlinear noise term. The self-consistent field theory [8] and the local energy transfer theory [9, 10] were developed independently, and shown to differ from the DIA only in the implementation of the fluctuation dissipation theorem [11, 12, 13]. Later a general theory of eddy viscosity for two- and three-dimensional homogeneous turbulence was developed in [14], which identified that in spectral space the eddy viscosity has a strong cusp-like form increasing with wavenumber up until the smallest resolved scale of motion. Additional statistical dynamical closure studies pertaining mainly to three-dimensional homogeneous turbulence include [15, 16, 17, 18, 19, 6, 20].

In the vast majority of studies, and in all of the aforementioned ones other than the DIA of [7], a physical hypothesis is proposed from which a subgrid model is developed. In contrast the following closure studies have made no assumptions on the physical relationship between the resolved and subgrid scales, with the model coefficients determined from higher resolution simulations. For homogeneous barotropic (two-dimensional) flow on a sphere, a self-consistent stochastic subgrid parameterisation was developed based on the DIA closure in [21], which comprised of a drain eddy viscosity and stochastic backscatter force. Backscatter is the physical process by which kinetic energy is transferred from small to large scales. A deterministic variant was also developed comprising of a net eddy viscosity representing the net effect of the drain and backscatter. These subgrid coefficients were determined from higher resolution closure calculations, and produced LESs that replicated the kinetic energy spectra of the benchmark simulation. For inhomogeneous turbulent flows, general expressions were later derived for the eddy viscosities, stochastic backscatter, meanfield shift, and eddy-topographic force based on a quasi-diagonal DIA (QDIA) closure in [22]. The various subgrid terms for atmospheric flows using the QDIA formalism were then evaluated in [23]. Each of the coefficients relating the subgrid tendency to the associated resolved quantities (mean field, fluctuating field or roughness elements) are referred to in [24] by the term self-energy (SE). More generally a self-energy is a linear operator that closes the effect of unresolved nonlinear processes on the resolved field [25, 26].

To broaden the applicability of the closure methods discussed above, a stochastic modelling approach was developed in [27] that determines the subgrid coefficients from the statistics of a higher resolution reference simulation solving for the instantaneous flow fields. The reference simulation is truncated back to a coarser grid and the eddy viscosities are then determined from the subgrid statistics, in a similar vein to the self-consistent closure calculations discussed above. We refer to this technique as the SE subgrid modelling approach, as it is essentially a numerical means of determining the SE closure coefficients. This approach has been successfully applied to quasi-geostrophic (QG) simulations of the atmosphere and ocean, comprising of sheared jets, Rossby waves, and baroclinic instability [28, 29, 30]. In addition, if the resolution of the LES is such that the smallest scales lie within a self-similar inertial range, then the subgrid coefficients may be governed by simple functions of resolution [31]. Using this approach eddy viscosity scaling laws have been developed for global QG simulations of the atmosphere [31] and ocean [32]. These scaling laws enable the subgrid parameterisations to be utilised more widely as they remove the need to generate the subgrid coefficients from a reference simulation for these flows.

Recently, a unifying SE closure theory for the subgrid modelling of two- and three-dimensional inhomogeneous flows was developed in [33, 24]. The fundamental functional forms were determined for the four main types of subgrid interactions: eddy-eddy; eddy-meanfield; eddy-topographic; and meanfield-meanfield. Subgrid parametrisations representing these different interactions are calculated in [23]. Eddy-eddy interactions represent the impact unresolved subgrid eddies have on the evolution of the resolved eddies. The eddy-meanfield interactions represent the impact unresolved subgrid eddies have on the evolution of the resolved meanfield. The eddy-topographic (or eddy-roughness) form represents the interactions between the subgrid eddies and the resolved topography (or equivalently surface roughness). The meanfield-meanfield type represent the interactions between the unresolved subgrid meanfield and the resolved meanfield.

In the current study we use the SE subgrid modelling approach of [27] to systematically determine the model coefficients governing the subgrid interactions in simulations of horizontally periodic turbulent channel flows with smooth walls. Truncations are made at resolutions applicable to research-focused LES. This approach requires a direct numerical simulation (DNS) to determine the subgrid coefficients and is initially computationally expensive. However, not only does it produce LESs in very good agreement with the high resolution DNS, but also illuminates the true relationship between the resolved and subgrid scales and the energy transfers between them. As the present channel flow has smooth walls there is no eddy-topographic force. Additionally only the mode with both a spanwise and streamwise wavenumber of zero has non-zero mean under sufficient sampling, which means there is no subgrid meanfield and hence no meanfield-meanfield interactions. This type of interaction can be important, however, for flow configurations with complicated geometry and topography, as is the case of the ocean. The eddy-meanfield interactions are non-zero, but negligible in comparison to the dominant eddy-eddy interactions, which will be the focus of the present study. In section 2 we present the details of the DNS Navier-Stokes solver used in the turbulent channel flow simulations, and also characterise the base flow. The details on how the subgrid parameterisation coefficients are determined from the DNS statistics are then presented in section 3. The subgrid model coefficients are illustrated in section 4 decomposed in three-dimensional scale space, and by horizontal scale and vertical position in section 5. The ability of the LES to replicate the statistics of the DNS is then illustrated in section 6. Concluding remarks are made in section 7.

2. Direct Numerical Simulation

In the present study we solve the incompressible isothermal Navier-Stokes equations. In physical space the momentum and continuity equations are given by

$$\frac{\partial u^a}{\partial t} + u^b \frac{\partial u^a}{\partial x^b} + \frac{1}{\rho}\frac{\partial p}{\partial x^b} - \nu \frac{\partial^2 u^a}{\partial x^b \partial x^b} = f_0^a \text{ , and} \tag{1}$$

$$\frac{\partial u^a}{\partial x^b} = 0 \, , \tag{2}$$

respectively, with the indices $a = 1,2,3$ and $b = 1,2,3$. The coordinates $(x^1, x^2, x^3) \equiv (x, y, z)$, where x is the streamwise direction, y the wall-normal, and z the spanwise. Likewise the velocity vector $(u^1, u^2, u^3) \equiv (u, v, w)$, where u is the streamwise direction, v the wall-normal, and w the spanwise velocity component. The force f_0^a symbolically represents the boundary conditions, ν is the kinematic viscosity, and ρ the fluid density. The general spectral form of the equations is given by

$$\frac{\partial}{\partial t} u_{\mathbf{k}}^a(t) + D_0^{ab}(\mathbf{k}) u_{\mathbf{k}}^b(t) = \sum_{\mathbf{k}'} \sum_{\mathbf{k}''} K^{abc}(\mathbf{k}, \mathbf{k}', \mathbf{k}'') u_{-\mathbf{k}'}^b(t) u_{-\mathbf{k}''}^c(t) + f_0^a(\mathbf{k}, t) \, . \tag{3}$$

Table 1. Numerical details of the channel simulations: friction-based Reynolds number ($Re_\tau \equiv u_\tau h/\nu$); centreline Reynolds number ($Re_0 \equiv u_0 h/\nu$); DNS truncation wavenumber (T); number of Chebyshev polynomials (N); domain sizes in streamwise, wall-normal and spanwise directions (L_x, L_y, L_z); in viscous units, the constant grid spacing in streamwise and spanwise directions ($\Delta x^+, \Delta z^+$) and the surface-normal grid spacing at the channel centreline (Δy^+_{cl}) and at the wall (Δy^+_{wall}); constant time-step size in viscous units (Δt^+); number of eddy turnover times over which the subgrid statistics were accumulated ($(t_f - t_i)u_0/L_x$); LES truncation levels (T_R).

Re_τ	Re_0	T	N	(L_x, L_y, L_z)	$(\Delta x^+, \Delta y^+_{cl}, \Delta y^+_{wall}, \Delta z^+)$	Δt^+	$(t_f - t_i)\frac{u_0}{L_x}$	T_R
180	3250	64	97	$(2\pi, 2, \pi)$	$(5.9, 6.1, 0.1, 2.9)$	0.13	45	32
550	11180	128	257	$(2\pi, 2, \pi)$	$(9.0, 6.7, 0.04, 4.5)$	0.04	12	64
950	20580	128	385	$(\pi, 2, \pi/2)$	$(7.8, 7.6, 0.03, 3.9)$	0.02	7	64

where $u^a_{\mathbf{k}}$ is the spectral component of wavenumber vector $\mathbf{k}$, for the velocity component of index a. The term $f^a_0(\mathbf{k}, t)$ are the spectral coefficients of f^a_0, the Navier-Stokes linear operator is given by $D^{ab}_0(\mathbf{k})$, and $K^{abc}(\mathbf{k}, \mathbf{k}', \mathbf{k}'')$ are the nonlinear interaction coefficients. The structure of (3) holds for any form of scale space decomposition, with only the specific form of the coefficients differing. The interaction coefficients for three-dimensional Fourier discretisation are detailed in [24]. In the present study the flow variables are discretised using a Fourier discretisation in x of wavenumber α, a collocated Chebyshev discretisation in y of polynomial index j, and a Fourier discretisation in z of wavenumber β. We define our pseudo wavenumber vector to be $\mathbf{k} \equiv (k_1, k_2, k_3) \equiv (\alpha, j, \beta)$. Note that $u^a_{-\mathbf{k}}(t) = u^{a*}_{\mathbf{k}}(t)$, where the superscript $*$ is the complex conjugate operation and $-\mathbf{k} = (-\alpha, j, -\beta)$. The specific form of $D^{ab}_0(\mathbf{k})$, $K^{abc}(\mathbf{k}, \mathbf{k}', \mathbf{k}'')$ and $f^a_0(\mathbf{k}, t)$ can be deduced from [34]. The DNS is solved over the rectangular wavenumber and Chebyshev polynomial set of maximum DNS truncation wavenumber T, and truncation Chebyshev polynomial index N. The summations immediately after the equals sign in (3) are over the wavenumber set

$$\mathbf{T} = [\ \mathbf{k}',\ \mathbf{k}''\ |\ 0 \le k'_1 \le T\ ,\ 0 \le k'_2 \le N\ ,\ -T \le k'_3 \le T \\ 0 \le k''_1 \le T\ ,\ 0 \le k''_2 \le N\ ,\ -T \le k''_3 \le T\ ]\ . \quad (4)$$

Note that the mechanics of the channel code are such that it solves for the prognostic variables: ζ, the wall-normal vorticity; and $\nabla^2 v$ where ∇^2 is the Laplacian operator [34]. The evolution of the velocity field, however, is still governed by (3). Finally unless otherwise stated, all quantities presented within are nondimensionalised by the channel half-height h, and the freestream centreline velocity u_0.

We undertake simulations at three friction-velocity-based Reynolds numbers of $Re_\tau \equiv u_\tau h/\nu = 180$, 550 and 950, where the friction velocity $u_\tau = \sqrt{\tau_{wall}/\rho}$ with τ_{wall} the magnitude of the wall shear stress. In the present simulations the channel flow is simulated in a rectangular box of streamwise extent L_x, wall-normal extent L_y, and spanwise extent L_z, with respective grid collocated points in each direction of N_x, $N_y = N$ and N_z. The number of grid points in the horizontal directions is related to the dealised truncation wavenumber T by $N_x = N_z = 3T$. The grid properties for each of the simulations are listed in Table 1, along with the LES truncation resolutions calculated from each DNS. Grid spacings are also presented in viscous units, denoted by the superscript $^+$, where they are nondimensionalised by the length scale ν/u_τ. The constant grid spacing in the streamwise and spanwise directions in viscous units is given by Δx^+ and Δz^+ respectively. In the wall-normal direction, the minimum grid spacing is located at the wall (Δy^+_{wall}), and the maximum grid spacing is located at the channel centreline (Δy^+_{cl}). For completeness Table 1 also includes the constant time-step size (Δt) used in the each of the

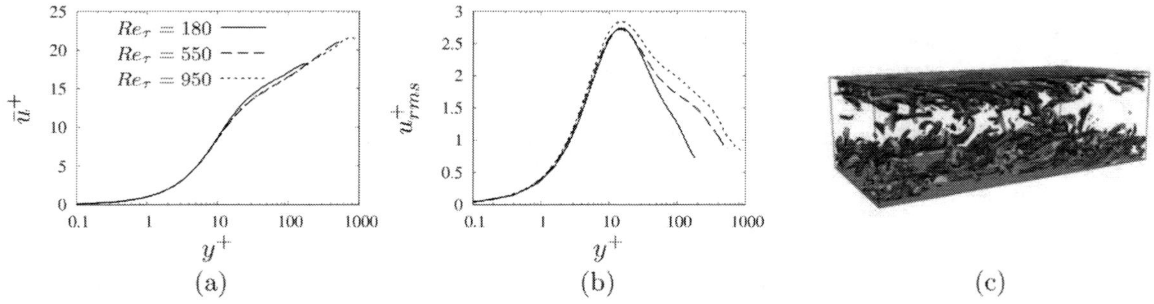

Figure 1. Characterisation of the channel flow. For all Re_τ the: (a) time-averaged streamwise velocity in viscous units $\bar{u}^+$, with key also applicable to (b); and (b) rms streamwise velocity in viscous units u^+_{rms}. (c) Instantaneous iso-surfaces of the discriminant of the velocity-gradient tensor equal to unity, coloured by streamwise vorticity with red positive rotation and blue negative rotation.

simulations presented in viscous units ($\Delta t^+ = \Delta t\ h/u_\tau$), and the number of flow-through times over which the statistics are accumulated given by $(t_f - t_i)L_x/u_0$, where t_f and t_i are the final and initial times respectively.

For all three Re_τ cases we plot the statistical velocity profiles in Fig. 1. Firstly we decompose the instantaneous velocity field, $\mathbf{u}$, into its time-averaged mean $\bar{\mathbf{u}}$, and its fluctuating component $\hat{\mathbf{u}}$. The velocity field is scaled into viscous units given by $\mathbf{u}^+ \equiv \mathbf{u}/u_\tau$. Time-averaged and root-mean-square (rms) streamwise velocity profiles are plotted in Fig. 1(a) and Fig. 1(b) respectively. All of the statistical profiles are consistent with previous DNS [35]. We also illustrate the instantaneous flow field of the $Re_\tau = 180$ case in Fig. 1(c), via iso-surfaces of the discriminant of the velocity gradient tensor equal to one. The iso-surfaces are coloured by streamwise vorticity, where red is positive rotation and blue negative rotation about the streamwise axis. The flow is left to right, with the walls indicated by the grey boundaries. We will now apply the SE subgrid modelling approach to these DNSs.

3. Stochastic self-energy modelling of the subgrid scales

The stochastic SE modelling approach developed in [27] is used to model the interactions between resolved and subgrid eddies and hence represent the transfer of energy between them. As discussed in section 2 the DNS is simulated over the complete wavenumber set $\mathbf{T}$ of truncation wavenumber T and truncation Chebyshev polynomial index N. In the present study we decompose the scales of motion using a rectangular truncation, where the resolved wavenumber set is defined by

$$\mathbf{R} = [\ \mathbf{k}',\ \mathbf{k}'' \mid 0 \le k'_1 \le T_R\ ,\ 0 \le k'_2 \le N\ ,\ -T_R \le k'_3 \le T_R \\ 0 \le k''_1 \le T_R\ ,\ 0 \le k''_2 \le N\ ,\ -T_R \le k''_3 \le T_R\ ]\ , \tag{5}$$

where $T_R < T$ is the LES truncation wavenumber. Recall that $\mathbf{k} \equiv (k_1, k_2, k_3) \equiv (\alpha, j, \beta)$. Note that we keep all of the Chebyshev polynomials, maintaining all of the vertical scales, and hence consider only the effect of the removal of the horizontal scales. The subgrid wavenumber set is then defined as $\mathbf{S} = \mathbf{T} - \mathbf{R}$.

To facilitate the following discussion on how the subgrid interactions are represented, we introduce a general state vector $\mathbf{q}$ of length N_q. There are several options for defining the specific form of $\mathbf{q}$, which will be discussed at the end of this section. Firstly the tendency (time derivative) of $\mathbf{q}$ is decomposed such that

$$\mathbf{q}_t(t) = \mathbf{q}^{\mathbf{R}}_t(t) + \mathbf{q}^{\mathbf{S}}_t(t)\ . \tag{6}$$

The tendency of the resolved scales is $\mathbf{q}_t^{\mathbf{R}}$, where all triadic interactions involve wavenumbers within the resolved set $\mathbf{R}$. The remainder is the subgrid tendency, $\mathbf{q}_t^{\mathbf{S}}$, which has at least one wavenumber within the subgrid set $\mathbf{S}$ involved in the triadic interactions. In its most primitive form, the subgrid modelling problem is the determination of the relationship between $\mathbf{q}_t^{\mathbf{S}}$ and $\mathbf{q}$ within the resolved set $\mathbf{R}$. The subgrid tendency is further decomposed by

$$\mathbf{q}_t^{\mathbf{S}}(t) = \overline{\mathbf{f}} + \hat{\mathbf{q}}_t^{\mathbf{S}}(t) \ , \qquad (7)$$

where $\overline{\mathbf{f}} \equiv \langle \mathbf{q}_t^{\mathbf{S}} \rangle$ is the time-averaged subgrid tendency, and $\hat{\mathbf{q}}_t^{\mathbf{S}}$ the fluctuating component. The $\hat{\mathbf{q}}_t^{\mathbf{S}}$ term represents the eddy-eddy interactions, and in the absence of surface roughness $\overline{\mathbf{f}}$ represents the summation of the eddy-meanfield and meanfield-meanfield interactions. One can decompose these interaction types using the process outlined in [36]. In the current work the values of $\langle \mathbf{q}_t^{\mathbf{S}} \rangle$ are determined directly from the DNS, and $\hat{\mathbf{q}}_t^{\mathbf{S}}$ is modelled.

The fluctuating component of the subgrid tendency, $\hat{\mathbf{q}}_t^{\mathbf{S}}$, is represented by the stochastic equation

$$\hat{\mathbf{q}}_t^{\mathbf{S}}(t) = -\mathbf{D_d}\ \hat{\mathbf{q}}(t) + \hat{\mathbf{f}}(t) \ , \qquad (8)$$

where $\mathbf{D_d}$ is the subgrid drain dissipation matrix, $\hat{\mathbf{q}}$ is the fluctuating component of $\mathbf{q}$, and $\hat{\mathbf{f}}$ is a random forcing vector. The matrix $\mathbf{D_d}$ is a time-independent $N_q \times N_q$ matrix, and $\hat{\mathbf{f}}$ is a time-dependent N_q element column vector. The drain dissipation operator is determined by post-multiplying both sides of (8) by $\hat{\mathbf{q}}^\dagger(t_0)$, integrating over the turbulent decorrelation period τ, and ensemble averaging. The expression for $\mathbf{D_d}$ is based on a generalisation of Gauss's theorem of least squares and given by

$$\mathbf{D_d} \ = \ -\left\langle \int_{t_0}^{t} \hat{\mathbf{q}}_t^{\mathbf{S}}(\sigma)\hat{\mathbf{q}}^\dagger(t_0)d\sigma \right\rangle \left\langle \int_{t_0}^{t} \hat{\mathbf{q}}(\sigma)\hat{\mathbf{q}}^\dagger(t_0)d\sigma \right\rangle^{-1} , \qquad (9)$$

where $\dagger$ denotes the Hermitian conjugate for vectors and matrices. The angled brackets denote ensemble averaging, with each ensemble member determined by shifting the initial time t_0 and the final time $t = t_0 + \tau$ forward by one time step. For the present data $\mathbf{D_d}$ increases and eventually plateaus with increasing τ, as previously observed for atmospheric and oceanic flows [31, 32]. We present $\mathbf{D_d}$ using a value of τ that maximises the dissipation, and hence captures the memory effects of the subgrid interactions. The value of τ at which $\mathbf{D_d}$ plateaus changes with resolution and Reynolds number, however, in all of the cases presented within τ is of the order of $100\Delta t$.

The model for $\hat{\mathbf{f}}$ is determined by calculating the non-linear noise covariance matrix $\boldsymbol{\mathcal{F}}_b = \mathbf{F_b} + \mathbf{F_b}^\dagger$, where $\mathbf{F_b} = \langle \hat{\mathbf{f}}(t)\ \hat{\mathbf{q}}^\dagger(t) \rangle$. By again post-multiplying both sides of (8) by $\hat{\mathbf{q}}^\dagger(t_0)$, and adding the conjugate transpose of (8) pre-multiplied by $\hat{\mathbf{q}}(t_0)$ yields the Lyapunov (or balance) equation

$$\left\langle \hat{\mathbf{q}}_t^{\mathbf{S}}(t)\hat{\mathbf{q}}^\dagger(t) \right\rangle + \left\langle \hat{\mathbf{q}}(t)\hat{\mathbf{q}}_t^{\mathbf{S}\dagger}(t) \right\rangle = -\mathbf{D_d} \left\langle \hat{\mathbf{q}}(t)\hat{\mathbf{q}}^\dagger(t) \right\rangle - \left\langle \hat{\mathbf{q}}(t)\hat{\mathbf{q}}^\dagger(t) \right\rangle \mathbf{D_d}^\dagger + \boldsymbol{\mathcal{F}}_b \ . \qquad (10)$$

Given that $\mathbf{D_d}$ is known, $\boldsymbol{\mathcal{F}}_b$ can now be calculated. From (10) it is clear that there is a balancing act between the deterministic ($\mathbf{D_d}$) and stochastic ($\boldsymbol{\mathcal{F}}_b$) components of the subgrid model. As $\mathbf{D_d}$ is dependent upon τ, it is τ that defines this balance. At this point the formulation is general, and $\hat{\mathbf{f}}$ is coloured noise. For the implementation of the stochastic subgrid parameterisation, $\hat{\mathbf{f}}$ is represented as a white noise process such that $\langle \hat{\mathbf{f}}(t)\ \hat{\mathbf{f}}^\dagger(t') \rangle = \boldsymbol{\mathcal{F}}_b\ \delta(t - t')$. An eigenvalue decomposition of $\boldsymbol{\mathcal{F}}_b$ is then used to produce a stochastic model for $\hat{\mathbf{f}}$, as detailed in [28].

As outlined in (8) the fundamental form of subgrid turbulence is a stochastic process, comprising of the deterministic drain dissipation and a stochastic backscatter force. However, one can approximate the subgrid interactions by a solely deterministic relationship $\hat{\mathbf{q}}_t^{\mathbf{S}}(t) = -\mathbf{D_{net}}\ \hat{\mathbf{q}}(t)$, where $\mathbf{D_{net}}$ is the net dissipation representing the net effect of the drain and backscatter [27]. The net dissipation is given by

$$\mathbf{D_{net}} = \mathbf{D_d} + \mathbf{D_b} = -\left\langle \hat{\mathbf{q}}_t^{\mathbf{S}}(t)\hat{\mathbf{q}}^{\dagger}(t)\right\rangle \left\langle \hat{\mathbf{q}}(t)\hat{\mathbf{q}}^{\dagger}(t)\right\rangle^{-1}, \text{ where} \tag{11}$$

$$\mathbf{D_b} = -\mathbf{F_b}\left\langle \hat{\mathbf{q}}(t)\ \hat{\mathbf{q}}^{\dagger}(t)\right\rangle^{-1}, \tag{12}$$

is the backscatter dissipation. From (11) it is clear that $\mathbf{D_{net}}$ is independent of τ. The drain, backscatter and net eddy viscosities are respectively given by $\boldsymbol{\nu}_\mathbf{d} \equiv \mathbf{D_d}\nabla^{-2}$, $\boldsymbol{\nu}_\mathbf{b} \equiv \mathbf{D_b}\nabla^{-2}$, and $\boldsymbol{\nu}_\mathbf{net} \equiv \mathbf{D_{net}}\nabla^{-2}$.

As all of the subgrid matrices discussed above have dimensions $N_q \times N_q$, it is advisable to minimise the size of $\mathbf{q}$. In the most general case, $\mathbf{q}$ would contain all of the spectral components of all field variables. For our present rectangular truncation, with the state vector containing all three velocity components, $N_q = 3T_R^2 N$, which at reasonable resolution is clearly too large for the SE subgrid approach to be tractable. One can reduce the dimension of the problem, however, by adopting the approach undertaken in the QDIA closure developed in [33]. This study formally showed that the subgrid tendency $\mathbf{q}_t^{\mathbf{S}}$ at a particular wavenumber (or pair, or triplet depending on the number of scale decomposed directions) can be renormalised to be a function of the resolved field $\mathbf{q}$ at only that same wavenumber. This approach was successfully applied to spherical QG atmospheric simulations in [28], producing resolution independent LES. The QDIA assumption was applied in the horizontal periodic directions, with the full matrix form used in the vertical. In this study there was one flow variable and two vertical levels, which meant that $N_q = 2$. Using this approach for the channel flow, for each wavenumber pair (α, β) we have a state vector $\mathbf{q}(\alpha, \beta, t)$ containing the Chebyshev coefficients of each of the three variables, with $N_q = 3N$. This may well be tractable but still introduces a significant issue concerning the sampling of the off-diagonal elements of the subgrid matrix statistics. We can take one step further, however, by also applying the QDIA approach in the wall-normal direction. We apply the approach in a pseudo-scale space by factorising the problem in terms of the Chebyshev polynomial indices j. Now for each $\mathbf{k} \equiv (\alpha, j, \beta)$ the state vector takes the form

$$\mathbf{q}(\mathbf{k}, t) \equiv [u(\mathbf{k}, t), v(\mathbf{k}, t), w(\mathbf{k}, t)]^T, \tag{13}$$

where $\mathbf{q}$ is now of length $N_q = 3$. This allows the subgrid scale model to be applied locally to each wavenumber pair and Chebyshev polynomial index, whilst still capturing the coupling between the three velocity components. The subgrid coefficients in this form are illustrated in section 4.

4. Subgrid eddy viscosities in scale space

In this section we present the subgrid coefficients for each of the $Re_\tau = 180$, 550 and 950 channel flows using a rectangular truncation in the two periodic horizontal directions, as outlined in (5). The subgrid interactions are assumed quasi-diagonal in both Fourier and Chebyshev polynomial space as presented in (13).

We initially present the subgrid coefficients for the $Re_\tau = 180$ truncated back to $T_R = T/2 = 32$. The subgrid net dissipation matrix elements are illustrated in Fig. 2. They are plotted in the horizontal wavenumber plane (α, β) for Chebyshev polynomial index $j = 41$, with an effective wall-normal scale of $\lambda_y^+ = 2$, where $\lambda_y^+ = \lambda_y Re_\tau$. We define λ_y to be the distance between the wall and the second turning point of the j-th polynomial index, given by $1 - \cos(2\pi/j)$. The matrix elements D_{net}^{uv} can be interpreted as the transfer function linking the resolved velocity field v to the subgrid tendency of velocity component u, and likewise for the other matrix

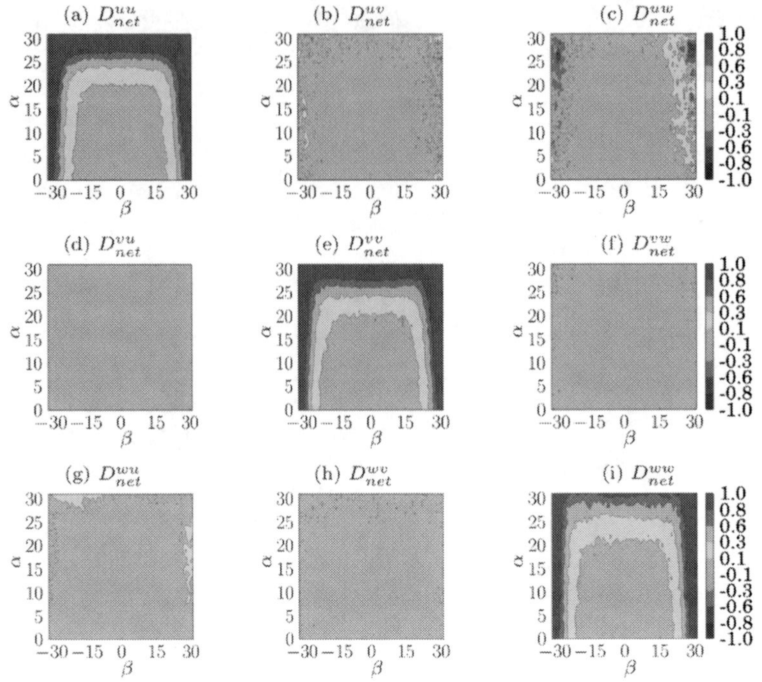

Figure 2. Subgrid net dissipation matrix elements for the $Re_\tau = 180$ case, illustrated in the horizontal wavenumber half-plane ($\alpha, \beta \geq 0$) at $\lambda_y^+ = 2$, using rectangular truncation with $T_R = T/2 = 32$: (a) D_{net}^{uu}; (b) D_{net}^{uv}; (c) D_{net}^{uw}; (d) D_{net}^{vu}; (e) D_{net}^{vv}; (f) D_{net}^{vw}; (g) D_{net}^{wu}; (h) D_{net}^{wv}; and (i) D_{net}^{ww}. The colour bars in the final column are the same and applicable to all figures.

components. It is clear from Fig. 2 that at this particular truncation level T_R, and for this Chebyshev polynomial j, the diagonal components are dominant. This means the subgrid tendency for a given velocity component is most strongly related to the resolved field of the same velocity component. The diagonal elements are also positive throughout the wavenumber plane which means energy is being drained out of the resolved scales and sent to the subgrid. The dissipations are also scale selective with the magnitude of the coefficients increasing as they approach the truncation boundaries. This means that the small resolved vortices near truncation have more significant interactions with the subgrid than the larger resolved structures. This is a typical phenomena and has previously been observed in the associated subgrid coefficients in atmospheric and oceanic simulations [31, 32].

The picture becomes somewhat more complicated, however, when inspecting the other Chebyshev polynomials. We concentrate for the moment on the D_{net}^{uu} term. This element is symmetric about the line $\beta = 0$, so we only plot these coefficients in the wavenumber half-plane ($\alpha, \beta \geq 0$). The D_{net}^{uu} coefficients are illustrated for $\lambda_y/h = 0.1$, $\lambda_y^+ = 10$, $\lambda_y^+ = 2$, and $\lambda_y^+ = 0.5$ in Fig. 3(a)-(d) respectively. The D_{net}^{uu} coefficients for $\lambda_y/h = 0.1$ and $\lambda_y^+ = 10$ have qualitatively the same properties for as $\lambda_y^+ = 2$ discussed above. For $\lambda_y^+ = 0.5$, however, the smallest resolved scales in the top right hand corner of Fig. 3(l) have negative net dissipations. This indicates that the subgrid interactions are conspiring to send energy from the subgrid to these wavenumber pairs, representative of a backscatter of energy. One can determine if this backscatter is coherent or incoherent by decomposing the net dissipation into its deterministic and stochastic components. The deterministic drain dissipation D_d^{uu} is illustrated for the same Chebyshev polynomial indices in Fig. 3(e)-(h), with the associated stochastic backscatter dissipation D_b^{uu} illustrated in Fig. 3(i)-(l). In all cases the drain dissipation is completely positive

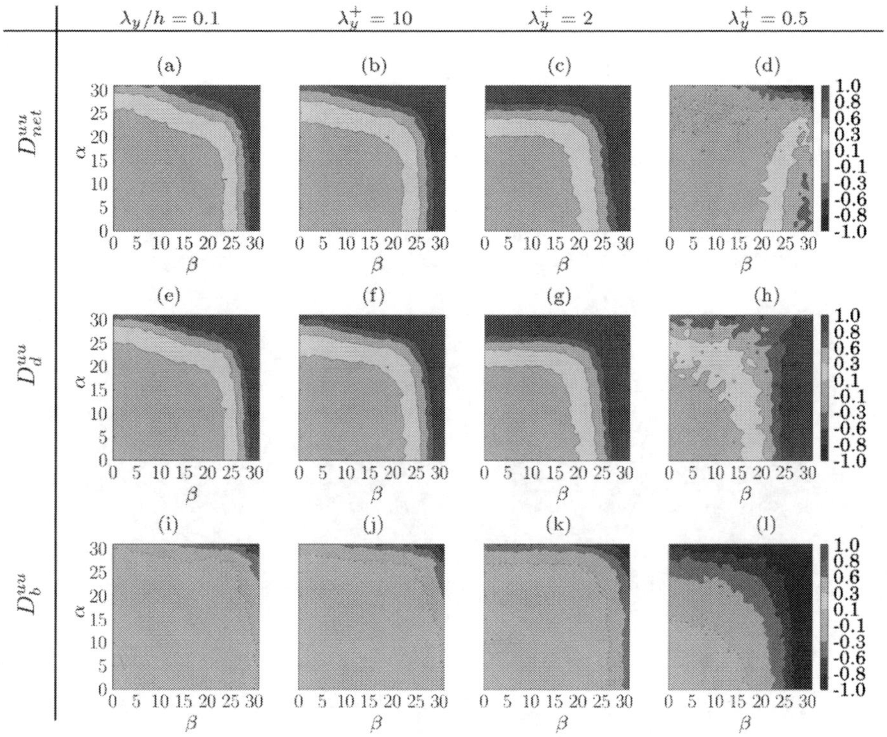

Figure 3. The uu component of the subgrid dissipation matrices of $Re_\tau = 180$ case illustrated in the horizontal wavenumber half-plane ($\alpha, \beta \geq 0$) using rectangular truncation $T_R = T/2$ for various Chebyshev polynomials (index j) with characteristic wall-normal scale $\lambda_y = 1 - \cos(2\pi/j)$. The first row (a)-(d) contains the net coefficients D_{net}^{uu}. The second row (e)-(h) contains the drain coefficients D_d^{uu}. The third row (i)-(l) contains the backscatter coefficients D_b^{uu}. In the first column (a),(e),(i) $\lambda_y/h = 0.1$. In the second column (b),(f),(j) $\lambda_y^+ = 10$. In the third column (c),(g),(k) $\lambda_y^+ = 2$. In the fourth column (d),(h),(l) $\lambda_y^+ = 0.5$. All coefficients are symmetric about the line $\beta = 0$. The colour bars in the final column are the same and applicable to all figures.

and the stochastic backscatter completely negative. This means that energy is sent from the resolved to the subgrid scales in a deterministic coherent manner, and energy is sent from the subgrid to the resolved in an incoherent stochastic manner. It is only at the smallest vertical scales that the stochastic backscatter overwhelms the drain dissipation resulting in negative net dissipation in Fig. 3(d) for certain wavenumber pairs.

The higher Re_τ cases have similar properties. The same D_{net}^{uu} are plotted for the $Re_\tau = 180$ case in Fig. 4(a)-(d). They are lined up with coefficients from $Re_\tau = 550$ in Fig. 4(e)-(h) and $Re_\tau = 950$ in Fig. 4(i)-(l) with the same vertical scales. These figures illustrate that as Re_τ increases so too does the magnitude of D_{net}^{uu}. For each Re_τ, the D_{net}^{uu} coefficients for $\lambda_y^+ = 0.5$ contain negative values for certain wavenumber pairs, indicating that energy is sent to these wavenumbers from the subgrid. Inspection of the associated drain and stochastic backscatter dissipations (not shown) indicates that the injection of energy is again stochastic in nature. Current work is looking to identify a potential universal scaling of these coefficients.

5. Subgrid eddy viscosities in horizontal scale and vertical physical space

So far we have presented the subgrid dissipations diagonalised based on horizontal and vertical scale, and identified that there is a net injection of energy into the smallest resolved vertical

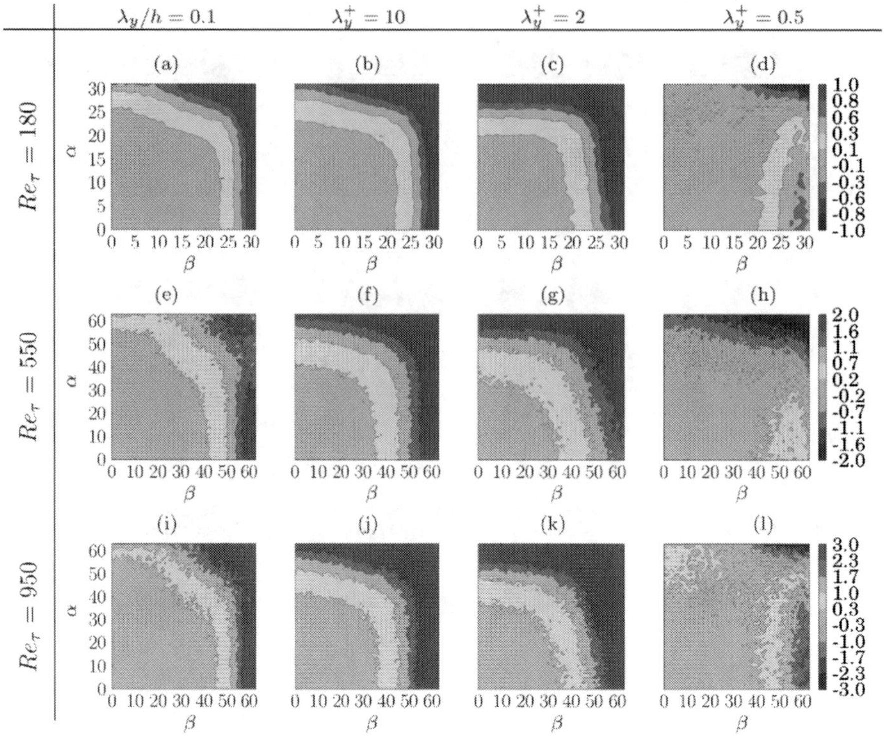

Figure 4. The net dissipation matrix element D^{uu}_{net} illustrated in the horizontal wavenumber half-plane ($\alpha, \beta \geq 0$) using rectangular truncation $T_R = T/2$ for various Chebyshev polynomials (index j) with characteristic wall-normal scale $\lambda_y = 1 - \cos(2\pi/j)$. The first row (a)-(d) contains the $Re_\tau = 180$ case. The second row (e)-(h) contains the $Re_\tau = 550$ case. The third row (i)-(l) contains the $Re_\tau = 950$ case. In the first column (a),(e),(i) $\lambda_y/h = 0.1$. In the second column (b),(f),(j) $\lambda_y^+ = 10$. In the third column (c),(g),(k) $\lambda_y^+ = 2$. In the fourth column (d),(h),(l) $\lambda_y^+ = 0.5$. All coefficients are symmetric about the line $\beta = 0$. The colour bars in the final column are applicable to all figures in that row.

scales (large Chebyshev polynomial index j). It is not clear, however, where in physical space this injection of energy is occurring. However, we can identify the regions of injection by looking at the subgrid interactions on the basis of vertical position instead of vertical scale. Whilst the subgrid interactions are quasi-diagonal in scale space, inspecting the structure of the subgrid dissipations diagonalised based on vertical position can also provide some insight.

The associated D^{uu}_{net} coefficients are plotted for each of the Re_τ cases in Fig. 5 at four different wall-normal positions of $y = 1$ (centreline), $y = 0.5$, $y^+ = 10$ and $y^+ = 1$. The net dissipations at each of these positions for the $Re_\tau = 180$ case are illustrated in Fig. 5(a)-(d), for $Re_\tau = 550$ in Fig. 5(e)-(h), and $Re_\tau = 950$ in Fig. 5(i)-(l). In all cases energy is drained out of the system in the body of the channel, with there being a net injection of energy at a wall-normal position of $y^+ = 1$. The magnitude of the dissipations again increase with Re_τ.

These observations are consistent with the physical view of wall-bounded turbulence. In the attached eddy hypothesis of [37] spanwise vorticity generated at the wall in the form of spanwise oriented vortex lines are lifted up and elongated in the streamwise direction by the mean flow. These wall-attached eddies of small vertical and horizontal scale, transfer energy to larger wall-attached eddies with their centreline located at a slightly greater distance from the wall, leading to a form of inverse energy cascade [38, 39, 40, 41, 42]. Truncating the DNS by removing the small horizontal scales removes the source of energy injection in the system (the

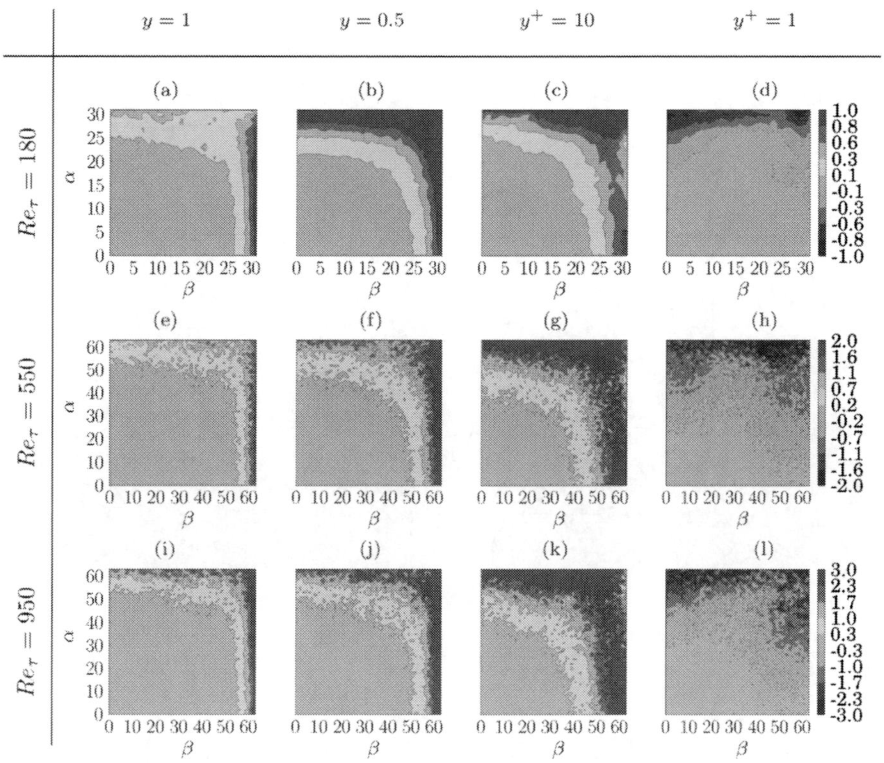

Figure 5. The net dissipation matrix element D^{uu}_{net} illustrated in the horizontal wavenumber half-plane ($\alpha, \beta \geq 0$) using rectangular truncation $T_R = T/2$ for various wall-normal positions y. The first row (a)-(d) contains the $Re_\tau = 180$ case. The second row (e)-(h) contains the $Re_\tau = 550$ case. The third row (i)-(l) contains the $Re_\tau = 950$ case. In the first column (a),(e),(i) $y = 1$. In the second column (b),(f),(j) $y = 0.5$. In the third column (c),(g),(k) $y^+ = 10$. In the fourth column (d),(h),(l) $y^+ = 1$. All coefficients are symmetric about the line $\beta = 0$. The colour bars in the final column are applicable to all figures in that row.

wall-attached eddies). This injection of energy is represented by the negative net coefficients of the small vertical scales illustrated in Fig. 4, and also at small wall-normal distance as illustrated in Fig. 5. At some point the initially wall-attached eddies detach themselves from the wall. The now detached eddy breaks down into smaller eddies giving rise to a forward energy cascade [38, 39, 40, 41, 42]. This dissipation of energy is represented by the positive net coefficients for larger vertical scales and wall-normal positions.

6. Large eddy simulation

In this section we will illustrate the ability of LESs of the $Re_\tau = 180$ case to replicate the spectra of the associated DNS. As discussed in section 3, LESs can be undertaken using either a stochastic or a deterministic subgrid parameterisation. The equations governing the LES are the same as for the DNS, but with $\mathbf{q}^{\mathbf{S}}_t(\mathbf{k})$ added to the right-hand-side of the prognostic equations, and solved over the wavenumber set $\mathbf{R}$ instead of $\mathbf{T}$. Recall that $\mathbf{k} \equiv (\alpha, j, \beta)$. In the stochastic representation $\mathbf{q}^{\mathbf{S}}_t(\mathbf{k})$ is given by the following matrix equation

$$\mathbf{q}^{\mathbf{S}}_t(\mathbf{k}, t) \quad = \quad -\mathbf{D_d}(\mathbf{k})\ \hat{\mathbf{q}}(\mathbf{k}, t) + \hat{\mathbf{f}}(\mathbf{k}, t) + \overline{\mathbf{f}}(\mathbf{k}) \ , \tag{14}$$

where the eddy-meanfield terms, $\overline{\mathbf{f}}$, are also those determined from the DNS. In the deterministic LES the drain dissipation and stochastic backscatter are replaced with the net dissipation, $\mathbf{D_{net}}$,

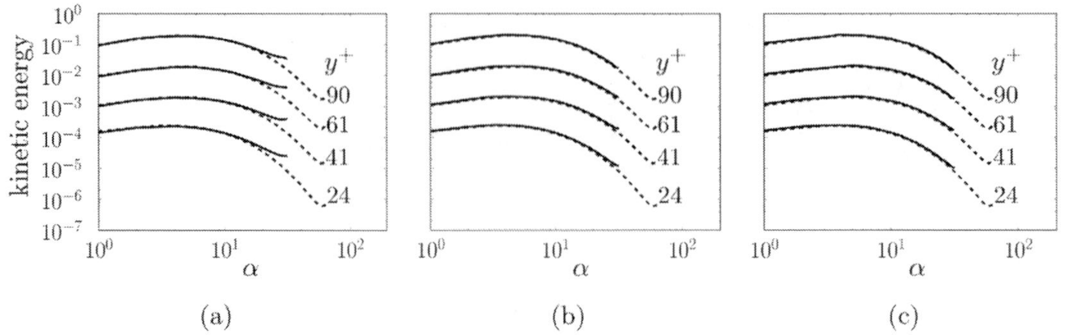

Figure 6. Comparison of the time-averaged spanwise-wavenumber-summed kinetic energy spectra of the $Re_\tau = 180$ DNS (dashed line) to LES (solid line) with: (a) no subgrid model; (b) deterministic subgrid model; and (c) stochastic subgrid model. Each pair of spectra are labelled by the associated value of y^+ and shifted down multiples of one decade for clarity.

such that

$$\mathbf{q}_t^{\mathbf{S}}(\mathbf{k},t) \quad = \quad -\mathbf{D}_{\mathbf{net}}(\mathbf{k})\ \hat{\mathbf{q}}(\mathbf{k},t) + \bar{\mathbf{f}}(\mathbf{k})\ . \tag{15}$$

On a technical matter, recall the channel code solves for the flow field on the basis of the state vector $(\zeta, \nabla^2 v)$, however, to facilitate physical interpretation of the results the subgrid parameterisations were developed on the basis of the velocity components. For the implementation of the LES, the same channel code is run at reduced resolution again solving for $(\zeta, \nabla^2 v)$. At each time step the Poisson equation is solved to convert the instantaneous $(\zeta, \nabla^2 v)$ to the velocity components (u, v, w). The subgrid model is then applied to the instantaneous resolved velocity field to produce the associated subgrid tendency in velocity space (u_t^S, v_t^S, w_t^S). The subgrid tendency is then converted back to $(\zeta_t^S, \nabla^2 v_t^S)$ and added to the right-hand-side of the DNS equations.

Comparisons between the DNS and LES are made at various wall-normal positions, on the basis of the time-averaged spanwise wavenumber summed kinetic energy spectra as a function of streamwise wavenumber. In all of the following plots the DNS is represented by the dashed lines, and the LES by the solid lines. The pairs of spectra are labelled by the associated value of y^+ with each pair shifted down multiples of one decade from the top pair for clarity. Firstly to serve as a baseline comparison, we truncate the DNS to the LES resolution of $T_R = T/2 = 31$ with no subgrid model prescribed. The resulting spectra are compared to the DNS spectra in Fig 6(a). This clearly illustrates an accumulation of energy at the smallest resolved scales of motion because the now unresolved subgrid interactions would have drained this energy out of the resolved scales and into the subgrid. This accumulation of energy can be alleviated by adopting either of the deterministic or stochastic SE subgrid models. The kinetic energy spectra of the deterministic LES, using the subgrid model defined in (15), is compared to the DNS spectra in Fig 6(b). Likewise the kinetic energy spectra of the stochastic LES, using the subgrid model defined in (14), is compared to the DNS spectra in Fig 6(c). Both the deterministic and stochastic implementations of the SE subgrid model are able to correct the kinetic energy spectra and reproduce the DNS statistics. This high level of agreement indicates that the deterministic and stochastic SE subgrid models accurately represent the subgrid interactions. It is, therefore, valid to make physical interpretations of the energy transfers from the sign, magnitude and spectral dependence of the SE subgrid model coefficients.

7. Concluding remarks and future applications

Self-energy subgrid models have been developed for turbulent channel flows. The stochastic parameterisation of the eddy-eddy interactions consists of a drain dissipation, and a stochastic backscatter noise term. The deterministic version is governed solely by the net dissipation, representing the net effect of the drain and backscatter. These subgrid matrices are functions of the streamwise and spanwise wavenumbers and the Chebyshev polynomial index, and explicitly capture the interactions between each of the three velocity components. These scale dependent matrices have been derived from the statistics of higher resolution DNSs. LESs adopting the stochastic and deterministic subgrid models diagonalised on the basis of horizontal and vertical scale were successful in replicating the statistics of the DNS.

Inspection of the three-dimensional scale-based net eddy viscosities indicates that the removal of small horizontal scales results in a net drain of energy out of the system for the large vertical scales, and an injection of energy to the smallest resolved vertical scales. Decomposing the net eddy viscosity into the drain and backscatter components, illustrates that the injection of energy is a stochastic one. Diagonalising the subgrid interactions on the basis of vertical position instead of vertical scale identified that the injection of energy was concentrated in the near wall region for all Reynolds numbers. This is consistent with the attached eddy hypothesis [37, 38] stating that small wall-attached eddies transfer energy to larger wall-attached eddies giving rise to an inverse energy cascade, and upon detachment break down into smaller eddies giving rise to a forward energy cascade [41, 42]. In the framework presented here, the inverse cascade is represented by the negative net dissipations, and the forward cascade is represented by the positive net dissipations.

These findings may also have implications for the development of wall models. First and foremost a wall model must apply an appropriate boundary condition at the first grid point away from the wall, as illustrated in [43, 5, 44, 45, 46, 47]. However, in the present study it has been identified that the removal of small wall-attached eddies results in a net injection of energy to the larger resolved eddies. Ideally, in addition to applying the correct boundary condition, wall models should also represent this injection of energy and the necessary dampening. The SE approach presented in this paper could be used to determine the complete structure of a wall model by decomposing the resolved and subgrid scales on the basis of vertical scale (or potentially position). Depending on the complexity of the resulting subgrid coefficients, the eddy viscosities may then be represented in a more simple form in order to construct a more general purpose wall model, complementing the application of the appropriate boundary conditions. The extent to which such an approach would improve current wall modelling efforts is an open question.

The SE subgrid modelling approach is initially computationally expensive as a DNS must be run to determine the subgrid coefficients. However, because no physical assumptions are made in the development of the model, physical interpretations can be made from the subgrid coefficients concerning the energy transfers between the resolved and subgrid scales. In the present work we decompose the resolved and subgrid horizontal scales in a rectangular manner in wavenumber space consistent with the spectral discretisation of the DNS. However, one could also truncate on the basis of an effective horizontal wavenumber $k_{2D} = \sqrt{\alpha\alpha^* + \beta\beta^*}$, and define the resolved scales as all wavenumber pairs with k_{2D} less than a threshold horizontal LES truncation wavenumber. Truncating the system in this form may be more amenable to representing the eddy viscosities by a set of broadly applicable scaling laws, as has been previously undertaken for atmospheric [31] and oceanic flows [32]. This would then remove the need to generate the eddy viscosities from a reference DNS. Additionally one could identify the effect of removing specific vortex structures defined by a certain wavenumber set – for example removing spectral components representative of streamwise oriented streaks, or any other type of flow structure for that matter. The work presented here paves the way for many research opportunities in a vast array of configurations, not only channel flows, and also not only fluid mechanics. In fact the approach is applicable to any classical nonlinear multi-scale system [33, 24].

Acknowledgments
The authors would like to acknowledge the co-funding of this research by the European Research Council via the 2013 Multiflow summer program, the Australian Research Council, and the Commonwealth Scientific and Industrial Research Organisation. The authors would also like to thank Dr Ricardo García-Mayoral for his feedback during the development of the manuscript.

References

[1] Smagorinsky J 1963 General circulation experiments with the primitive equations: I. The basic experiment *Mon. Wea. Rev.* **91** 99–164
[2] Germano M, Piomelli U, Moin P and Cabot WH 1991 A dynamic subgrid-scale eddy viscosity model *Phys. Fluids A* **3** 1760–1765
[3] Lilly DK 1992 A proposed modification of the Germano subgrid-scale closure method *Phys. Fluids A* **4** 633–635
[4] Misra A and Pullin DI 1997 A vortex-based subgrid stress model for large-eddy simulation *Phys. Fluids* **9** 2443–2454
[5] Chung D and Pullin DI 2009 Large-eddy simulation and wall modelling of turbulent channel flow *J. Fluid Mech.* **631** 281–309
[6] McComb WD, Hunter A and Johnson C 2001 Conditional mode elimination and the subgrid-modelling problem for isotropic turbulence *Phys. Fluids* **13** 2030–2044
[7] Kraichnan RH 1959 The structure of isotropic turbulence at very high Reynolds numbers *J. Fluid Mech.* **5** 497–543
[8] Herring JR 1965 Self consistent field approach to turbulence theory *Phys. Fluids* **8** 2219–2225
[9] McComb WD 1974 A local energy-transfer theory of isotropic turbulence *J. Phys. A: Math. Gen.* **7** 632–649
[10] McComb WD 1990 *The Physics of Fluid Turbulence* (Clarendon Press, Oxford)
[11] Carnevale GF and Frederiksen JS 1983 Viscosity renormalization based on direct-interaction closure *J. Fluid. Mech.* **131** 289–303
[12] Frederiksen JS, Davies AG and Bell RC 1994 Closure theories with non-Gaussian restarts for truncated two-dimensional turbulence *Phys. Fluids* **6** 3153–3163
[13] Kiyani K and McComb WD 2004 Time-ordered fluctuation-dissipation relation for incompressible isotropic turbulence *Phys. Rev. E* **70** 066303
[14] Kraichnan RH 1976 Eddy viscosity in two and three dimensions *J. Atmos. Sci.* **33** 1521–1536
[15] Leslie DC and Quarini GL 1979 The application of turbulence theory to the formulation of subgrid modelling procedures *J. Fluid Mech.* **91** 65–91
[16] Chollet JP and Lesieur M 1981 Parameterization of small scales of three-dimensional isotropic turbulence utilizing spectral closures *J. Atmos. Sci.* **38** 2747–2757
[17] Leith CE 1990 Stochastic backscatter in a subgrid-scale model: Plane shear mixing layer *Phys. Fluids* **2** 297–299
[18] Chasnov JR 1991 Simulation of the Kolmogorov inertial subrange using an improved subgrid model *Phys. Fluids A* **3** 188–200
[19] Domaradzki JA, Liu W and Brachet ME 1993 An analysis of subgrid scale interactions in numerically simulated isotropic turbulence *Phys. Fluids A* **5** 1747–1759
[20] McComb WD and Johnson C 2001 Conditional mode elimination and scale-invariant dissipation in isotropic turbulence *Physica A* **292** 346–382
[21] Frederiksen JS and Davies AG 1997 Eddy viscosity and stochastic backscatter parameterizations on the sphere for atmospheric circulation models *J. Atmos. Sci* **54** 2475–2492
[22] Frederiksen JS 1999 Subgrid-scale parameterizations of eddy-topographic force, eddy viscosity and stochastic backscatter for flow over topography *J. Atmos. Sci* **56** 1481–1493
[23] O'Kane TJ and Frederiksen JS 2008 Statistical dynamical subgrid-scale parameterizations for geophysical flows *Phys. Scr.* **T132** 014033 (11pp)
[24] Frederiksen JS 2012 Self-energy closure for inhomogeneous turbulence and subgrid modeling *Entropy* **14** 769–799
[25] Schwinger J 1948 Quantum electrodynamics. I: A covariant formulation *Phys. Rev.* **74** 1439–1461
[26] Berera A, Salewski M and McComb WD 2013 Eulerian field-theoretic closure formalisms for fluid turbulence *Phys. Rev. E* **87** 013007

[27] Frederiksen JS and Kepert SM 2006 Dynamical subgrid-scale parameterizations from direct numerical simulations *J. Atmos. Sci.* **63** 3006–3019
[28] Zidikheri MJ and Frederiksen JS 2009 Stochastic subgrid parameterizations for simulations of atmospheric baroclinic flows *J. Atmos. Sci.* **66** 2844–2858
[29] Zidikheri MJ and Frederiksen JS 2010 Stochastic modelling of unresolved eddy fluxes *Geophys. Astrophys. Fluid Dyn.* **104** 323–348
[30] Zidikheri MJ and Frederiksen JS 2010 Stochastic subgrid-scale modelling for non-equilibrium geophysical flows *Phil. Trans. Royal Soc. A* **368** 145–160
[31] Kitsios V, Frederiksen JS and Zidikheri MJ 2012 Subgrid model with scaling laws for atmospheric simulations *J. Atmos. Sci.* **69** 1427–1445
[32] Kitsios V, Frederiksen JS and Zidikheri MJ 2013 Scaling laws for parameterisations of subgrid eddy-eddy interactions in simulations of oceanic circulations *Ocean Modelling* **68** 88–105
[33] Frederiksen JS 2012 Statistical dynamical closures and subgrid modeling for inhomogeneous QG and 3D turbulence *Entropy* **14** 32–57
[34] Kim J, Moin P and Moser R 1987 Turbulence statistics in fully developed channel flow at low Reynolds number *J. Fluid Mech.* **177** 133–166
[35] Del Álamo JC and Jiménez J 2003 Spectra of the very large anisotropic scales in turbulent channels *Phys. Fluids* **15** L41–L44
[36] Kitsios V, Frederiksen JS and Zidikheri MJ 2013 Subgrid parameterisation of the eddy-meanfield interactions in a baroclinic quasi-geostrophic ocean *ANZIAM J.* **54** C394–C410
[37] Townsend AA 1961 Equilibrium layers and wall turbulence *J. Fluid. Mech.* **11** 97–120
[38] Perry AE and Chong MS 1982 On the mechanism of wall turbulence *J. Fluid. Mech.* **119** 173–217
[39] Lozano-Durán A and Jiménez J 2011 Time-resolved evolution of the wall-bounded vorticity cascade *J. Phys.: Conf. Series* **318** 062016
[40] Lozano-Durán A, Flores O and Jiménez J 2012 The three-dimensional structure of momentum transfer in turbulent channels *J. Fluid. Mech.* **694** 100–130
[41] Cimarelli A and de Angelis E 2012 Anisotropic dynamics and sub-grid energy transfer in wall-turbulence *Phys. Fluids* **24** 015102
[42] Cimarelli A, de Angelis E and Casciola CM 2013 Paths of energy in turbulent channel flows *J. Fluid. Mech.* **715** 436–451
[43] Piomelli U and Balaras E 2002 Wall-layer models for large eddy simulation *Annu. Rev. Fluid Mech.* **34** 349–74
[44] Marusic I, Mathis R and Hutchins N 2010 Predictive model for wall-bounded turbulent flow *Science* **329** 193–196
[45] Podvin B and Fraigneau Y 2011 Synthetic wall boundary conditions for the direct numerical simulation of wall-bounded turbulence *J. Turb.* **12** 1–26
[46] García-Mayoral R, Pierce B and Wallace J 2012 Off-wall boundary conditions for turbulent simulations from minimal flow units in transitional boundary layers *CTR Summer Prog.* (Stanford Univ.) pp 87–96
[47] Mizuno Y and Jiménez J 2013 Wall turbulence without walls *J. Fluid. Mech.* **723** 429–455

Journal of Physics: Conference Series **506** (2014) 012002 doi:10.1088/1742-6596/506/1/012002

Scaling of velocity fluctuations in off-wall boundary conditions for turbulent flows

M P Encinar[1], **R García-Mayoral**[1,2,3] **and J Jiménez**[1]

[1]School of Aeronautics, Universidad Politécnica de Madrid, 28040, Spain

[2]Center for Turbulence Research, Stanford University, 94305 CA, USA

[3]Department of Engineering, University of Cambridge, CB2 1PZ, UK

E-mail: r.gmayoral@eng.cam.ac.uk

Abstract.
A model for off-wall boundary conditions for turbulent flow is investigated. The objective of such a model is to circumvent the need to resolve the buffer layer near the wall, by providing conditions in the logarithmic layer for the overlying flow. The model is based on the self-similarity of the flow at different heights in the logarithmic layer. It was first proposed by Mizuno and Jiménez (2013), imposing at the boundary plane a velocity field obtained on-the-fly from an overlying region. The key feature of the model was that the lengthscales of the field were rescaled to account for the self-similarity law. The model was successful at sustaining a turbulent logarithmic layer, but resulted in some disagreements in the flow statistics, compared to fully-resolved flows. These disagreements needed to be addressed for the model to be of practical application. In the present paper, a more refined, wavelength-dependent rescaling law is proposed, based on the wavelength-dependent dynamics in fully-resolved flows. Results for channel flow show that the new model eliminates the large artificial pressure fluctuations found in the previous one, and a better agreement is obtained in the bulk properties, the flow fluctuations, and their spectral distribution across the whole domain.

1. Introduction

Wall-bounded turbulent flows are ubiquitous both in nature and in industry, and their simulation is often necessary to predict their behaviour and/or performance. As the Reynolds number of the flow increases, so does the range of scales at play, and the resolution of all of them becomes inaccessible with the computing power available at present – as well as that predicted in the next few decades. To overcome this difficulty, large-eddy simulation (LES) solves only the largest scales, modelling the effect on them of the smaller ones. This approach can be applied with satisfactory results, provided some conditions are met. One is that the largest and smallest scales must be separated by a broad range of scales, called the 'inertial' range. Another is that the smallest, dissipative scales must be reasonably isotropic, so that some degree of universality across different flows can be assumed for them. Both assumptions fail in turbulence near walls. The distance y to the wall imposes a constraint on all the turbulent eddies, which could in principle be of any length in the wall-parallel plane, but whose wall-normal extent must vanish as it approaches the wall, and cannot exceed $O(y)$. This constraint not only imposes anisotropy on the eddies, but also a y-dependent distribution of sizes. Figure 1, taken from Jiménez (2012), shows that while the size of the smaller, dissipative eddies varies slowly with y, the size of large,

Journal of Physics: Conference Series **506** (2014) 012002 doi:10.1088/1742-6596/506/1/012002

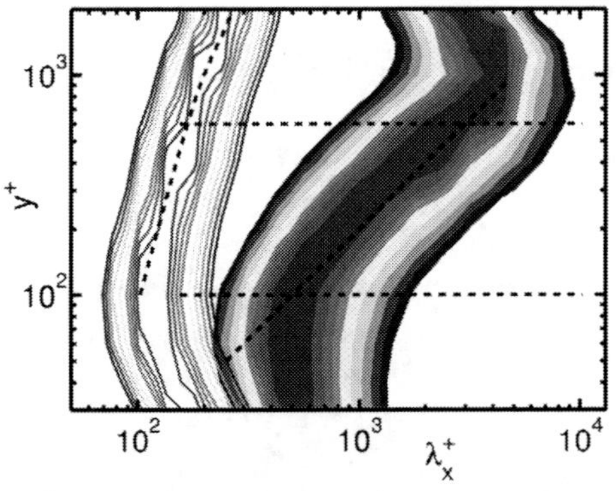

Figure 1. Spectral densities in a channel at $\mathrm{Re}_\tau \approx 2000$, as functions of the streamwise wavelength λ_x^+ and of the wall distance y^+ in viscous units, from Jiménez (2012). The shaded contours correspond to the premultiplied spectrum of the velocity fluctuations, $k_x E_{uu}(k_x)$, and the line contours to the surrogate dissipation, $\nu k_x E_{\omega\omega}(k_x)$, where ω is the vorticity magnitude. The lowest contour is 0.86 times the local maximum at each y. The horizontal lines, $y^+ = 100$ and $y/\delta = 0.4$, are the approximate limits in which the energy length scale grows linearly with y. The diagonal line through $k_x E_{uu}(k_x)$ is $\lambda_x = 5y$, and the one through $\nu k_x E_{\omega\omega}(k_x)$ is $\lambda_x = 40\eta$.

energy-carrying eddies varies linearly as they separate from the wall. As a consequence, the range of separation between the two increases with y. Close to the wall, they essentially overlap, in the so-called buffer region, which on these grounds can be defined as $y^+ \lesssim 100$. If we define the logarithmic layer as that in which the larger eddies scale linearly with y, it would span from the upper limit of the buffer layer to $y \approx 0.4\delta$, where δ is the flow thickness (the half-height in the case of channels). Far enough from the wall, e.g. $y \approx 0.1\delta$, and as long as $0.1y^+ \gtrsim 100$, the large and small scales are sufficiently separated, and the distance to the wall is large enough for the small ones to be reasonably isotropic. It is at this height where the assumptions for LES begin to hold.

Below that height, as the larger scales become increasingly small, the resolution required to resolve them becomes comparable to that for the direct numerical simulation (DNS) of all the scales. Resolving all of them is the most straightforward approach to LES of wall-bounded flows, and is called wall-resolved LES (Piomelli, 1999). However, the computational cost of such approach makes it impractical (Piomelli and Balaras, 2002; Choi and Moin, 2012), and models for the near-wall flow have to be used instead. The flow in the wall layer is often represented only in the average sense, using the Reynolds-averaged Navier–Stokes (RANS) equations, or approximated by the thin-turbulent-boundary-layer equations. Many of these wall models attempt, in various ways, to relate the wall stress, which LES cannot obtain accurately because of its insufficient grid resolution, to the outer flow, in order to obtain boundary conditions for the computation. This subclass of wall models are called wall stress models (Cabot and Moin, 1999).

An alternative also described by Cabot and Moin (1999) is to use off-wall boundary conditions. In this case, the outer flow is computed through LES, with a grid providing sufficient resolution only down to a chosen distance from the wall, where approximate boundary conditions are provided. In their review, Cabot and Moin (1999) cite attempts of this type by Bagwell et al. (1993), Balaras et al. (1996) and Nicoud et al. (1998), which they characterize as being largely unsuccessful, due to a lack of accurate representation of the relative phases and time scales of the flow at the artificial boundary. They also cite Jiménez and Vasco (1998), who observed that the wall layer flow is quite sensitive to the transpiration of the vertical velocity across the off-wall plane in order to maintain continuity.

A recent attempt to develop off-wall models is that of Chung and Pullin (2009), in an LES of a turbulent channel flow up to very high Reynolds numbers. They determine the slip velocity at an off-wall plane in the logarithmic region, with the Kármán constant calculated dynamically. This is done by relating the slip velocity to the shear stress at each location on the wall which, in turn, is calculated from an ODE obtained by wall-parallel filtering and wall-normal averaging of the streamwise momentum equation. The model is used to calculate a logarithmic relation at the off-wall location, and with it the slip velocity. The original work has been further extended by Inoue and Pullin (2011) and Inoue et al. (2012).

Pascarelli et al. (2000) addressed the need for greater resolution in the wall layer by using a multi-block LES. The outer flow was computed with a lower-resolution grid and the near-wall flow was computed with a higher-resolution one, in a domain periodic in the wall-parallel directions, but smaller than the outer-flow domain. This near-wall block was repeated in a lattice to cover the full streamwise and spanwise lengths of the overlying, low-resolution domain. The length and width of the wall-layer blocks was somewhat larger than those determined to be necessary to sustain turbulence (Jiménez and Moin, 1991). Although the flow at the interface had an artificial wavelength set by the inner-flow grid, the authors found that longer wavelengths occurred within a few grid points from the interface. At much higher Reynolds numbers, where the lengthscale separation within the inner and outer flows becomes much larger, many repeated wall-layer blocks would need to be used in this periodically repeated fashion. When the wall layer block had its upper surface at $y^+ \approx 100$, reaching into the logarithmic layer, first- and second-order statistics showed good agreement with single-block calculations. However, when the top of their near-wall blocks was placed at $y^+ \approx 30$, within the buffer layer, the Reynolds stresses were underpredicted, and spurious pressure fluctuations occurred.

The idea of simulating the overlying flow separately from the buffer layer, suggested by the work of Pascarelli et al. (2000) and some of the investigations cited above, can be pushed further by removing the buffer layer completely and replacing it by a dynamic boundary condition at its interface with the logarithmic layer, or well within the logarithmic layer itself. This approach has been tested in three recent studies. Podvin and Fraigneau (2011) generated synthetic boundary conditions from proper-orthogonal-decomposition eigenfunctions, which needed to be obtained a priori. Mizuno and Jiménez (2013) constructed boundary conditions dynamically from information in the overlying flow, by capitalizing on the self-similarity of the turbulent fluctuations across the logarithmic layer, and the independence, to a large extent, of this layer from the dynamics beneath it. García-Mayoral et al. (2012) used the idea of 'minimal' units of buffer-layer dynamics (Jiménez and Moin, 1991), and the universality of these dynamics across different flows, to construct a boundary condition at $y^+ \approx 100$, which could be imposed in a lattice in the spirit of Pascarelli et al. (2000). They synthesized conditions from different flows, including early stages of transition in boundary layers and channel flow, and tested them successfully on turbulent channels.

Compared to Mizuno and Jiménez (2013), the strategy of implementing a boundary condition just above the buffer layer, as proposed by García-Mayoral et al. (2012), has the advantage of not requiring a logarithmic layer to draw information from, so it can be used at low Reynolds numbers, when the logarithmic region is relatively thin. Conversely, it is ill-suited for large Reynolds numbers, in which resolving the flow at $y^+ \approx 100$, or any other $y/\delta \ll 1$, would be very expensive. In that case, the approach of Mizuno and Jiménez (2013), which can set the boundary condition at a fixed y/δ within the logarithmic layer, say $y/\delta = 0.1$, would be much more efficient. In that sense, both approaches can be viewed as complementary. For the full simulation of a boundary layer, the strategy of García-Mayoral et al. (2012) could be used in the transitional and early-turbulence regions, while that of Mizuno and Jiménez (2013) could be implemented further downstream.

The present work takes up from that of Mizuno and Jiménez (2013), and attempts to address

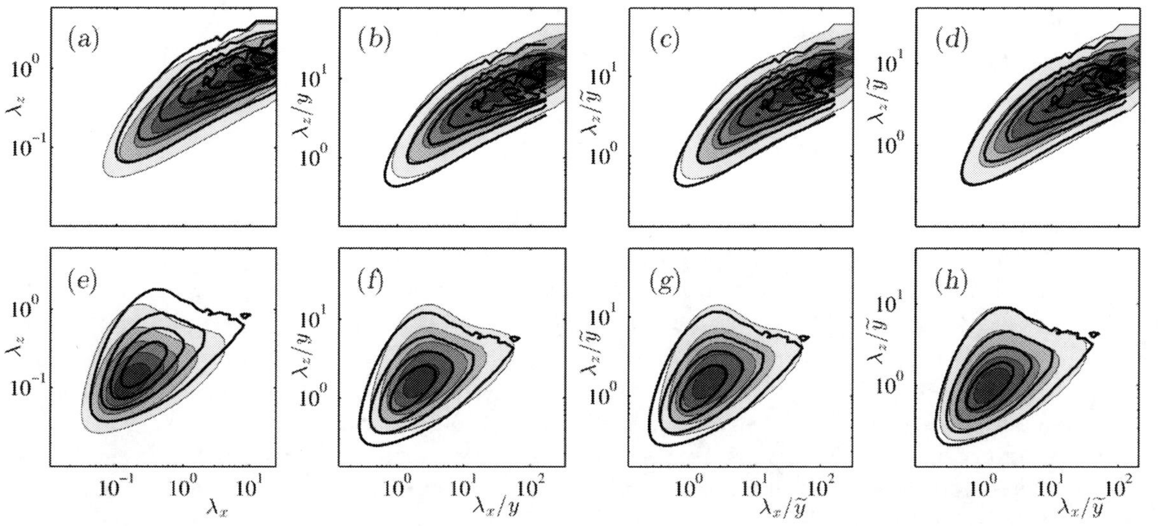

Figure 2. Comparison of the spectral energy densities of the streamwise velocity $k_x k_z E_{uu}$ (top) and the spanwise one $k_x k_z E_{vv}$ (bottom), at $y^+ \approx 150$ and $y^+ \approx 300$, from the channel at $\mathrm{Re}_\tau \approx 2000$ of Hoyas and Jiménez (2006). The shaded contours represent the spectra at $y^+ \approx 150$, and the solid ones the spectra at $y^+ \approx 300$. In (*a*) and (*e*), the stream- and span-wise wavelengths, λ_x and λ_z, have been normalised with the channel half-height δ. In (*b*) and (*f*), they are normalised with the distance to the wall, y. In (*c*) and (*g*), the wavelengths are normalised with the distance $\widetilde{y} = y - y_0$ to a plane $y_0^+ = -12$, following Mizuno and Jiménez (2011), and in (*d*) and (*h*) using $y_0^+ = -100$, as suggested by García-Mayoral et al. (2012).

some of the caveats encountered in that first approach. Mizuno and Jiménez (2013) based their off-wall model in the fact that the length-scales of the fluctuations in the logarithmic layer are proportional to the distance to the wall (Millikan, 1938; Townsend, 1961; Jiménez, 2012). They constructed a boundary condition by rescaling the lengthscales of the velocity fluctuations at a reference plane, and imposed the resulting velocities in a plane closer to the wall, with the ratio of heights of both planes being the rescaling factor. The condition was shifted in the streamwise direction with respect to the reference plane, to enforce a mean advection equal to the mean velocity at the boundary plane. The resulting simulations successfully proved that a logarithmic layer can exist independently of a supporting buffer layer, or even a physical wall, so long as the self-similarity of the lengthscales of the fluctuations is enforced. However, a few shortcomings needed to be addressed for such a method to be accurate enough for general LES applications.

From a physical perspective, the brute-force rescaling of all the flow structures by the same factor is crude. In full wall-bounded flows, the smaller eddies scale linearly with the height, but the larger ones extend all the way to the wall with their stream- and spanwise scales unmodified, except in the the viscous sublayer, $y^+ \leq 5$ where they are finally dissipated before reaching the wall (Jiménez, 2012). This is particularly evident for the structures formed by the streamwise u-velocity fluctuations, which can have very large lengthscales (Hutchins and Marusic, 2007), as portrayed in Figure 2. The figure shows statistically how the smaller eddies scale linearly with the distance to the wall, while the larger scales remain roughly the same at different heights. A further refinement was suggested in García-Mayoral et al. (2012), who analysed the spectral densities of the fluctuations observed in DNSs, and reported that their linear scaling with y does not have its origin exactly at the wall, but roughly at $y_0^+ = -100$ below it, independently of the Reynolds number of the flow. That scaling is also shown in Figure 2. In contrast, Mizuno and

Jiménez (2011) analysed the scaling with y of the mixing length, defined as

$$\ell(y) = \left(\frac{1}{u_\tau}\frac{\mathrm{d}U}{\mathrm{d}y}\right)^{-1}, \tag{1}$$

and found that it also has an origin y_0^+ below the wall, although it is Re_τ-dependent, tending to vanish with increasing Re_τ, and already of order ~ -10 for $\mathrm{Re}_\tau \approx 2000$. When a similar y_0^+ was calculated a posteriori for the wall-less channels of Mizuno and Jiménez (2013), values of $y_0^+ \approx +30$ for $\mathrm{Re}_\tau \approx 850$ and $\approx +80$ for $\mathrm{Re}_\tau \approx 2600$ were obtained. Mizuno and Jiménez used those values to rescale their results with $\widetilde{y} = y - y_0$ and $\widetilde{u}_\tau = u_\tau(\widetilde{\delta}/\delta)^{1/2}$, obtaining a better agreement with the full channel data. The scaling using the virtual origin obtained by Mizuno and Jiménez for the channel of Hoyas and Jiménez (2006) is also shown in Figure 2.

Another important problem in Mizuno and Jiménez (2013) was the appearance of very large pressure fluctuations near the boundary plane. The authors traced this problem to the vanishing correlation between the flow field imposed at the boundary and that immediately above. The large pressure gradients would arise in order to enforce the incompressibility of the flow. Finally, the resulting Reynolds stresses $-uv$ were somewhat below those of the full channel, and for the flow to be in equilibrium the defect needed to be compensated by a larger mean shear stress. As a result, the mean profile had a higher slope than for a full DNS, and showed a slightly rounder shape.

In this paper we present some modifications to the method of Mizuno and Jiménez (2013), aimed at improving both its accuracy and dynamical resemblance to fully-resolved wall-turbulence. In Section 2, we briefly outline the method of Mizuno and Jiménez (2013), and describe the modifications we have introduced. In Section 3, we present and discuss the results of our simulations. Conclusions and directions for future work are summarised in the final Section 4.

2. Numerical method

We conduct our simulations using the numerics of Mizuno and Jiménez (2013), modifying only the boundary conditions. The code solves the time-dependent, incompressible flow in a doubly-periodic channel of height 2δ and streamwise and spanwise dimensions L_x and L_z. We denote the streamwise, wall-normal and spanwise coordinates by x, y, and z, and the corresponding velocities by u, v, and w. The density is assumed to be $\rho = 1$, and the viscosity is set to $\nu = 1/86000$ to obtain the target friction Reynolds number, $\mathrm{Re}_\tau = u_\tau\delta/\nu \approx 2000$. Wall units, scaled with ν and the friction velocity u_τ, are used throughout the paper, and indicated by the conventional '+' superscript. The code is pseudo-spectral in the two wall-parallel directions, with 2/3 dealiasing, and uses fourth-order compact finite differences on a non-uniform, staggered grid along y (Lele, 1992). The integration in time uses a fractional-step, third-order RungeKutta (Spalart et al., 1991). The simulation is run at constant, zero flow rate, with the velocity scale set by the difference between the mean u at the centre plane and the mean u at the boundary planes, $\Delta U = 0.3$. We have applied these method on a simulation with DNS-like resolution (DNS for brevity from here on, even if the boundary conditions are artificial, and in no way 'direct'), and on a coarse-mesh simulation with the same LES treatment of Mizuno and Jiménez (2013). The parameters of the simulations are summarised in Table 1, together with those of the original DNS-like simulation of Mizuno and Jiménez (2013) and the DNS of Hoyas and Jiménez (2006), used for comparison.

In Mizuno and Jiménez (2013), the boundary conditions for the velocities at $y_{bc}^+ \approx 100$ were constructed from the velocity fields at a reference plane $y_{ref}^+ \approx 200$. The fields were 'shrunk' by a lengthscale ratio $r = y_{ref}/y_{bc}$, maintaining the original amplitudes, and shifting the result in the streamwise direction to account for the different mean advection velocities at both planes,

Table 1. Simulations parameters. L_x, L_z, and δ are the channel length, width, and half-height, and N_x, N_z, and N_y, the number of collocation points in the three respective directions directions. Re_τ is the resulting friction Reynolds number. C2000 is the DNS of Hoyas and Jiménez (2006), WL2000 that of Mizuno and Jiménez (2013), and DNS and LES refer to the present simulations.

Case	L_x/δ	L_z/δ	N_x	N_z	N_y	Re_τ
C2000	8π	3π	6144	4608	633	2003
WL2600	4π	2π	3072	3072	569	2570
DNS	4π	2π	3072	3072	569	1980
LES	4π	2π	384	384	307	2040

which were assumed to be the local means. This procedure is relatively simple in Fourier space, so long as $1/r$ is an integer, and is thoroughly described in Mizuno and Jiménez (2013). Here we have refined the procedure to include the effect of an origin $y_0 = -100$ for the self-similarity law, and to restrict the 'shrinking' process to all but the largest scales. For the former refinement, we have chosen $y_{bc}^+ \approx 100$ and $y_{ref}^+ \approx 300$, so that $r = (y_{ref} - y_0)/(y_{bc} - y_0) = 1/2$. For the latter we have applied a wavelength-selective rescaling. Take for instance the streamwise velocity field at y_{ref}^+ in the Fourier plane, $\widehat{u}_{ref}(k_x, k_z)$, where k_x and k_z are the wavenumbers associated to the wavelengths λ_x and λ_z. In Mizuno and Jiménez (2013), a rescaled field was obtained through a rescaling function $\Re$, so that $\widehat{u}_{bc} = \Re(\widehat{u}_{ref})$. Here, we obtain the same field, but multiply it by a filter that leaves out the resulting large scales, and replace them by the original ones from $\widehat{u}_{ref}$. The filter is constructed from two one-dimensional ones, $\Phi = \Phi_x(\lambda_x) \times \Phi_z(\lambda_z)$, where Φ_x is zero for $\lambda_x > 3h$ and unity for $\lambda_x < 17.5 y_{ref}$, and Φ_z is zero for $\lambda_z > 1.2h$ and unity for $\lambda_z < 7 y_{ref}$, with smooth fifth-order-spline transitions in-between. The inverse filter $(1 - \Phi)$ is applied to the unrescaled $\widehat{u}_{ref}$, to obtain the large scales to be added. In summary, we have

$$\widehat{u}_{bc} = \Phi \times \Re(\widehat{u}_{ref}) + (1 - \Phi) \times \widehat{u}_{ref}. \tag{2}$$

The thresholds given above are selected so that the 'local' scales at y_{ref} are not filtered by Φ, while those larger –which would be 'local' when viewed at some plane higher than y_{ref}, are filtered out. According to Flores and Jiménez (2010), 'healthy' turbulence can be obtained at a given height y when the largest wavelengths considered are no less than $\lambda_x \approx 5y$, $\lambda_z \approx 2y$, giving a suitable estimate for the size of the 'local' turbulence at each height y. The criterion that we have imposed is that anything larger than ≈ 6 times the 'local' minimum unit at $y = y_{ref}$ is filtered out completely, and anything smaller than ≈ 3.5 times that minimum unit is left completely unfiltered.

The filtering process is illustrated in Figure 3. Note that according to Eq. (2), Φ is applied on the rescaled field and $(1 - \Phi)$ on the unrescaled one, so that some intermediate wavelengths are included in both contributions, and they are in a sense added twice. This is done intentionally to energize the intermediate scales. If both filters were applied on the original, unrescaled signal, a zero-energy gap would appear in the spectral region separating the rescaled and un-rescaled wavelengths. In real wall-turbulence, that gap is filled by a continuous cascade of scales, which are 'local' to the planes between y_{ref} and y_{bc}. Since information on those scales is missing at y_{ref}, it is actually convenient to artificially synthesize a signal to fill that spectral gap. Note also that the streamwise shifting, which accounts for the different advection velocities at y_{ref} and y_{bc}, is applied through the operator $\Re$, so it does not affect the large scales. This is again intentional, since in real flows the small scales are advected with the local mean velocity, but the large ones travel with a velocity roughly uniform along their wall-normal span, and approximately equal to the bulk velocity (del Álamo and Jiménez, 2009). By making copies of the larger structures

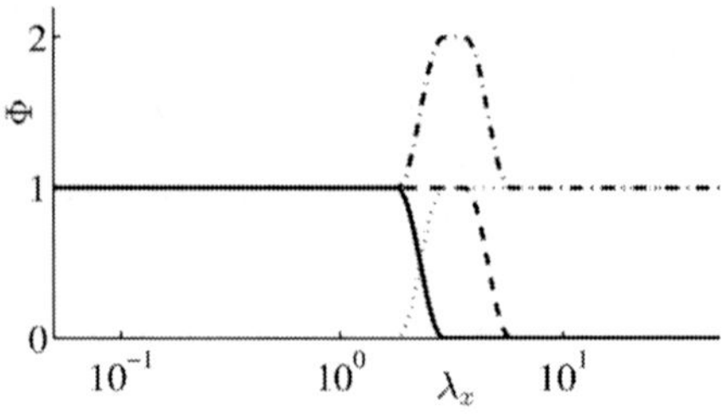

Figure 3. Sketch of the filtering method in the streamwise direction. ——, filter Φ_x applied on the rescaled signal, $\Re(\widehat{u}_{ref})$; ----, equivalent filter $\widehat{\Phi_x}$ for the un-rescaled u_{ref}, $\Re(\widehat{\Phi_x} \times \widehat{u}_{ref}) = \Phi_x \times \Re(\widehat{u}_{ref})$; ·······, filter $1 - \Phi_x$ applied on the unrescaled signal, to extract the large scales; – · –, amplification that would be obtained on each mode of a white-noise signal by applying eq. (2).

exactly below them, we force them to be advected at the same velocity in both planes.

The direct copy of the large scales works well for the wall-parallel velocities, but fails for v. Simulations which tested an implementation of eq. (2) for the three velocities soon developed large pressures near the boundaries and became unstable. However, the large scales of real flows have little v, since they reach the wall and cannot pass beyond it. That impermeability is transmitted through continuity to the regions farther from the wall, where the large scales therefore experience only very weak wall-normal motions. This suggested that we could simply remove all the large scales from v, so that

$$\widehat{v}_{bc} = \Phi \times \Re(\widehat{v}_{ref}). \tag{3}$$

The condition on large scales derived from continuity is actually more subtle, and this direct approach has negative implications, particularly on the Reynolds stress, which will be discussed in the following section. A finer, more precise implementation has been left for future work.

3. Results

In this section we present and discuss the results of our simulations. Before that, a word of caution must be said on the DNS case. This simulation is very demanding in computational resources, both in terms of machine allocation and actual running time. Because of that, we have been unable to run the case for a time longer than $u_\tau t/\delta \approx 1$. That implies that the results are not statistically converged, and should be taken with caution. On the other hand, the LES results are fully converged, but lack information on how the boundary conditions affect the smaller scales, which are in this case modelled instead of resolved. Fortunately, the characteristic time for those scales is of order y/u_τ, so the DNS results are somewhat more significant, when restricted to the smaller scales.

The refinements to the boundary condition of Mizuno and Jiménez (2013) proposed here produce significant improvements on the statistics of the resulting simulations. Bulk and fluctuation results are summarised in Figure 4. The most dramatic improvement is the substantial reduction of the spike in the pressure fluctuation p'^+ as the boundary is approached, which is now essentially confined to the first plane. Besides that, the new simulations show slightly larger kinks in the velocity fluctuations near the boundary, especially for u. These kinks are ubiquitous in off-wall models (Mizuno and Jiménez, 2013; Podvin and Fraigneau, 2011; García-Mayoral et al., 2012). In contrast, the new results are more accurate away from the wall, and match almost exactly the real fluctuations for $y \gtrsim 0.2\delta$. It was argued in Mizuno and Jiménez (2013) that the effect of the large pressures near the boundary acted on relatively large lengthscales, and its influence reached well into the channel. Now that those have been

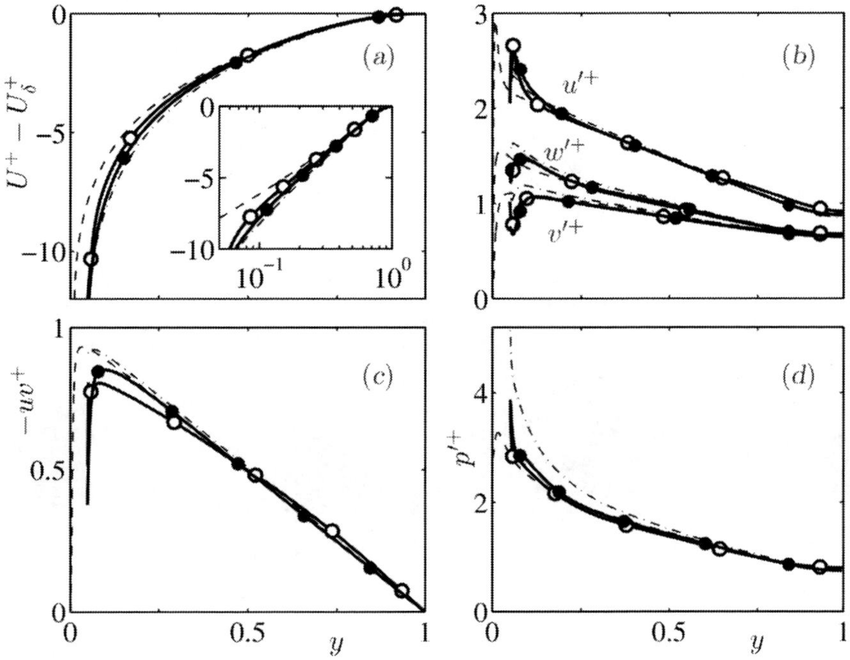

Figure 4. Flow statistics from the present channel DNS and LES, compared to those of Mizuno and Jiménez (2013) and Hoyas and Jiménez (2006). (*a*) mean velocity profile in defect form; (*b*) rms velocity fluctuations; (*c*) Reynolds shear stress; (*d*) pressure fluctuations. -○-, present DNS; -●-, present LES; –·– DNS of Mizuno and Jiménez (2013); ----, DNS of Hoyas and Jiménez (2006).

eliminated, the perturbations induced by the boundary conditions vanish soon away from their plane.

The better agreement away from the boundary also translates in a better behaviour of the mean profile. To obtain an accurate profile, it is essential that the Reynolds stresses are well represented across the whole channel. A thorough discussion on the delicate equilibrium between the local turbulence and the shear of the mean profile can be found in Tuerke and Jiménez (2013). In essence, any defect in $-uv$ needs to be compensated by a larger mean shear $\nu dU/dy$. Since most of the total stress away from the walls is Reynolds stress, small relative errors in it may lead to large relative errors in dU/dy. Those errors accumulate along the U-profile, and can lead to substantial errors in U itself. The effect is particularly grievous if it is not confined to a region near the boundary, where the simulation can always be expected to be 'wrong,' and the region discarded (Kawai and Larsson, 2012). Even if it shows an improved mean profile compared to Mizuno and Jiménez (2013), the Reynolds stress of the present DNS departs significantly from that of the full-channel DNS of Hoyas and Jiménez (2006). Furthermore, the total stress, not shown, but of which the dominant contribution away from the wall is $-uv$, also departs significantly from the linear trend, even in the core region of the channel. This implies that the mean flow is, in the time span considered, not in equilibrium, and is a clear indicator of the failure to reach statistical convergence, mentioned above. The LES, which is in contrast fully converged, exhibits a sharp dip in $-uv$ at the boundary (as does the DNS), but it nevertheless follows better the results of the full direct simulation C2000 above $y \gtrsim 0.2\delta$, and thus results in a somewhat improved U-profile for most of the channel, although the error is still comparable to that of Mizuno and Jiménez (2013).

The fluctuations provide information on the flow in the broad picture, but to analyse whether those fluctuations are correctly distributed among small and large scales, we need to inspect their

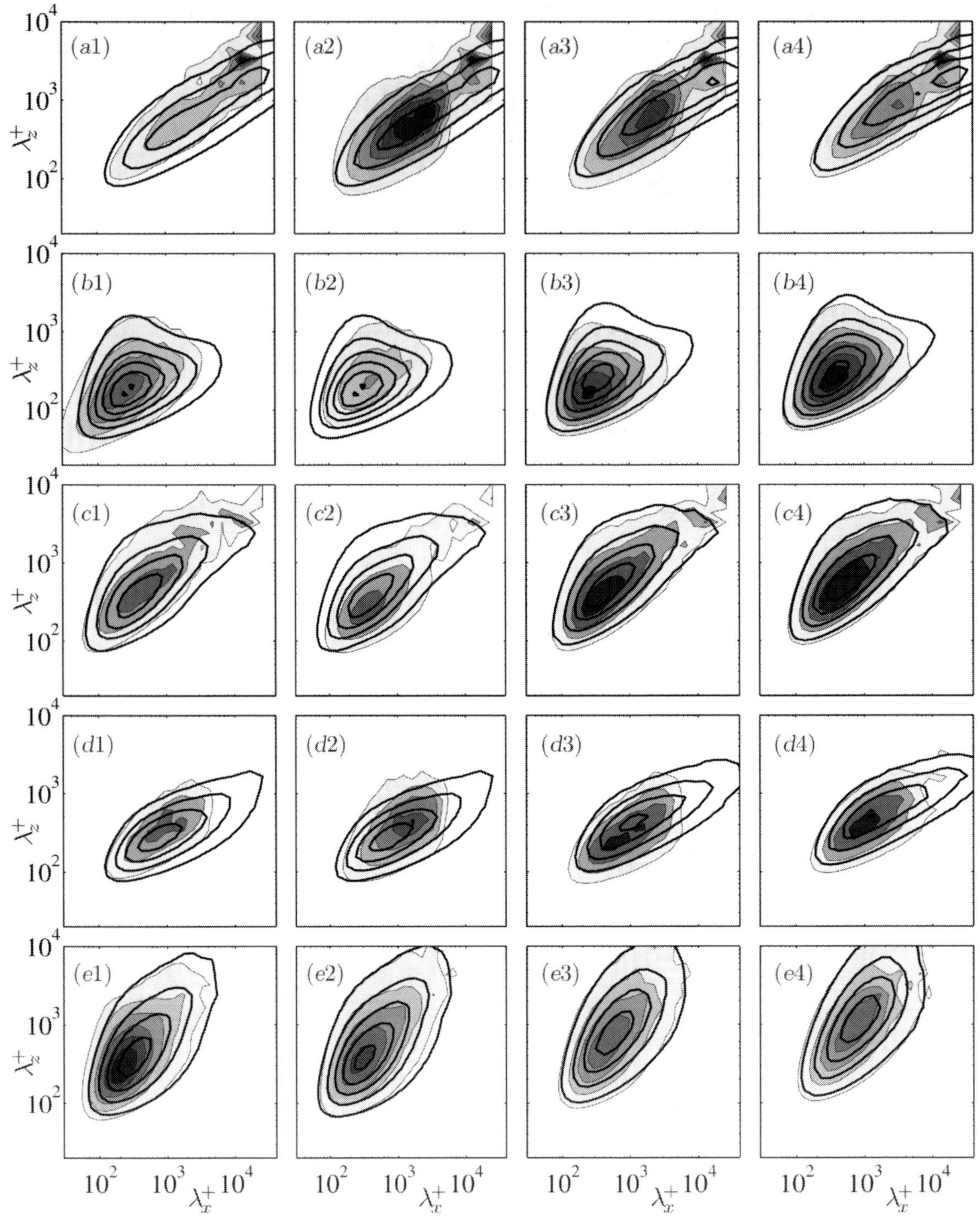

Figure 5. Two-dimensional spectral energy densities, from left to right, at $y^+ \approx 100$, 105, 150 and 200. Shaded contours represent the present DNS simulation, and solid lines the reference full-channel DNS (Hoyas and Jiménez, 2006). Premultiplied spectra of the (a) streamwise, (b) wall-normal, and (c) spanwise velocities, $k_x k_z E_{uu}$, $k_x k_z E_{vv}$, and $k_x k_z E_{ww}$, with contours every $0.025\, u'^{+2}$, $0.025\, v'^{+2}$ and $0.025\, w'^{+2}$. (d) Cospectrum of the Reynolds stress, $-k_x k_z E_{uv}$, with contours every $0.035\, \overline{uv}^+$. (e) Spectrum of the pressure fluctuations, $k_x k_z E_{pp}$, with contours every $0.025\, p'^{+2}$.

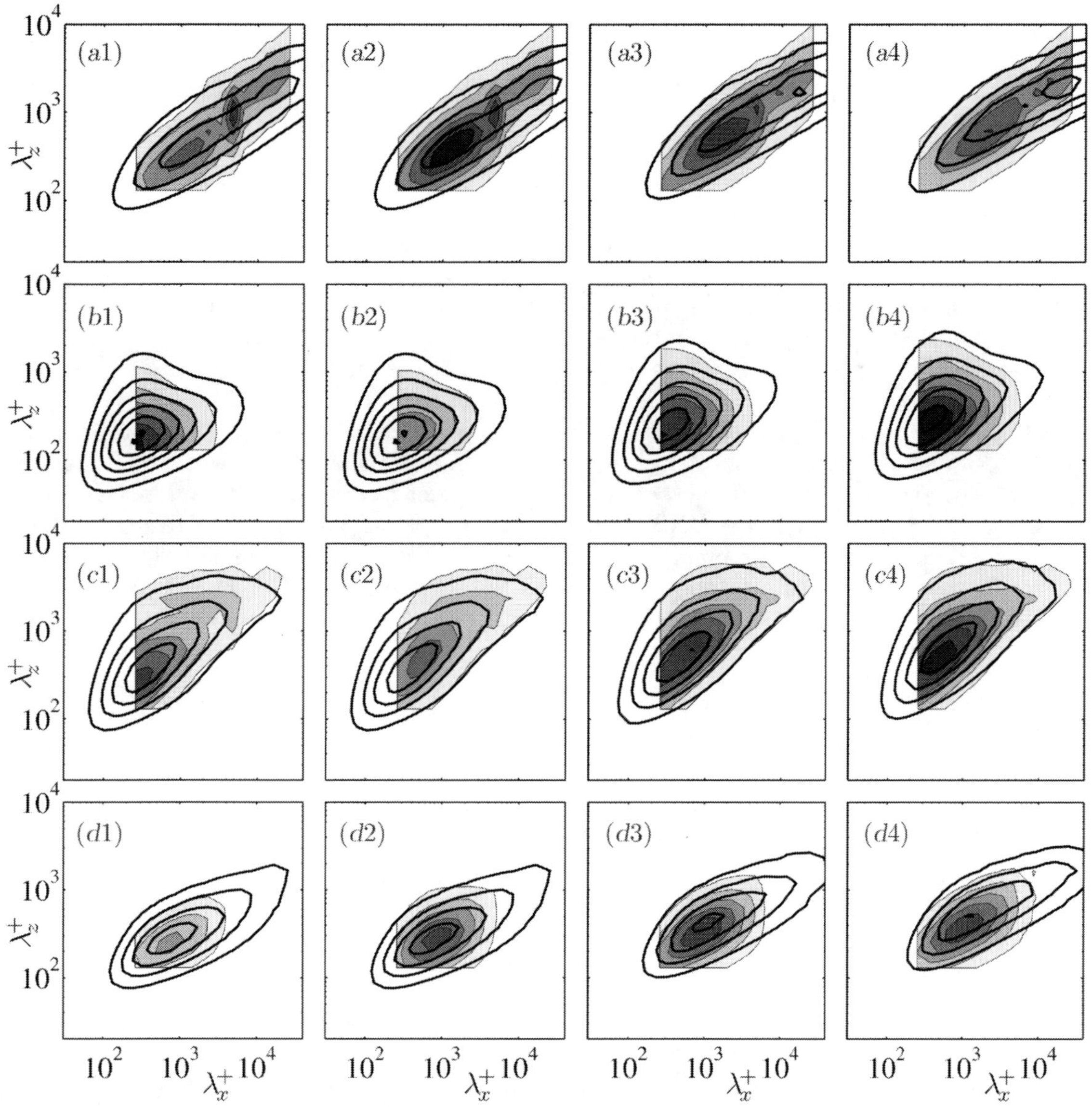

Figure 6. Two-dimensional spectral energy densities, from left to right, at $y^+ \approx 100$, 150, 200 and 300. Shaded contours represent the present LES simulation, and solid lines the reference full-channel DNS (Hoyas and Jiménez, 2006). Premultiplied spectra of the (*a*) streamwise, (*b*) wall-normal, and (*c*) spanwise velocities, $k_x k_z E_{uu}$, $k_x k_z E_{vv}$, and $k_x k_z E_{ww}$, with contours every $0.025\,u'^{+2}$, $0.025\,v'^{+2}$ and $0.025\,w'^{+2}$. (*d*) Cospectrum of the Reynolds stress, $-k_x k_z E_{uv}$, with contours every $0.035\,\overline{uv}^+$.

spectral energy distributions, both along the stream- and spanwise wavelengths, and along the wall-normal coordinate y. Energy densities are portrayed in Figure 5 for our DNS, and in Figure 6 for our LES, and compared with the true densities of Hoyas and Jiménez (2006). In Figure 5, they are shown at the boundary plane, one plane immediately above it, and 50 and 100 wall units above. At that height, $y^+ = 200$ or $y = 0.1\delta$, the flow is already beginning to relax to a state that in general terms resembles that of the full, reference simulation C2000 quite closely. The most significant differences occur below that height. Figure 6 includes results at the boundary $y^+ \approx 100$ and at $y^+ = 150$, 200 and 300, already showing that the flow resembles quite closely the full DNS at that height. Of the three velocities, the spanwise w exhibits a quite accurate behaviour through the full collection of y planes, and the energy in v is essentially distributed among the correct lengthscales, although its values are somewhat low immediately away from the boundary. That translates in the dip in v' observed in Figure 4(b) for y values just above the boundary. The removal of the large v-scales, very clear in the absence of the tail for $\lambda_x^+ \gtrsim 5000$ in Figure 5($b4$), and to a lesser extent in Figure 6($b4$), is likely responsible of the slight shift of the full v-spectrum toward smaller scales.

On the other hand, the modifications in the distribution of u are more profound. The boundary condition is distributing the energy correctly in a statistical sense, as shown in Figures 5($a1$) and 6($a1$). However, there is a sharp transition between the behaviour at the boundary and one plane immediately above, shown in Figure 5($a2$). In the second plane, the intensity of E_{uu} is unusually large, and spreads into regions to the sides of the true high-density spectral region. As a result, the distance y taken by u to relax is somewhat larger than for the other two velocities. The only possible source for the sharp change in u at the wall is large pressure fluctuations. We have already discussed the existence of a residual spike in p', compared to Mizuno and Jiménez (2013), which now exists only at the boundary plane. The spectral distribution of that peak is revealed in Figure 5($e1$) for the DNS (the computation of the pressure spectra for the LES simulation was under way while preparing this manuscript but, given the similarity in all the other results, there is no reason to believe it should be substantially different from that of the DNS). At that first plane, the pressure spectrum increases in magnitude, and is slightly shifted towards smaller wavelengths. The shape of the spectral distribution is somewhat reminiscent of that of the u spectrum in the second plane, particularly in the amplified region with $\lambda_z^+ \approx 1000$-2000 (figure 5$a2$), supporting the connection of the sudden increase in u fluctuations immediately above the boundary with the large pressure fluctuations at the boundary itself. Nevertheless, those large pressures have already vanished at the second plane, and the spectral distribution of p closely resembles that of Hoyas and Jiménez (2006) from that second panel on. This strongly suggests that, save for the boundary plane, our modifications to Mizuno and Jiménez (2013) have solved the deviations experienced by the pressure.

There is, however, a price to be paid for removing completely the large scales of v, which is made clear in the spectral distribution of the Reynolds stresses in Figures 5(d) and 6(d). While the large-scale tail of E_{vv} is negligible, its contribution, together with E_{uu}, to populating the cospectral region of Reynolds stresses with $\lambda_x^+ \gtrsim 5000$ is not. In real flows at moderate or high Reynolds numbers, that tail actually contributes a significant amount to the total $-uv$, as shown in the figures. That is because, even if the corresponding values of v are quite low, those of u can be fairly large. Unlike the rest of the variables, the spectral distribution of the Reynolds stress takes longer wall-normal distances to relax into a 'healthy'-turbulence state. In fact, the tail for the long streamwise wavelengths has not yet recovered at $y^+ \approx 300$. Solving this defect in $-uv$, which is now identified as a large-sale issue, should be the next step.

By introducing a variable rescaling between the flow velocities at $y = y_{ref}$ and $y = y_{bc}$, we have identified a potential problem in leaving the large scales of the three velocities unrescaled, which in fact led to the destabilization of the simulations. Setting the large scales of the wall-normal velocity to zero has solved that issue, but it is probably not the most accurate solution, even from a physics perspective, and it is easy to see that it should only strictly hold for infinitely long and infinitely wide structures, $k_x = k_z = 0$. More accurate procedures are currently under study.

4. Conclusions and future work

The direct simulation of turbulent flows at moderate to high Reynolds numbers is inaccessible for any existing computing facility. Alternatively, LES can successfully reproduce physical flows, but it requires a sufficiently large separation between the scales at which energy is generated and those at which it is dissipated. Unfortunately, that assumption does not hold near walls, since the production scale approaches linearly the viscous one as the distance to the wall decreases.

This forces the resolution requirements of LES to approach those of DNS near walls, so the advantage of the latter is lost, unless the near-wall layer is substituted by a model.

In the present work, we have reassessed the modelling strategy of Mizuno and Jiménez (2013), which constructs boundary conditions at an off-wall plane. This is done in real time during the simulation, by rescaling the lengthscales of the flow at some reference, overlying plane. The rescaling assumes, and imposes that the length scales of the fluctuations are proportional to the wall-normal distance. We have modified this model to introduce several concepts, based mainly on our physical understanding of the logarithmic layer. From the analysis of the spectral densities of the fluctuations in full-channel DNSs, a virtual origin can be derived for the lengthscales, which is essentially independent of the Reynolds number, and 100 wall-units below the wall. Other virtual origins have also been proposed (Mizuno and Jiménez, 2011), but the effect is in any case small, and negligible at large Reynolds numbers, once the distances from the wall to the reference and boundary planes are much larger than the distance from the wall to the virtual origin. More importantly, we have introduced a method to incorporate the dependence with wavelength of the rescaling factor. While the lengthscales of smaller eddies are proportional to the height –as in the original model of Mizuno and Jiménez (2013), larger structures maintain their size constant at different distances from the wall. In our new model, we have proposed to leave the large scales unrescaled. If that is done simultaneously on the three velocities, the code becomes unstable, but this can be sorted by setting the large-scale v to zero. This procedure is actually reasonable from a physical perspective, since very large eddies extend all the way to the wall, where v is always zero. Because of mass conservation, their v should also be very small at any other height.

We have presented results obtained with the model thus refined, showing significant improvement over the original model. In particular, the large artificial pressure fluctuations that appeared at the boundary, and extended well into the flow, have now been removed. Only a much smaller residual peak remains, now affecting only the first plane of the simulation. Nevertheless, that peak seems to perturb intensely the fluctuations of the streamwise velocity immediately above, but those perturbations extend, with milder intensity, not farther than $y \approx 0.1\delta$. A more important problem is that the slope of the velocity profile is overpredicted, leading to errors in the profile itself. This occurs because of a defect in the Reynolds stress, which needs to be compensated by the viscous one. The defect is mainly located in the large scales, $\lambda_x^+ \gtrsim 5000$, for which the contribution to the Reynolds stress has been artificially forced to vanish when setting the wall-normal velocity to zero. Strategies to mitigate this effect, while keeping the large-scale v small, are currently under investigation.

Acknowledgments
This work was partially supported by the European Research Council through the 2013 Multiflow Summer Workshop.

References

Bagwell T, Adrian R, Moser R and Kim J 1993 Improved approximation of wall shear stress boundary conditions for large eddy simulation *Near-Wall Turbulent Flows* ed R So *et al* (New York: Elsevier) pp. 265–76.

Balaras E, Benocci C and Piomelli U 1996 Two-layer approximate boundary conditions for large-eddy simulations *AIAA J.* **34**, 1111–19.

Cabot W and Moin P 1999 Approximate wall boundary conditions in the large eddy simulation of high Reynolds number flow. *Flow, Turb. and Comb.* **63**, 269–91.

Choi H and Moin P 2012 Grid-point requirements for large-eddy simulation: Chapman's estimates revisited *Phys. Fluids* **24**, 011702.

Chung D and Pullin D 2009 Large-eddy simulation and wall modelling of turbulent channel flow *J. Fluid Mech.* **631**, 281–309.

del Álamo J C and Jiménez J 2009 Estimation of turbulent convection velocities and corrections to Taylor's approximation *J. Fluid Mech.* **640**, 5–26.

Flores O and Jiménez J 2010 Hierarchy of minimal flow units in the logarithmic layer *Phys. Fluids* **22**, 071704.

García-Mayoral R, Pierce B and Wallace J 2012 Off-wall boundary conditions for turbulent simulations *Annu. Research Briefs* ed P Moin and J Nichols (Stanford: Center for Turbulence Research) pp. 275–88.

Hoyas S and Jiménez J 2006 Scaling of the velocity fluctuations in turbulent channels up to $\mathrm{Re}_\tau = 2003$ *Phys. Fluids* **18**, 011702.

Hutchins N and Marusic I 2007 Evidence of very long meandering streamwise structures in the logarithmic region of turbulent boundary layers *J. Fluid Mech.* **579**, 1–28.

Inoue M, Mathis R, Marusic I and Pullin D 2012 Inner-layer intensities for the flat-plate turbulent boundary layer combining a predictive wall-model with large-eddy simulations *Phys. Fluids* **24**, 075102.

Inoue M and Pullin D 2011 Large-eddy simulation of the zero-pressure-gradient turbulent boundary layer up to $\mathrm{Re}_\theta = O(10^{12})$ *J. Fluid Mech.* **686**, 507–33.

Jiménez J 2012 Cascades in wall-bounded turbulence *Ann. Rev. Fluid Mech.* **44**, 27–45.

Jiménez J and Moin P 1991 The minimal flow unit in near-wall turbulence *J. Fluid Mech.* **225**, 213–40.

Jiménez J and Vasco C 1998 Approximate lateral boundary conditions for turbulent simulations *Studying turbulence using numerical simulation* ed P Moin and W Reynolds (Stanford: Center for Turbulence Research) pp. 399–412.

Kawai S and Larsson J 2012 Wall-modeling in large eddy simulation: length scales, grid resolution and accuracy *Phys. Fluids* **24**, 015105.

Lele S 1992 Compact finite difference schemes with spectral-like resolution *J. Comput. Phys.* **103**, 16–42.

Millikan C B 1938 A critical discussion of turbulent flows in channels and circular tubes *Proc. 5th Int. Conf. on Applied Mechanics* (New York: Wiley) pp. 386–92.

Mizuno Y and Jiménez J 2011 Mean velocity and length-scales in the overlap region of wall-bounded turbulent flows *Phys. Fluids* **23**, 085112.

Mizuno Y and Jiménez J 2013 Wall turbulence without walls *J. Fluid Mech.* **723**, 429–55.

Nicoud F, Winklemans G, Carati D, Baggett J and Cabot W 1998 Boundary conditions for LES away from the wall *Studying turbulence using numerical simulation* ed P Moin and W Reynolds (Stanford: Center for Turbulence Research) pp. 413–22.

Pascarelli A, Piomelli U and Candler G 2000 Multi-block large-eddy simulations of turbulent boundary layers *J. Comp. Phys.* **157**, 256–79.

Piomelli U 1999 Large-eddy simulations: achievements and challenges *Prog. Aerosp. Sci.* **35**, 335–62.

Piomelli U and Balaras E 2002 Wall-layer models for large-eddy simulations *Annu. Rev. Fluid Mech.* **34**, 349–74.

Podvin B and Fraigneau Y 2011 Synthetic wall boundary conditions for the direct numerical simulation of wall-bounded turbulence *J. Turbulence* **12**, 1–26.

Spalart P R, Moser R D and Rogers M M 1991 Spectral methods for the Navier-Stokes equations with one infinite and two periodic directions *J. Comput. Phys.* **96**, 297–324.

Townsend A A 1961 Equilibrium layers and wall turbulence *J. Fluid Mech.* **11**, 97–120.

Tuerke F and Jiménez J 2013 Simulations of turbulent channels with prescribed velocity profiles *J. Fluid Mech.* **723**, 587–603.

Numerical issues in Lagrangian tracking and topological evolution of fluid particles in wall-bounded turbulent flows

C Atkinson[1], J Hackl[2], P Stegeman[1], G Borrell[2] and J Soria[1,3]

[1] Laboratory for Turbulence Research in Aerospace and Combustion, Dept. Mechanical and Aerospace Engineering, Monash University, Victoria, 3800, Australia
[2] School of Aeronautics, Universidad Politécnica de Madrid, 28040 Madrid, Spain
[3] Department of Aeronautical Engineering, King Abdulaziz University, Jeddah, Kingdom of Saudi Arabia

E-mail: callum.atkinson@monash.edu

Abstract. The determination of the local Lagrangian evolution of the flow topology in wall-bounded turbulence, and of the Lagrangian evolution associated with entrainment across the turbulent / non-turbulent interface into a turbulent boundary layer, require accurate tracking of a fluid particle and its local velocity gradients. This paper addresses the implementation of fluid-particle tracking in both a turbulent boundary layer direct numerical simulation and in a fully developed channel flow simulation. Determination of the sub-grid particle velocity is performed using both cubic B-spline, four-point Hermite spline and higher-order Hermite spline interpolation. Both wall-bounded flows show similar oscillations in the Lagrangian tracers of both velocity and velocity gradients, corresponding to the movement of particles across the boundaries of computational cells. While these oscillation in the particle velocity are relatively small and have negligible effect on the particle trajectories for time-steps of the order of CFL = 0.1, they appear to be the cause of significant oscillations in the evolution of the invariants of the velocity gradient tensor.

1. Introduction

The study of the temporal evolution of wall-bounded flows is complicated by the range of motion experienced by the near-wall and outer flow, and the dependence of many fluid properties on the relative motion of the observer. In order to overcome this challenge many studies have focused on Galilean invariant quantities such as the velocity gradient tensor A_{ij} (VGT) which governs the formation of intense vortical filaments and whose invariants (Q_A, R_A) can be used to classify the local topology at any point in the flow, within the framework of critical-point theory [1], as indicated in Figure 1. Following this methodology a number of researchers have calculated spiralling conditional mean trajectories (CMT) in Q_A-R_A space similar to that shown in Figure 2, with associated time-scales, which represent the non-local mean topological evolution [2]. Typically these calculations involve the solution of the Lagrangian evolution equations for the Q_A, R_A invariants using the right-hand side of Navier–Stokes equations [3] in an Eulerian frame of reference, evaluated from direct numerical simulations (DNS) [2, 4, 5]. Alternatively, experimental measurements of the Eulerian evolution of the invariants can be used, and the convective terms estimated [6]. This enables an investigation of the mean evolution of the invariants, conditioned by their location in the Q_A-R_A space, for a given fluid domain, but does not allow the investigation of individual fluid particle traces, or account for the movement of particles throughout the domain. Given that the evolution of the invariants and their associated time-scales have been shown to be strongly dependent on the wall-normal position and, in the outer region of a turbulent boundary layer at $\mathrm{Re}_\tau \approx 1000$, can have evolution times of the order of 2.5 eddy-turnover times and a convected distance of approximately 20 boundary layer thickness (δ) [5], it remains to be seen how representative these CMT are of the true mean Lagrangian evolution of the flow, and the extent to which individual fluid particles follow this mean.

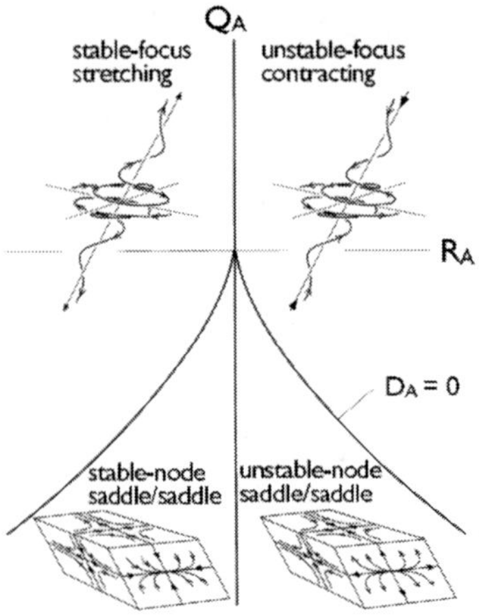

Figure 1. Three-dimensional local topologies (R_A, Q_A)-plane for incompressible flow. Taken from Soria et al. [7]

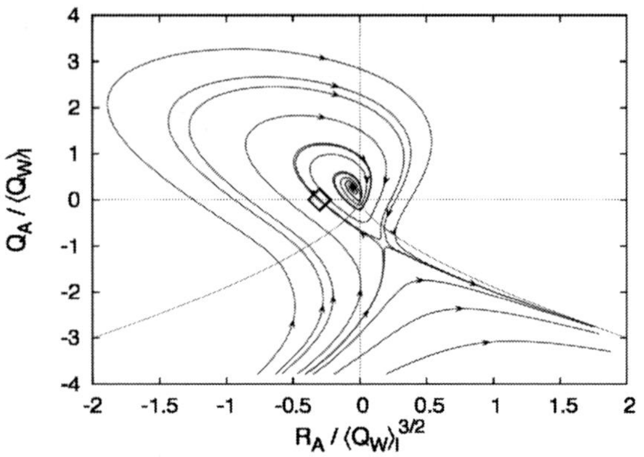

Figure 2. Conditional mean trajectories in the (R_A, Q_A)-plane over a domain corresponding to the most probable gradients in the logarithmic layer and wake region of a turbulent boundary layer from $Re_\theta = 730$ to 1954. Taken from Atkinson et al. 2012 [5].

The Lagrangian evolution of individual fluid particles is also of interest when it comes to the study of the entrainment of irrotational fluid into a turbulent flow, such as the interface between turbulent and non-turbulent (T/NT) regions of a turbulent boundary layer, which have been shown to extend far into the core of the boundary layer [8]. The dynamics of this interface play an important role in the entrainment of mass, momentum and passive scalars into a turbulent flow and in the growth of the boundary layer. Advances in experimental measurement techniques and access to the high-resolution data of direct numerical simulations has lead to an increasing interest in the geometry, role of coherent structures and mechanisms associated with T/NT interfaces in both free and wall-bounded shear flows [9–13]. To date, most of the research regarding the T/NT interface has focused on the investigation and the classification of the interface and the surrounding flow, with less attention given to the Lagrangian view of the entrainment process. For instance, what happens to fluid particles in an irrotational region? how do they interact with the rotational flow and gain enstrophy? what is the topological evolution and time-scale associated with this process?

In order to answer these questions about the true topological evolution experienced by fluid particles in various regions of a turbulent wall-bounded flow, or the evolution and dynamics of a fluid particle as it is entrained across the T/NT interface, it is necessary to implement fluid-particle tracking and thereby determine the Lagrangian evolution of the VGT as it follows a particle at each time step of a DNS of the wall-bounded flow. Fluid-particle tracking has received significant attention in DNS of homogeneous isotropic turbulence [14, 15] but, with the exception of [16], presumably owing to the higher computational demands, it has received relatively little attention in DNS of wall-bounded flows. In this paper we investigate the implementation of particle tracking in both a turbulent boundary layer and a fully developed channel flow, and examine the influence of different interpolation and particle time-marching approaches on the calculated particle paths, and the Lagrangian evolution of the VGT.

2. Particle Tracking in Direct Numerical Simulations of Wall-bounded flows

Codes that perform direct numeric simulations of fluids flow can be modified to enable the calculation of the Lagrangian evolution of the fluid by treating element or packets of the fluid as if they were particles that exactly follow the fluid dynamics, so that

$$\mathrm{d}\vec{x}/\mathrm{d}t = \vec{u}(\vec{x}), \tag{1}$$

where $\vec{x}$ is the particle's position and $\vec{u}$ is the fluid flow velocity [14]. Given the simplicity of the above equation there should be no need to modify the time-marching of the principle code, assuming that it resolves all the relevant time-scales of the fluid flow. However, if during the time-marching the code uses an implicit or predictor step, a decision must be made as to whether or not to increment particle position at each intermediate step, assuming continuity is enforced at each step, and then re-calculate the particle velocity, or to only update the particle position at the end of the time-step. This later option only involves the use of the higher-order-accurate velocity fields, but will likely require a small overall time-step and more computation in order to achieve an accurate particle trajectory. Ultimately the accuracy of the fluid particle dynamics and the calculated particle trajectory will depend on both the magnitude of the time-step and the accuracy with which the velocity $\vec{u}(\vec{x})$ at the particle location can be determined from the computational cells of the fluid simulation. Assuming that the velocity components at the faces of each cell are calculated to a sufficient accuracy to capture the dynamics of the fluid, the accuracy of the particle velocity will still depend on the numerical interpolation technique used to compute the sub-cell velocity components. Owing to the expense of direct three-dimensional interpolation, such interpolation is typically determined via successive one-dimensional interpolations of the required grid points along each axis, which can also be advantageous when it comes to the interpolation of particle velocity from a computational domain that is split across a series of processors nodes.

Optimal implementation of particle tracking depends on the DNS code in which the tracking is to be performed. In the present case, fluid-particle tracking is to be implemented both in the high-resolution incompressible boundary layer code, and in the fully developed channel code developed by the School of Aeronautics at the Universidad Politécnica de Madrid. Simulation of the turbulent boundary layer is performed using a fractional-step method with a three-step Runge–Kutta scheme for time-marching that is second-order accurate for velocity, and enables the use of variable pressure gradients and time-dependent boundary conditions [17]. Convective and viscous terms in the streamwise (x) and wall-normal (y) directions are calculated using staggered three-point compact finite differences. The computational domain is periodic in the span-wise (z) direction with velocity and pressure components expanded in a Fourier series so that the span-wise gradient can be calculated using spectral methods with the 2/3 rule to prevent aliasing. A pressure correction is applied using the Poisson equation at each sub-step of the Runge–Kutta. Velocity gradients and divergence for the Poisson correction are calculated by centred second-order finite differences in the x, y directions and spectral in the z direction. The present version of the code uses a combination of multi-processing (OMP) and message passing (MPI) for hybrid multi-thread and multi-core parallel processing [18] with the decomposition of the computational domain alternating between cross-stream y, z slabs and stream-wise pencils. Tracking is carried out in the slab decomposition, so that communication is only required in the streamwise direction.

The channel-flow simulations are performed using the established pseudospectral method of Kim *et al* [19] with particle tracking incorporated into the parallel code of Hoyas & Jiménez [20]. Homogeneity in both the stream-wise and span-wise directions allows the pseudospectral solution of the Navier-Stokes equations rewritten in terms of wall-normal vorticity ω_y and the Laplacian of v, hence eliminating the need for pressure. While computationally efficient in terms of the fluid phase, this formulation requires the generation of a realisable velocity field $u_i(x, y, z)$ in order to interpolate for the velocity at each particle. As in the case of the boundary layer, particle positions and velocities are updated at each substage of a semi-implicit third-order Runge–Kutta time marching scheme, hence requiring the generation of the fluid velocity field at each sub-step. In this formulation, derivatives in both x and z directions are handled spectrally, while wall-normal derivatives are computed via a fourth-order seven-point compact finite difference.

3. Sub-grid Interpolation methods

For structured grids, a variety of techniques can be used for sub-cell interpolation. If the parent simulation code involves periodic dimensions in an incompressible flow field, a spectral interpolation can be performed with no interpolation error for all scales resolved by the grid. For non-periodic dimensions compact (spectral-like) interpolation schemes such as those found in [21] and [22] may be used. It is important to note that, while providing excellent interpolation accuracy, both the spectral and compact methods involve significant computational cost that must be repeated for every particle in the domain at each time-step or Runge–Kutta sub-step. Previous applications of fluid particle tracking in DNS of turbulent flows have typically used more computationally efficient and scalable explicit Basis- or Hermite-spline based interpolation schemes [14, 16] with Choi et al. [16] demonstrating that a four-point Hermite interpolation (degree = 8) provides results similar to the far more expensive spectral interpolation, albeit with derivatives at each grid point evaluated spectrally. Recent work of Stegeman et al. [23] indicates that even higher-order Hermite interpolation is required if the interpolation error is to be limited to only a few percent at the smallest wavenumber of the computation. Only the use of Basis and Hermite splines will be consider in this paper.

While both Basis- and Hermite-spline interpolation involves fitting and evaluating a local piecewise polynomial, Hermite splines require the value of the function (in this case velocity) and its derivatives at the two grid points nearest to the target location of the interpolation. In contrast Basis or B-splines represent a recursion of only the target quantity at multiple grid points.

Consider a scalar function (e.g. a single component of the fluid velocity) $f(x)$ sampled with N points over a uniform grid $\{\{x_i\} : i \in [0, N)\}$ with spacing Δ such that $f_i = f(x_i)$. A Hermite spline interpolation operator of degree M is defined as

$$f_I(\check{x}) = \sum_{p=0}^{M-1} a_p \check{x}^p, \tag{2}$$

where the parametric position $\check{x}$ between two consecutive nodes is defined from the spatial position x as

$$\check{x} = \frac{x - x_n}{\Delta} \quad \text{where} \quad x_n \leq x < x_{n+1}. \tag{3}$$

The polynomial coefficients a_p are determined from the boundary conditions of the scalar and its derivatives at $\check{x} = 0$ and 1:

$$\frac{\partial^\alpha f_I(\check{x}_b)}{\partial \check{x}^\alpha} = \frac{\partial^\alpha f(x_n + \Delta \check{x}_b)}{\partial \check{x}^\alpha} \quad \text{where} \quad \check{x}_b = \{0, 1\}, \ \alpha = \{0, ..., K-1\}. \tag{4}$$

This ensures that the derivatives of the fitted spline correspond to the derivative of $f(x)$ and are continuous across grid points up to an order $K = M/2$. The boundary derivatives of the sampled function are determined by a central differencing method with an order K such that the spline coefficients a_p are a linear function of the set of sampled points $\left\{f_{i-K/2}, ..., f_{i+K/2+1}\right\}$,

$$a_p = \sum_{q=0}^{K+1} c_{p,q} f_{n+q-K/2}. \tag{5}$$

Substituting the above expression for the spline coefficients into Equation (2) the spline can be presented as

$$f_I(\check{x}) = \sum_{p=0}^{M-1} \left(\sum_{q=0}^{K+1} c_{p,q} f_{n+q-K/2} \right) \check{x}^p, \tag{6}$$

so that each spline of degree M has an array of $(M-1) \times (K+1) = (M^2 + M - 2)/2$ unique coefficients $c_{p,q}$. Interpolation of a three-dimensional function can then be performed via successive one-dimensional interpolations of the required grid points along each axis. Given that the spline represents a piecewise polynomial, it can also be used to determine the gradient of a scalar at a given interpolated location, albeit at lower order of accuracy. Both this method and the interpolation of the velocity gradient from scalar fields of each component of $\partial u_i/\partial x_j$ are examined in section 5.

From the above definition, a cubic Hermite spline consists of order M=4, and requires the calculation to second-order accuracy of derivatives up to first order at the nearest grid points. This can be obtained from a second-order central difference. Following the formulation of Equation (6) a cubic B-spline will have pre-determined coefficients

$$c_{p,q} = \frac{1}{6} \begin{bmatrix} -1 & 3 & -3 & 1 \\ 3 & -6 & 3 & 0 \\ -3 & 0 & 3 & 0 \\ 1 & 4 & 1 & 0 \end{bmatrix}.$$

An alternative form to this Hermite interpolation was considered in the turbulent channel flow, where rather than using higher-order derivative boundary conditions at the two nearest grid points, a higher-order interpolation was instead performed using the zero- and first-order derivatives at the four nearest grid points [24]. In one dimension, a four-point Hermite spline can be represented as

$$f_I(\check{x}) = \sum_{p=1}^{4} H_p(\check{x}) f_q + \sum_{p=1}^{4} G_p(\check{x}) \frac{df}{d\check{x}}|_p, \tag{7}$$

where H and G are basis functions and f and all its first-order derivatives are evaluated at the grid points.

The relative error of both Basis- and Hermite-spline interpolation is due to the truncation error of the polynomial approximation as well as the error in the finite-difference approximation of the boundary derivatives in the case of the Hermite spline. This error can be explored using a plane-wave error analysis which provides the transfer function of the spline interpolation operator. A continuous plane-wave with a biased phase can be expressed in terms of the parametric coordinate system as

$$f(\check{x}, k\Delta, \phi) = Ae^{ik(x_n + \Delta\check{x}) + \phi} \quad \text{where} \quad \phi \in [-\pi, \pi). \tag{8}$$

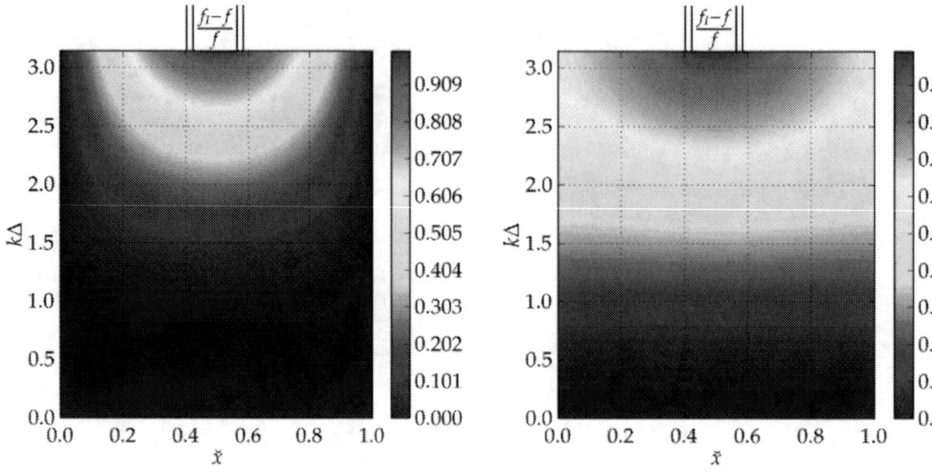

(a) Hermite (b) Basis

Figure 3. Relative magnitude of interpolation error for Cubic (Degree = 4) Hermite and B-spline interpolation.

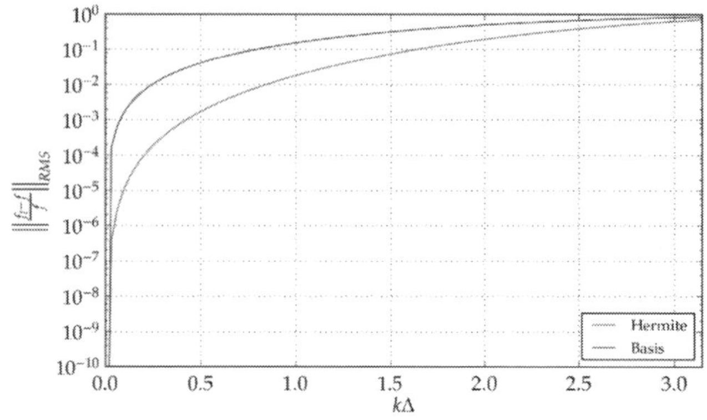

Figure 4. Root-mean-squared relative error over the range of interpolation positions for Cubic (Degree = 4) Hermite and B-spline interpolation.

Substituting Equation (8) into Equation (6), the interpolated value can be expressed as

$$f_I\left(\check{x}, k\Delta, \phi\right) = Ae^{\mathrm{i}k(x_n+\Delta\check{x})}e^{\mathrm{i}\phi}\sum_{p=0}^{M-1}\left(\sum_{q=0}^{K+1} c_{p,q}e^{\mathrm{i}k\Delta(q-K/2)}\right)\check{x}^p. \tag{9}$$

The difference between the true value of the plane wave and its interpolated value can be determined from a Monte Carlo simulation in the parameter space consisting of the parametric wavenumber $k\Delta$, the phase ϕ, and the location of the interpolated point $\check{x}$ relative to the nearest grid point. Figure 3 shows the magnitude of the interpolation error, defined as the magnitude of the difference between the interpolated value $f_I(\check{x})$ and the true value of the scalar $f(\check{x})$ at the same location with respect to the grid points. Not surprisingly the error increases as the distance between the interpolation point and the grid points increases, for both Hermite and B-splines, particular at higher wavenumbers. This change is far more gradual in the case of the B-spline but, as shown in Figure 4, the total error in the Hermite spline is lower than that of the B-spine for the same wavenumber.

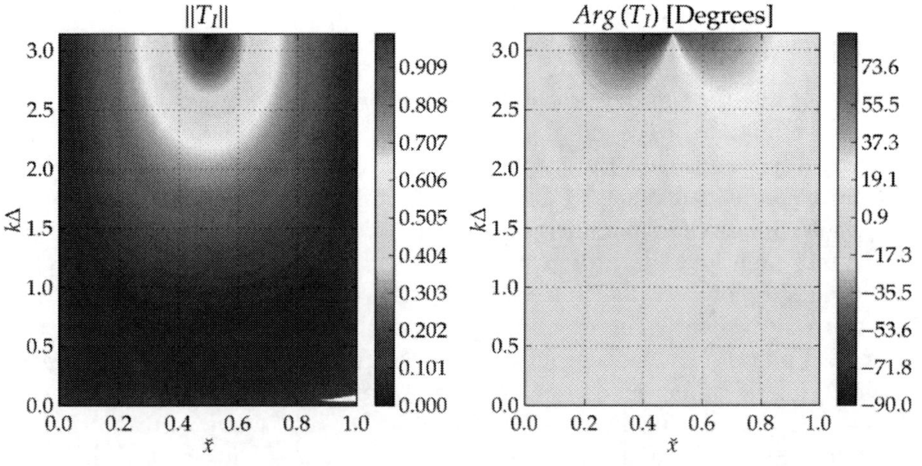

Figure 5. Transfer Function for a Cubic (degree = 4) Hermite Spline

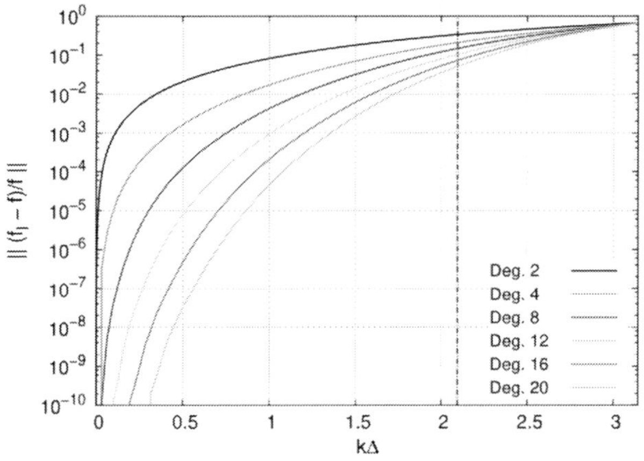

Figure 6. Root mean square relative error over the range of interpolation positions for different degree Hermite splines.

From Equation 9, a complex transfer function for this interpolation can be defined as

$$T_I\left(\check{x}, k\Delta\right) = \frac{f_I\left(\check{x}, \phi\right)}{f\left(\check{x}, \phi\right)} = \sum_{p=0}^{M-1}\left(\sum_{q=0}^{K+1} c_{p,q} e^{\mathrm{i}k\Delta(q-K/2)}\right)\check{x}^p, \tag{10}$$

which is only a function of the interpolation position $\check{x}$ and the relative wavenumber $k\Delta$. The modulus and argument of the transfer function defines the relative difference in magnitude and absolute difference in phase between the interpolated plane-wave and the original plane-wave, respectively. The contour plots in Figure 5 show the modulus and argument for the Hermite spline. The transfer function of the B-spline is similar, but shows less variation in interpolation position, as observed by the relative magnitude of the interpolation error. In both cases the argument of this transfer function suggests that both interpolation schemes may introduce oscillations in the higher wavenumbers.

The absolute magnitude of these interpolation errors and the effect on the particle trajectories and the evolution the of VGT and its invariants will depend on the energy content of the scalar (the fluid-velocity field) at different wavenumbers, the grid spacing, and the chosen degree of the fitted spline. Figure 6 shows the effect of varying the degree of Hermite interpolation. Assuming negligible error in the scalar function at the grid points, a higher-order spline reduces the error at all wavenumbers. While the reduction in error from a linear (degree = 2) to a cubic (degree = 4) spline is significant, the incremental reduction in error decreases with higher order. If the effective cut-off wavenumber of the simulation is taken at $k\delta = 2/3$, following the 2/3 rule for de-aliasing, Figure 6 indicates that the error at this wavenumber should be approximately 22% for a cubic Hermite spline, decreasing to 5% for a Hermite spline of degree 20.

Higher-order interpolation requires the use of higher-order gradients, either directly in the case of Hermite interpolation, or through a larger coefficient matrix in the case of B-splines. Both cases ultimately require a wider interpolation kernel of size $N_P = M$, equal to the degree of the spline. This significantly increases the computation required for each particle. Given that the entire fluid phase must be computed for each time-step and that Lagrangian statistics require the calculation of numerous particles, it is therefore desirable to run the DNS with as many particles as possible. Naturally, this requires that the additional computation imposed by the particles be kept to a minimum. The benefits of higher-order interpolation are assessed in the context of the particle tracers in section 5.

4. Fluid-Particle Tracking Implementation

In order to test the benefits and possible need for high-order interpolation methods and to enable the processing of high-resolution boundary-layer simulations, the DNS code was modified to enable the use of Hermite spline interpolation of any specified order for both uniform and irregular grid spacing. Following the formulation for Hermite splines discussed in section 3, derivatives up to the required order are calculated at grid points nearest to the particle location using central or forward- or backward-biased finite-difference schemes, depending on the location of the particle with respect to the domain boundaries. Coefficients for these finite-difference schemes were computed for different-order schemes during runtime, following [25].

The velocity of each particle was determined at their initial location before the first Runge–Kutta time-step by converting one velocity component into physical space, and then performing three-dimensional interpolation of that velocity component at each particle location, before considering the next component. Interpolation was performed by first fitting one-dimensional Hermite splines of order M in the y, z cross-stream directions over a three-dimensional interpolation kernel of $M \times M \times M$ points around each particle, with final interpolation performed in the x direction. Structuring the interpolation in this format has the advantage that when a particle's interpolation kernel extends across the boundaries between computational nodes, the cross-stream interpolation can be handled by separate nodes. This reduces the communication between nodes to at most $M/2$ quantities for each velocity component.

Particle positions where updated at the start of each Runge–Kutta sub-step using their velocity from the end of the previous step. Velocities were then re-interpolated at this updated particle location after the fluid velocity was updated by the right-hand side of the governing equation, and after the Poisson correction for the divergence. The positions of all master (centre of particle is contained in the node) and slave particles (part of the interpolation kernel of a particle from an adjacent node) are stored for each computational domain, as done by Uhlmann [26]. With each update of the particle position, a test is performed to see whether a particle or its interpolation kernel crosses the boundary of its parent node, forcing the transition of a slave to a master or the formation of a new slave, respectively.

The velocity gradient tensor (VGT) at each particle was evaluated by first determining the velocity gradients at the centre of each cell using second-order central finite differences, consistent with the methodology used for the Poisson correction. It was assumed that the error associated with this scheme would be insignificant in comparison to the error introduction by the interpolation (at least for lower-order schemes), although this assumption has yet to be tested. Each component of the VGT was then interpolated following the same procedure used for the

Table 1. Parameters of the zero-pressure gradient boundary layer DNS. N_x, N_y and N_z are the grid sizes along the three axes, expressed for z in terms of collocation points, with Δ representing the corresponding resolutions at their coarsest points.

Re_τ	$(L_x, L_y, L_z)/\delta$	$\Delta x^+, \Delta y^+, \Delta z^+$	N_x, N_y, N_z	Particles
$126 - 190$	$7.6 \times 9.0 \times 10.5$	$13.4 \times 0.33 \times 6.2$	$257 \times 315 \times 768$	1200
$137 - 800$	$20.7 \times 15.7 \times 55.4$	$13.6 \times 0.18 \times 12.5$	$4097 \times 315 \times 768$	28800

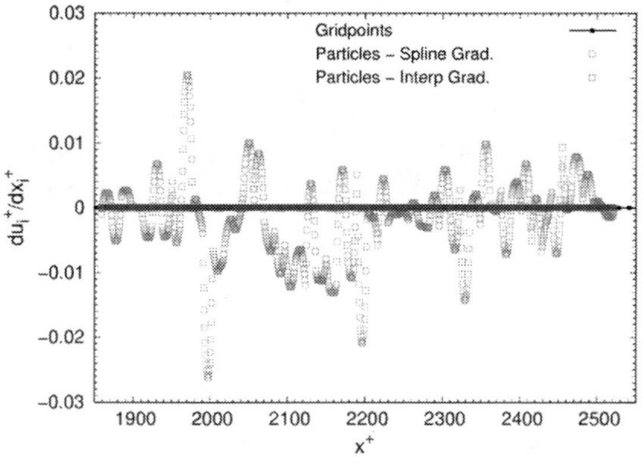

Figure 7. Comparison of the divergence in the particle velocity calculated from the gradient of a degree-4 Hermite spline fitted to the velocity field or by fitting Hermite splines to the gradients of the velocity field at each grid point. Values are shown for a series of particles distributed along a stream-wise line though the boundary layer simulation at $y^+ = 53$.

velocity field. Tests were also performed to see if the velocity gradients could be determined from differentiation of the spline fits as suggested by Yeung and Pope [14], which does not require the differentiation of the velocity field or additional interpolation, and is therefore far more efficient.

A similar process was performed in the channel flow using both cubic (degree = 4) B-splines and the four-point Hermite interpolation discussed in section 3. Particle positions and velocities were similarly computed at each Runge–Kutta sub-step. Pseudo-spectral methods were used to compute the velocity gradient at each grid point, from which each component of the velocity gradients tensor was calculated at each particle location via the same spline interpolation.

5. Results

To test the effect of the Hermite spline order and the size of the time-step on the particle trajectories and on the Lagrangian evolution of the invariants of the VGT, a turbulent boundary layer DNS was run over two moderate resolution grids whose dimensions and grid spacing are listed in Table 1. Initial testing was performed on the smaller grid in order to test the basic implementation of the code and the interpolation before extending the code to more realistic simulations.

In the absence of known particle trajectories, the accuracy of a given interpolation scheme was tested based on how much the particle position differed from that produced by a higher-order interpolation or a shorter time-step Δt, given that position error is a function of both the interpolation error and the time-step. The time-step from this point on will be quoted in terms of Courant-Friedrichs-Lewy number $CFL = U\Delta t/\Delta x$, where U is the free-stream velocity and Δx is the length of the computational cell. In addition to this, for an incompressible flow, the divergence $\partial u_i/\partial x_i$ measured at the particle location can be used as an indication of the accuracy of the velocity gradients calculated for each particle. Figure 7 shows the difference in the divergence obtained when calculating the velocity gradients at each particle from the gradient of a spline fitted to the velocity field (as suggested by [14]) compared to fitting a

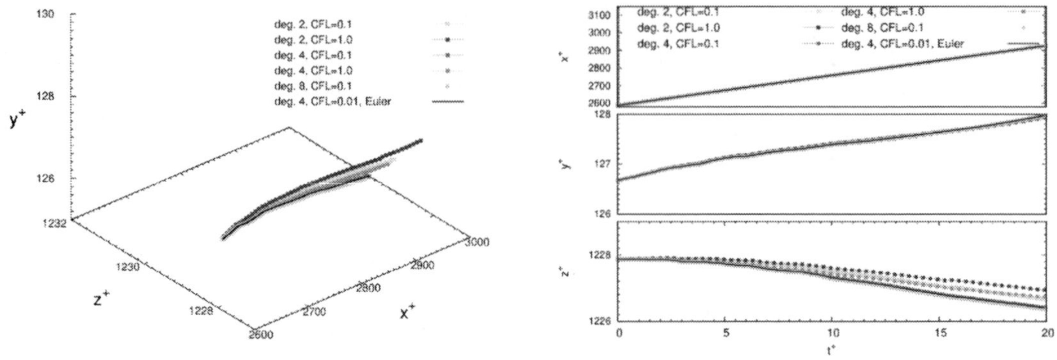

Figure 8. Particle trajectories calculated in the boundary layer simulation using Hermite splines of varying order over a series of different time-steps.

separate spline to the gradients of the velocity field at each grid point. Separate interpolation of each component of the VGT provides a divergence that is typically five orders of magnitude smaller than that obtained by differentiating the spline. Interpolating each component of the VGT is considerable more expensive, but provides a far more accurate estimation of the sub-grid particle velocity gradients.

The behaviour for the different interpolation and time-steps can be illustrated by considering the behaviour of an individual particle trajectory. Similar patterns were observed for most particles although, naturally, the degree to which a given particle trajectory diverges for different interpolation schemes will ultimately depend on the dynamics of the flow field at that time and position. Figure 8 shows the trajectory followed by a particle started at a wall-normal position of $y^+ = 126.7$ using a series of Hermite interpolations. Particle positions and velocities are updated for each Runge–Kutta time-step with the exception of the final trajectory, where the particle positions are instead updated using a first-order Euler time marching with CFL $= 0.01$. This final trajectory was generated in order to ensure that the use of the velocity fields inside the Runge–Kutta time-step did not have an adverse effect on the calculation of the particle's velocity. Comparison of the trajectories indicates that, for the same interpolation order, updating the particle position within the Runge–Kutta substeps enables time-steps at least an order of magnitude longer than for the Euler method, with negligible influence on the particle trajectory.

Figure 9 shows the variation in particle velocity as a function of time for each of the interpolation methods discussed above, demonstrating that both velocity and the consequent position are highly sensitive to the order of interpolation scheme and to the size of the time-step used in the simulation. The use of linear (degree $= 2$) interpolation results in significant oscillations in each component of the particle velocity, which is ultimately responsible for the difference in the particle trajectory. The use of cubic (degree $= 4$) interpolation reduces the amplitude of these oscillations, but the benefit of moving to a degree $= 8$ is less clear. In all cases these oscillations maintain approximately the same period and are strongly correlated with the particle position relative to the Eulerian simulation grid points.

The Lagrangian evolution of the Q_A, R_A invariants of the VGT following the same particle are shown in Figure 10 where the trajectories originate in the top left corner of the domain, which corresponds to the particle being in a region of stable focus stretching (see Figure 1). The invariants shown have been normalised by the mean of the second invariant of the rate of rotation tensor Q_W, which is directly proportional to the enstrophy [5]. While it is not clear what the

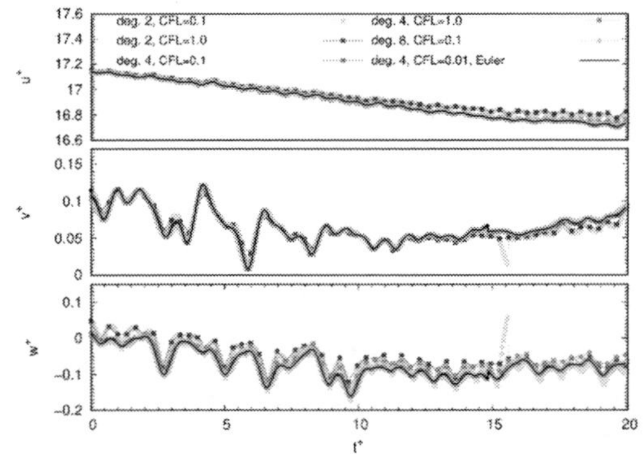

Figure 9. Particle velocity calculated in the boundary layer simulation from Hermite splines of varying order over a series of different time-steps.

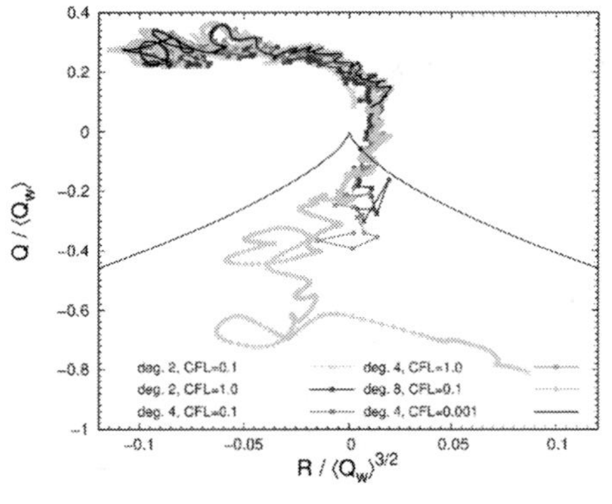

Figure 10. Lagrangian evolution of Q_A, R_A invariants of the velocity gradient tensor in the boundary layer simulation calculated for a single particle using Hermite splines of varying order over a series of different time-steps.

Re_τ	$(L_x, L_y, L_z)/\delta$	$\Delta x^+, \Delta y^+, \Delta z^+$	N_x, N_y, N_z	*Particles*
2000	$2\pi \times 2 \times \pi$	$10 \times 9 \times 5$	$1536 \times 633 \times 1536$	7.296×10^5

Table 2. Parameters for channel flow tests. Nomenclature as given in table 1.

Lagrangian trajectory of the invariants of an individual particle should look like, results shows a similar pattern of oscillation to the particle velocity. This oscillation is again reduced by using a higher-order interpolation although, unlike in the case of the particle velocity, these oscillations continue to reduce up to an interpolation degree of at least 8. This decreased oscillation for higher-degree splines suggests that the oscillations are at least partly due to the interpolation method.

The channel flow has been tested most extensively in the configuration given in Table 2. Figure 11 illustrates oscillations in the time series of quantities interpolated using both cubic B-splines and four-point Hermite splines. As in the case of the boundary layer, these oscillations do not correspond to non-monotonic behaviour of the unsteady flow in the surrounding grid points. As shown by Figure 12, oscillations in the velocity gradients are more noticeable. It should be noted that the magnitude of the oscillations is typically small but, as shown in Figure 10, they can have a significant effect on the observed Lagrangian evolution of the invariants, and will have consequences when it comes to the calculation of Lagrangian accelerations.

Despite significant differences between the resolution and the computation of velocity gradients in both the boundary layer and channel flow simulation, the similarity in the computed Lagrangian velocity and velocity gradient tracers suggests that the accuracy of the Lagrangian statistics is highly sensitive to the interpolation scheme. Given the shape of the transfer function of the interpolation method (see Figure 5) and the spacing of these oscillation with respect to the grid points, it is likely that these oscillations are purely the result of the sub-grid velocity interpolations. It may therefore be possible to low-pass filter these oscillations without effecting the Lagrangian statistics, assuming that the time-step of the simulation is small enough that these oscillations do not significantly influence the calculated particle trajectory, and given that there should be no appreciable energy in the fluid phase for parametric wavenumbers $k\Delta > 2\pi/3$ under the 2/3 de-aliasing of the DNS. Technically, this filtering should be a function

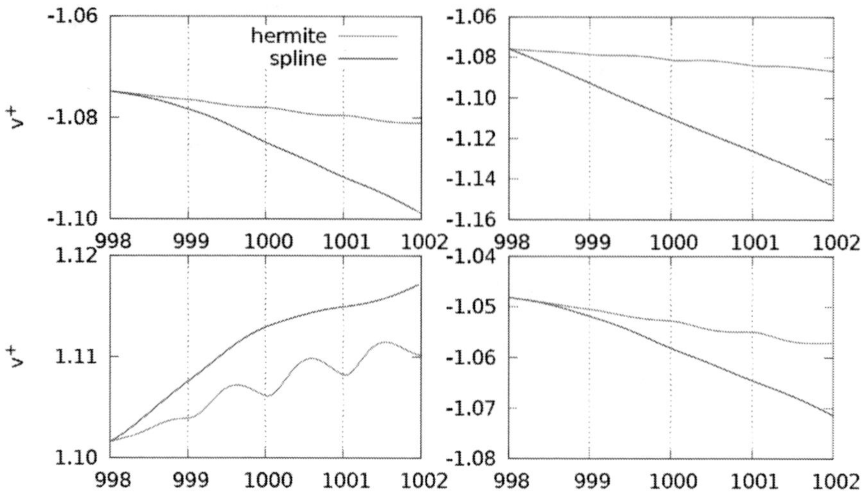

Figure 11. Time series of wall-normal velocity v for representative particles initialized at grid points on a cross-stream plane of the channel at $Re_\tau = 2000$. CFL = 0.3

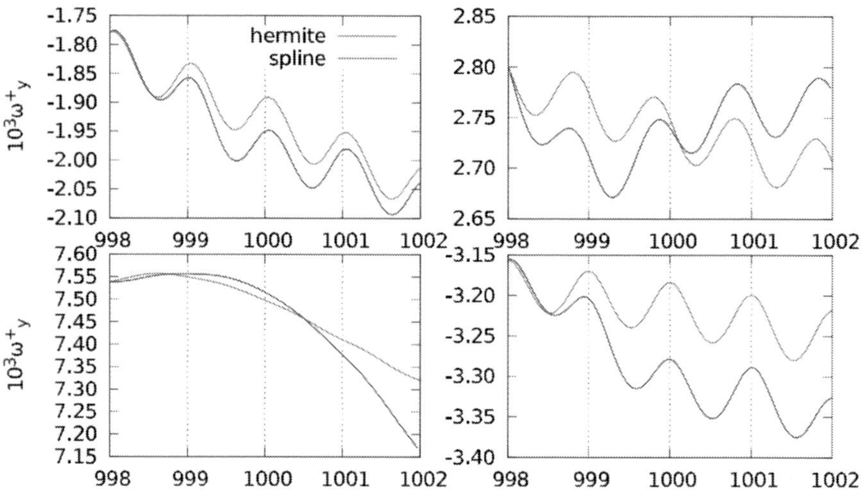

Figure 12. Time series of wall-normal vorticity ω_y for representative particles initialized at grid points on a cross-stream plane of the channel at $Re_\tau = 2000$. CFL=0.3

of positional variation in all three-directions and not a function of time. Given that the size of the computational cell varies with the wall-normal position, and that oscillations occur as the particle moves with respect to each face of the computational cell, the implementation of this filtering is non-trivial.

6. Conclusions

In order to enable the calculation of Lagrangian statistics and time-scale in wall-bounded flows, the Lagrangian tracking of fluid particles has been implemented in direct numerical simulation codes for both a boundary layer and for a fully developed channel flow. Sub-grid interpolation was implemented using both cubic B-spline, four-point Hermite and higher-order Hermite interpolation schemes. Despite significant differences in cell size, the computation of velocity gradients, and the implementation of the interpolation, the particle velocities and gradients in both the boundary layer and the channel flow showed similar oscillations in their Lagrangian tracers. These oscillations are highly correlated with the movement of particles with respect to the computation grid and, given the similar pattern in the phase error of the interpolation method, appear to be due to the sub-grid interpolation method. While these oscillations in the particle velocity are relatively small and have negligible effect on the particle trajectories for time-steps on the order of CFL = 0.1, they appear to be the cause of more significant oscillations in the evolution of the invariants of the velocity gradient tensor. The extent to which these interpolation errors may influence the Lagrangian statistics is not yet clear.

Acknowledgments
This research was instigated as part of, and partially funded by, the Multiflow program of the European Research Council, and was partially supported by a grant from the Australian Research Council, both of which are gratefully acknowledged.

References

[1] Chong MS, Perry AE and Cantwell BJ 1990 A general classification of three-dimensional flow fields *Phys. Fluids* **2** 765–77
[2] Ooi A, Martín J, Soria J and Chong MS 1999 A study of the evolution and characteristics of the invariants of the velocity-gradient tensor in isotropic turbulence *J. Fluid Mech.* **381** 141–74
[3] Cantwell BJ 1992 Exact solution of a restricted Euler equation for the velocity gradient tensor *Phys. Fluids A* **4** 782–93
[4] Martín J, Ooi A, Chong MS and Soria J 1998 Dynamics of the velocity gradient tensor invariants in isotropic turbulence *Phys. Fluids* **10** 2336–46
[5] Atkinson C, Chumakov S, Bermejo-Moreno I and Soria J 2012 Lagrangian evolution of the invariants of the velocity gradient tensor in a turbulent boundary layer *Phys. Fluids* **24** 105104
[6] Elsinga GE and Marusic I 2010 Evolution and lifetimes of flow topology in a turbulent boundary layer *Phys. Fluids* **22** 015102
[7] Soria J, Sondergaard R, Cantwell BJ, Chong MS and Perry AE 1994 A study of the fine-scale motions of incompressible time-developing mixing layers *Phys. Fluids* **6** 871–84
[8] Jiménez J, Hoyas S, Simens MP and Mizuno Y 2010 Turbulent boundary layers and channels at moderate Reynolds numbers *J. Fluid Mech.* **657** 335–60
[9] Bisset DK, Hunt JCR and Rogers MM 2002 The turbulent/non-turbulent interface bounding a far wake *J. Fluid Mech.* **451** 383–410
[10] Westerweel J, Hofmann T, Fukushima C and Hunt JCR 2002 The turbulent/non-turbulent interface at the outer boundary of a self-similar turbulent jet *Exp. Fluids* **33** 873–78
[11] Mizuno Y, Atkinson C and Soria J 2011 Effect of the outer laminar flow on zero-pressure-gradient turbulent boundary layers *Bull. Am. Phys. Soc.* **56** G74
[12] Da Silva CB and Dos Reis RJN 2011 The role of coherent vortices near the turbulent/non-turbulent interface in a planar jet *Phil. Trans. Roy. Soc. A* **369** 738–53
[13] Da Silva CB, Dos Reis RJN and Pereira JC 2011 The intense vorticity structures near the turbulent/non-turbulent interface in a jet *J. Fluid Mech.* **685** 165–90
[14] Yeung P and Pope S 1989 Lagrangian statistics from direct numerical simulations of isotropic turbulence *J. Fluid Mech.* **207** 531–86
[15] Squires KD and Eaton JK 1991 Lagrangian and Eulerian statistics obtained from direct numerical simulations of homogeneous turbulence *Phys. Fluids A* **3** 130–43
[16] Choi JI, Yeo K and Lee C 2004 Lagrangian statistics in turbulent channel flow *Phys. Fluids* **16** 779–93
[17] Simens MP, Jiménez J, Hoyas S and Mizuno Y 2009 A high-resolution code for turbulent boundary layers *J. Comput. Phys.* **228** 4218–31
[18] Sillero J, Borrell G, Jiménez J and Moser RD 2011 *Recent Advances in the Message Passing Interface* (Berlin: Springer) pp 218–27
[19] Kim J, Moin P and Moser R 1987 Turbulence statistics in fully developed channel flow at low Reynolds number *J. Fluid Mech.* **177** 133–66
[20] Hoyas S and Jiménez J 2006 Scaling of the velocity fluctuations in turbulent channels up to $Re_\tau = 2003$ *Phys. Fluids* **18** 011702
[21] Denga X and Maekawab H 1997 Compact high-order accurate nonlinear schemes *J. Comput. Phys.* **130** 77–91

[22] Nagarajan S, Lele SK and Ferziger JH 2003 A robust high-order compact method for large-eddy simulation *J. Comput. Phys.* **191** 392–419

[23] Stegeman PC, Soria J and Ooi A 2012 Comparison of the dynamics of particles in a flow field with Reynolds and Favre filtered flow velocities *Australasian Fluid Mech. Conf.* eds. P.A. Brandner and B.W. Pearce (http://www.afms.org.au) **320**

[24] Hildebrand F 1974 Introduction to numerical analysis *New York: McGraw-Hill, 2nd ed.*

[25] Fornberg B 1988 Generation of finite difference formulas on arbitrarily spaced grids *Math. of Comput.* **51** 699–706

[26] Uhlmann M 2004 *Simulation of particulate flows on multi-processor machines with distributed memory* (Madrid: Ciemat)

Journal of Physics: Conference Series **506** (2014) 012004 doi:10.1088/1742-6596/506/1/012004

Turbulence in the highly restricted dynamics of a closure at second order: comparison with DNS

N C Constantinou[1], A Lozano-Durán[2], M-A Nikolaidis[1], B F Farrell[3], P J Ioannou[1], J Jiménez[2]

[1]University of Athens, Department of Physics, Panepistimiopolis, Zografos, 15784 Athens, Greece

[2]School of Aeronautics, Universidad Politécnica de Madrid, 28040 Madrid, Spain

[3]Department of Earth and Planetary Sciences, Harvard University, Cambridge, MA 02138, USA

E-mail: navidcon@phys.uoa.gr

Abstract. S3T (Stochastic Structural Stability Theory) employs a closure at second order to obtain the dynamics of the statistical mean turbulent state. When S3T is implemented as a coupled set of equations for the streamwise mean and perturbation states, nonlinearity in the dynamics is restricted to interaction between the mean and perturbations. The S3T statistical mean state dynamics can be approximately implemented by similarly restricting the dynamics used in a direct numerical simulation (DNS) of the full Navier–Stokes equations (referred to as the NS system). Although this restricted nonlinear system (referred to as the RNL system) is greatly simplified in its dynamics in comparison to the associated NS, it nevertheless self-sustains a turbulent state in wall-bounded shear flow with structures and dynamics comparable to those observed in turbulence. Moreover, RNL turbulence can be analysed effectively using theoretical methods developed to study the closely related S3T system. In order to better understand RNL turbulence and its relation to NS turbulence, an extensive comparison is made of diagnostics of structure and dynamics in these systems. Although quantitative differences are found, the results show that turbulence in the RNL system closely parallels that in NS and suggest that the S3T/RNL system provides a promising reduced complexity model for studying turbulence in wall-bounded shear flows.

1. Introduction

The Navier–Stokes equations (NS), while comprising the complete dynamics of turbulence, have at least two disadvantages for theoretical investigation of the physics of turbulence: NS lacks analytical solution for the case of the fully turbulent state, and the nonlinear advection term results in turbulent states of high complexity that tend to obscure the fundamental mechanisms underlying the turbulence. One approach to overcoming these impediments has been the search for simplifications of NS that retain essential features of the turbulence dynamics. The Linearized Navier–Stokes equations (LNS) provide one example of the successful application of this approach in which the power of linear systems theory is made available to the study of turbulence [1, 2]. The LNS system captures the non-normal mechanism responsible for perturbation growth in NS [3]. This linear mechanism retained in LNS underlies both the process of subcritical transition to turbulence and the maintenance of the turbulent state [4–7]. However, linear models are unable to capture other essential phenomena in turbulence that are

Journal of Physics: Conference Series **506** (2014) 012004 doi:10.1088/1742-6596/506/1/012004

intrinsically nonlinear, including the establishment of the turbulent mean velocity profile and the maintenance of turbulence in a statistically steady state in the absence of extrinsic excitation, the mechanism of which is referred to as the self-sustaining process (SSP). While there remain many further aspects of the turbulent state to be addressed, such as the structures and spectrum of the inertial subrange, a basis of understanding of wall turbulence comprises these fundamental aspects. One approach to studying these phenomena that has proven successful is to impose simplifications that isolate certain mechanisms of the turbulence dynamics. An example is restricting the size of the channel [8, 9]. The existence of the minimal channel for sustaining turbulence isolated the streak structure component of the SSP,[1] and showed that the SSP cycle comprises interactions among the streak, a pair of streamwise vortices of opposite sign (referred to as rolls) and the perturbation field. In minimal channel turbulence a single streak fits in the channel and the streamwise extent of the channel is of such length that the perturbations can adopt a structure allowing them to extract energy from the mean flow through a time dependent interaction [8, 10, 11]. Another example is imposing simplification of the dynamics by isolating specific regions of the turbulent flow. For example, it was found that when the inner and outer regions of the flow are isolated, turbulence was independently maintained in each region, arguing that an independent SSP is operating in these regions [11–13].

The minimal channel flow studies referred to above focused attention on the effort to obtain a self-consistent dynamical description of the interaction among the streak, the roll and the perturbation field. The streaks in minimal simulations are associated with the zero streamwise harmonic ($k_x = 0$). This observation motivates constructing a reduced complexity turbulence model based on the simplest closure of the LNS equations, which govern interaction between the streamwise mean flow (with the streak structure included) and the streamwise varying perturbation field (characterized by the $k_x \neq 0$ flow field). This system is obtained by augmenting LNS by only the nonlinearity resulting from including the feedback of the $k_x \neq 0$ perturbations on the $k_x = 0$ streamwise mean flow. This dynamics, which greatly restricts the nonlinearity of the NS, will be referred to as the restricted nonlinear system (RNL). Remarkably, as will be demonstrated in this work, the RNL self-sustains a realistic turbulent state, not only in minimal channels at low Reynolds numbers, but also in larger channels and at moderate Reynolds numbers.

The RNL system was introduced here as a simple extension of LNS that supports self-sustained turbulence. However, the RNL has deeper roots: it approximates the second-order cumulant closure of the NS, which is the basis of the Stochastic Structural Stability Theory (S3T). S3T defines a statistical mean state dynamical system, and implementations of this system have recently been used to develop theories of turbulence [14–20]. S3T employs an ensemble closure which produces autonomous statistical mean-state dynamics in which turbulent mean states exist as statistical equilibria. This makes the turbulent state available as an object for stability study, extending classical hydrodynamic stability theory, which addresses only the stability of stationary sample state solutions of the NS. In contrast, S3T can be used to determine the structural stability of statistical mean state entities such as attractors (characterised by a specific probability density function), so that for instance, when such an attractor becomes S3T-unstable the fluid state bifurcates to a new attractor characterized by a different probability density function. An example application of S3T is to a constant shear flow subjected to homogeneous turbulence excitation in which it is found that a bifurcation occurs when the Reynolds number exceeds a critical value resulting in a new stationary state comprising the mean flow with a roll-streak structure and a perturbation field supporting it in a new statistical steady state [7]. At an even higher Reynolds number a saddle-node bifurcation occurs in the S3T system, and the flow transitions to a time-dependent state that self-sustains, which is identified with transition to the

[1] In this context the term "streak" describes well-defined elongated regions of spanwise alternating bands of low and high speed fluid superimposed on the mean shear.

turbulent state. In the self-sustaining turbulent state the statistical mean state dynamics of S3T approaches the deterministic dynamics of the RNL. Recent work has exploited S3T dynamics to explore the role of streamwise coherent structures in turbulence, including the dynamics of the roll and streak structures [21, 22]. In this paper, we verify predictions of S3T for turbulence structure at high Reynolds numbers in pressure-driven channel flow by comparing flow statistics, structures, and dynamical diagnostics obtained from the RNL system to results obtained in the S3T system and to direct numerical simulations (DNS) of the full NS equations.

2. Modeling framework

Consider a pressure-driven plane Poiseuille flow maintained by application of the time-dependent pressure, $-G(t)x$, where x is the streamwise coordinate. The wall-normal direction is y and the spanwise direction is z. The lengths of the channel in the streamwise, wall-normal and spanwise direction are respectively L_x, $2h$ and L_z. The channel walls are at $y/h = 0$ and 2. Streamwise mean, spanwise mean and time-mean quantities are denoted respectively by an overbar, $\overline{\bullet} = L_x^{-1}\int_0^{L_x} \bullet \, dx$, square brackets, $[\bullet] = L_z^{-1}\int_0^{L_z} \bullet \, dz$ and a wide hat $\widehat{\bullet} = T^{-1}\int_0^{T} \bullet \, dt$, with T sufficiently long. The velocity, $\mathbf{u}$, is decomposed into its streamwise mean value, denoted $\mathbf{U}(y,z,t)$, and the deviation from the mean (the perturbation), $\mathbf{u}'(x,y,z,t)$, so that the flow velocity is $\mathbf{u} = \mathbf{U} + \mathbf{u}'$. The pressure gradient is similarly written as $\nabla p = \nabla\left(-G(t)x + P(y,z,t) + p'(x,y,z,t)\right)$. The NS can be then decomposed into an equation for the mean and an equation for the perturbation as follows:

$$\partial_t \mathbf{U} + \mathbf{U}\cdot\nabla\mathbf{U} - G(t)\hat{\mathbf{x}} + \nabla P - \nu\Delta\mathbf{U} = -\overline{\mathbf{u}'\cdot\nabla\mathbf{u}'} \ , \quad (1a)$$

$$\partial_t \mathbf{u}' + \mathbf{U}\cdot\nabla\mathbf{u}' + \mathbf{u}'\cdot\nabla\mathbf{U} + \nabla\mathbf{p}' - \nu\Delta\mathbf{u}' = -\left(\mathbf{u}'\cdot\nabla\mathbf{u}' - \overline{\mathbf{u}'\cdot\nabla\mathbf{u}'}\right) \ , \quad (1b)$$

$$\nabla\cdot\mathbf{U} = 0 \ , \quad \nabla\cdot\mathbf{u}' = 0 \ , \quad (1c)$$

where ν is the coefficient of kinematic viscosity. The x, y, z components of $\mathbf{U}$ are (U, V, W) and the corresponding components of $\mathbf{u}'$ are (u', v', w'). The streak component of the streamwise mean flow is denoted with U_s and defined as $U_s = U - [U]$. The V and W are the streamwise mean velocities of the roll vortices. Streamwise mean perturbation Reynolds stress components are therefore written as e.g. $\overline{u'u'}$, $\overline{u'v'}$.

The RNL approximation is obtained by neglecting the perturbation-perturbation interaction terms in Eq. (1b). The RNL system is:

$$\partial_t \mathbf{U} + \mathbf{U}\cdot\nabla\mathbf{U} - G(t)\hat{\mathbf{x}} + \nabla P - \nu\Delta\mathbf{U} = -\overline{\mathbf{u}'\cdot\nabla\mathbf{u}'} \ , \quad (2a)$$

$$\partial_t \mathbf{u}' + \mathbf{U}\cdot\nabla\mathbf{u}' + \mathbf{u}'\cdot\nabla\mathbf{U} + \nabla\mathbf{p}' - \nu\Delta\mathbf{u}' = 0 \ , \quad (2b)$$

$$\nabla\cdot\mathbf{U} = 0 \ , \quad \nabla\cdot\mathbf{u}' = 0 \ . \quad (2c)$$

Equation (2a) describes the dynamics of the streamwise mean flow, $\mathbf{U}$, which is driven by the divergence of the streamwise mean Reynolds stresses. These Reynolds stresses are obtained from Eq. (2b) which incorporates the influence of the time dependent streamwise mean flow $\mathbf{U}(y,z,t)$ on the streamwise varying perturbations $\mathbf{u}'$, but not the nonlinear interaction among the perturbations. Only the interaction of the perturbations directly on the streamwise mean flow, $\mathbf{U}$, is retained in the r.h.s. of Eq. (2a). Remarkably, RNL self-sustains turbulence by incorporating this one essential nonlinear interaction, in the absence of which a self-sustained turbulent state cannot be established [23, 24].

Table 1. Simulation parameters. $[L_x, L_z]/h$ is the domain size in the streamwise, spanwise direction. N_x, N_z are the number of Fourier components after dealiasing and N_y is the number of Chebyshev components. Re_τ is the Reynolds number of the simulation based on the friction velocity and $[L_x^+, L_z^+]$ is the channel size in wall units.

abbreviation	$[L_x, L_z]/h$	$N_x \times N_z \times N_y$	Re_τ	$[L_x^+, L_z^+]$
NS350	$[\pi\ ,\ \pi]$	$128 \times 255 \times 193$	357.1	$[1122, 1122]$
RNL350	$[\pi\ ,\ \pi]$	$128 \times 255 \times 193$	353.5	$[1111, 1111]$
NS950	$[\pi\ ,\ \pi/2]$	$256 \times 255 \times 385$	939.9	$[2953, 1476]$
RNL950	$[\pi\ ,\ \pi/2]$	$256 \times 255 \times 385$	882.4	$[2772, 1386]$

3. Numerical approach and simulation parameters

The data were obtained from a DNS of Eqs. (1) and from the RNL that is directly associated with the DNS. Both the DNS and its directly associated RNL are integrated with no-slip boundary conditions in the wall-normal direction and periodic boundary conditions in the streamwise and spanwise directions. The dynamics were expressed in the form of evolution equations for the wall-normal vorticity and the Laplacian of the wall-normal velocity, with spectral discretization and Fourier dealiasing in the two wall-parallel directions, and Chebychev polynomials in the wall-normal direction [25]. Time stepping was implemented using the third-order semi-implicit Runge-Kutta method.

Quantities reported in outer-units lengths are scaled by the channel half-width, h, and time by h/u_τ, and the corresponding Reynolds number is $\mathrm{Re}_\tau = u_\tau h/\nu$ where $u_\tau = \sqrt{\nu\, \mathrm{d}U/\mathrm{d}y|_\mathrm{w}}$ ($\mathrm{d}U/\mathrm{d}y|_\mathrm{w}$ is the shear at the wall) is the friction velocity. Inner-units lengths are scaled by $h_\tau = \mathrm{Re}_\tau^{-1} h$ and time by $\mathrm{Re}_\tau^{-1} h/u_\tau$. Velocities scaled by the friction velocity u_τ will be denoted with the superscript $+$, which indicates inner-unit scaling.

4. Comparison of turbulence structure and dynamics diagnostics between NS and RNL

In this section we compare turbulence diagnostics obtained from self-sustained turbulence in the RNL system, Eqs. (2), to diagnostics obtained from the parallel DNS of the NS, Eqs. (1). The geometry and resolution of the NS and RNL cases are given in Table 1. Results are reported for Poiseuille turbulence at either $\mathrm{Re}_\tau = 350$ or $\mathrm{Re}_\tau = 950$. The RNL simulations were initialized with an NS state and run until a steady state was established. The RNL simulations produce self-sustained turbulence with the time mean estimated Re_τ values reported in Table 1, which are close to the Re_τ values of the NS turbulent state. Henceforth, the RNL simulations will be identified with the Re_τ value of the corresponding NS.

The turbulent mean profiles for the NS and RNL simulations with $\mathrm{Re}_\tau = 350$ and $\mathrm{Re}_\tau = 950$ are shown in outer variables in Fig. 1(a,c), while Fig. 1(b,d) shows the same data in inner-wall units. Previous simulations in Couette turbulence [22] at lower Reynolds numbers ($\mathrm{Re}_\tau = 65$) showed very small differences between the mean turbulent profile in NS and RNL. These simulations at larger Reynolds numbers show significant differences in the mean turbulent profiles sustained by NS and RNL. This is especially pronounced in the outer regions, where RNL produces a mean turbulent profile with substantially smaller shear. Both profiles produce a logarithmic layer. However, the shear in these logarithmic regions are different: the von Kármán constant of NS at $\mathrm{Re}_\tau = 950$ is $\kappa = 0.4$, while for RNL it is $\kappa = 0.77$. Formation of a logarithmic layer indicates that the underlying dynamics producing the logarithmic layer are retained in RNL. Because RNL maintains essentially the same stress and variance as NS in the logarithmic layer with a smaller shear, RNL is in this sense more efficient than NS in transferring

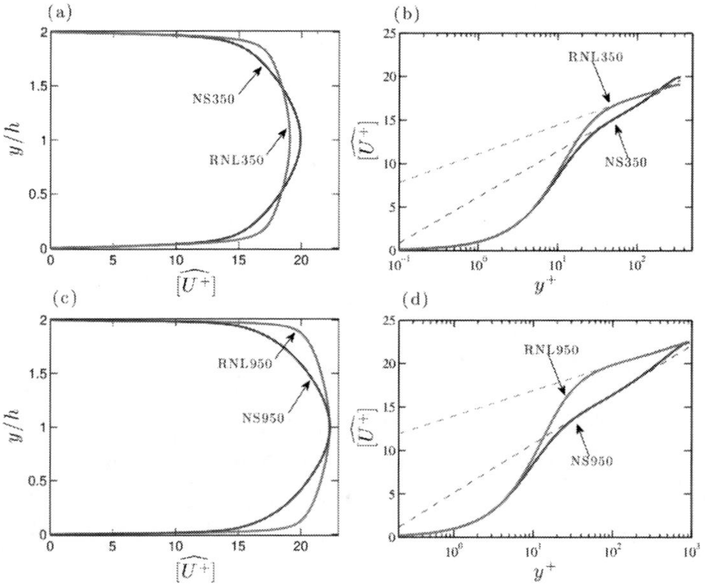

Figure 1. Streamwise velocity $\widehat{[U^+(y)]}$ for the simulations in Table 1. (a) NS350 and RNL350 simulations, (c) NS950 and RNL950 simulations. The corresponding profiles are shown in (b) and (d) in wall units. The dashed lines indicate the best fit to the law of the wall, $\widehat{[U^+]} = (1/\kappa)\log(y^+) + C$, with coefficients: NS350: $\kappa = 0.44$, $C = 6.1$, RNL350: $\kappa = 0.71$, $C = 11.1$, NS950: $\kappa = 0.40$, $C = 5.1$, RNL950: $\kappa = 0.77$, $C = 14.0$.

energy from the mean to the perturbations.

A comparison of perturbation statistics in RNL and NS is shown in Fig. 2 for $\mathrm{Re}_\tau = 950$. The streamwise perturbation velocity fluctuations are significantly more pronounced in RNL, and the magnitude of the streak in RNL exceeds significantly the streak magnitude in NS in the inner wall region (cf. Fig. 2a). In contrast, the wall-normal and spanwise fluctuations in RNL are less pronounced than in NS (cf. Fig. 2b,c) and similarly the streak fluctuations in the outer

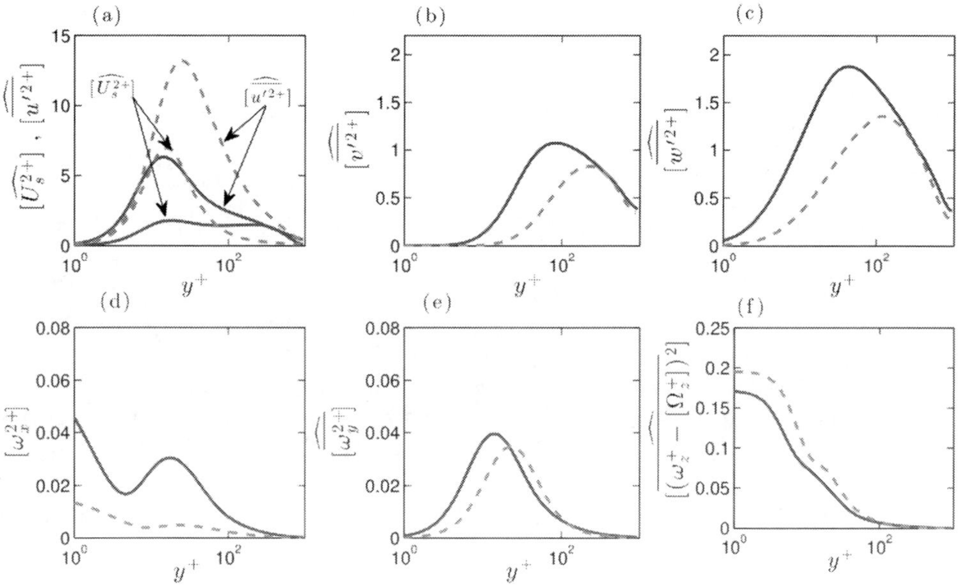

Figure 2. Comparison of velocity and vorticity fluctuations between NS950 and RNL950. Shown are (a) $\widehat{[U_s^{2+}]}$, $\widehat{[u'^{2+}]}$, (b) $\widehat{[v'^{2+}]}$, (c) $\widehat{[w'^{2+}]}$, (d) $\widehat{[\omega_x^{+2}]}$, (e) $\widehat{[\omega_y^{+2}]}$ and (f) $\widehat{[(\omega_z^+ - [\Omega_z^+])^2]}$, where $\Omega_z^+ = -\partial U^+/\partial y^+$ for the NS950 (solid) and RNL950 (dashed).

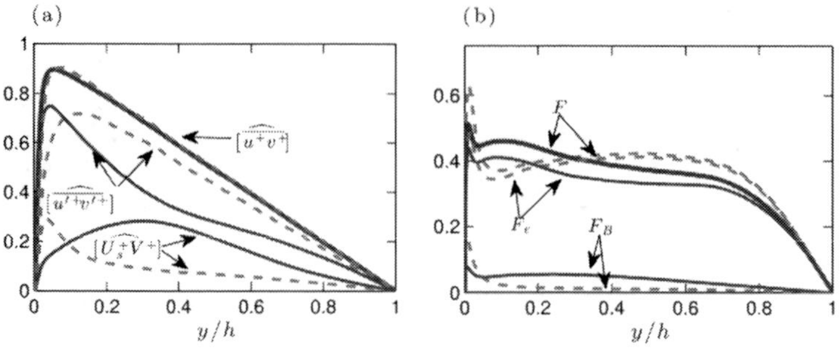

Figure 3. (a) The Reynolds stress component, $\widehat{[\overline{uv}]}(y)$ in NS950 (solid) and in RNL950 (dashed). Also shown are each of the terms, $\widehat{[U_s V]}$ and $\widehat{[\overline{u'v'}]}$ that sum to $\widehat{[\overline{uv}]}$. Although the NS and RNL values of the total $\widehat{[\overline{uv}]}$ are almost identical, the contribution of $\widehat{[U_s V]}$ and $\widehat{[\overline{u'v'}]}$ differ in NS and RNL. (b) Structure function, F, in NS950 (solid) and RNL950 (dashed). Shown are $F_B = -\widehat{[UV]}\,(\widehat{[U]^2}\widehat{[V]^2})^{-1/2}$, $F_e = -\widehat{[\overline{u'v'}]}\,(\widehat{[\overline{u'^2}]}\widehat{[\overline{v'^2}]})^{-1/2}$ and $F = -\widehat{[\overline{uv}]}\,(\widehat{[\overline{u^2}]}\widehat{[\overline{v^2}]})^{-1/2}$.

region are also less pronounced in RNL (cf. Fig. 2a). The structure of the fluctuations of the vorticity components as a function of y is shown in Fig. 2(d,e,f). The ω_z and ω_y fluctuations are similar in NS and RNL. The large ω_z fluctuations are associated with the shear of the streamwise velocity, while the ω_y fluctuations are associated with the streamwise streak structure. Only the ω_x fluctuations differ appreciably in amplitude between RNL and NS. This vorticity component is primarily associated with the fluctuations in the streamwise roll circulations, which are responsible for maintaining the streak, and which is central to sustaining the turbulence.

Despite this difference in the r.m.s. values of the velocity fluctuations, both RNL and NS produce very similar uv Reynolds stress. The increased amplitude of the velocity fluctuations in RNL is consistent with the fact that RNL and NS produce nearly the same energy dissipation rate. The Reynolds stress $\widehat{[\overline{uv}]}$ is the sum of $\widehat{[U_s V]}$ and $\widehat{[\overline{u'v'}]}$. Comparison of the wall-normal distribution of the time mean of these two components of the Reynolds stress is shown in Fig. 3(a). Because the turbulence in NS and RNL is sustained with essentially the same pressure gradient, the sum of these Reynolds stresses is the same linear function of y outside the viscous layer. The Reynolds stress is dominated by the perturbation Reynolds stress $\widehat{[\overline{u'v'}]}$ in both simulations, with the RNL stress penetrating farther from the wall. This is consistent with the fact that the perturbation structure in RNL has larger scale. This can be seen in a comparison of the NS and RNL perturbation structure shown in Fig. 4. Note that the Reynolds stress $\widehat{[U_s V]}$ associated with the streak and roll in the outer region of the NS simulation is larger than that in RNL. Further, the average correlation between the perturbation u' and v' fields are almost the same in both simulations, while the correlation between the U_s and V in the RNL is much smaller than that in NS in the outer layer. This is seen in a plot of the structure function (cf. [26]) shown in Fig. 3(b).

Turning now to the NS950 and RNL950 simulations, (y, z)-plane snapshots of the streamwise mean flow component (corresponding to $k_x = 0$ streamwise wavenumber) are shown in Fig. 5. Contours of the streamwise streak flow field, U_s, are shown together with vectors of the streamwise mean (V, W) field, which indicates the velocity components of the large-scale roll structure. The presence of organized streaks and associated rolls is evident both in the inner-wall and in the outer-wall region. Figure 6 shows a three-dimensional perspective of the flow of the NS and RNL simulations, in which all of the k_x components of the velocity field are included. Note that in the RNL there is no visual evidence of the $k_x = 0$ roll/streak structure which is required

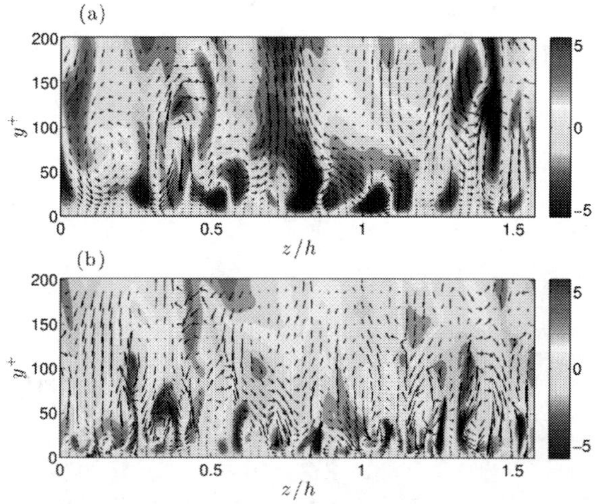

Figure 4. Perturbation structure, $\mathbf{u}'^+$ in (y, z) plane cross-section for (a) RNL950 and (b) NS950 in the inner wall region, $0 \leq y^+ \leq 200$. Both panels show contours of the u'^+ field, superimposed with components the (v'^+, w'^+) velocities.

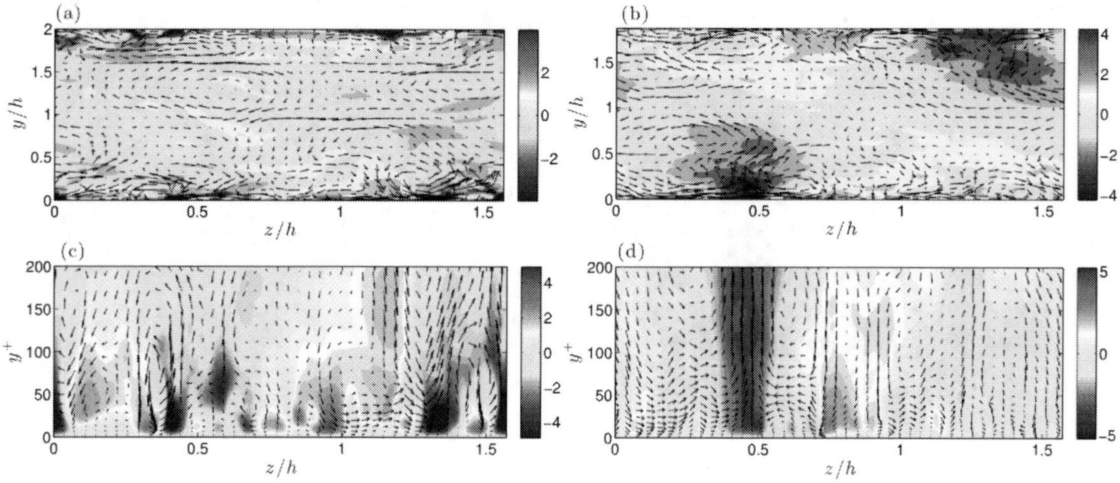

Figure 5. Instantaneous streamwise average flow velocity $\mathbf{U}^+$ (the $k_x = 0$ component of the flow) shown as a (y, z) plane cross-section for (a), (c) RNL950 and (b), (d) NS950. All panels show contours of the streak velocity, $U_s^+ = U^+ - [U^+]$, superimposed with the components of the (V^+, W^+) velocities. The top panels show the whole channel, while the bottom panels show the inner-wall region, $0 \leq y^+ \leq 200$.

by the restrictions of RNL dynamics to be the primary structure responsible for maintaining the RNL self-sustained turbulent state. Rather, the most energetic streamwise harmonic ($hk_x = 2$ for this channel) is most prominent and dominates the perturbation structure.

A comparison of the spectral energy densities of the velocity fields as a function of streamwise and spanwise wavenumber, (k_x, k_z), provides an alternative view of the turbulent structure. The premultiplied spectral energy densities of each of the three contributions to the kinetic energy, E_{uu}, E_{vv} and E_{ww} at channel heights $y^+ = 20$, representative of the inner-wall region, and $y/h = 0.65$, representative of the outer-wall region, are shown in Fig. 7. Near the wall, RNL produces spanwise streak spacing and rolls similar to those in NS. The tendency of RNL to favor longer structure is also evident in these figures. The spectra for the outer region indicate a similar large-scale structure and good agreement in the spanwise spacing between RNL and NS. This figure establishes the presence of large-scale structure in the outer region in both RNL and NS. It has been noted that in NS, while the dominant large-scale structures scale linearly with distance from the wall in the inner-wall region, structures having the longest possible streamwise

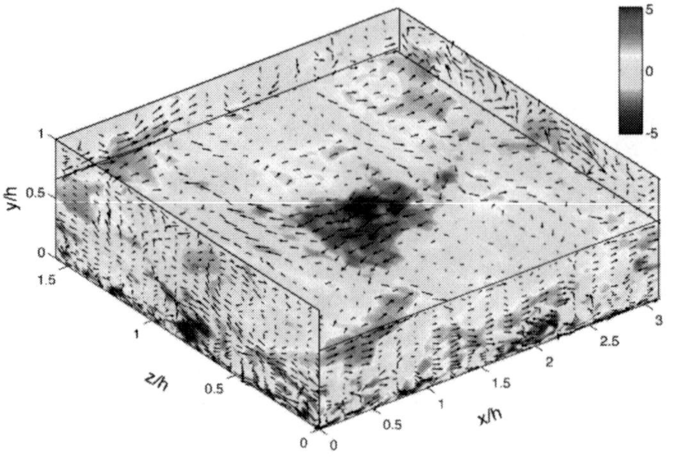

(a) **NS**

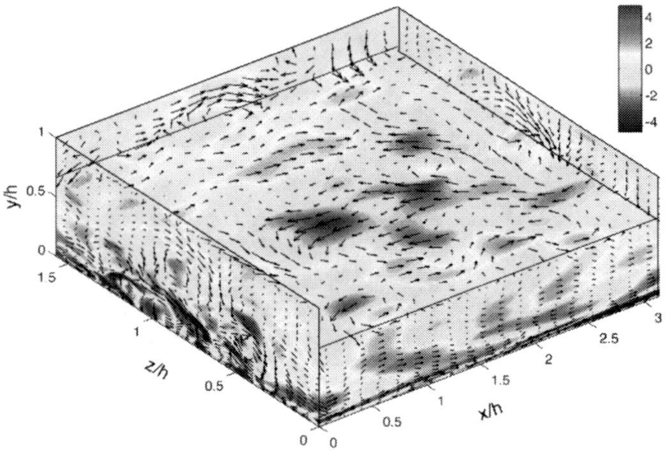

(b) **RNL**

Figure 6. Three-dimensional perspective plots of the flow at a single time for (a) NS950, and (b) RNL950, for the lower half of the channel, $0 \leq y/h \leq 1$. Both images show contours of the streak component plus streamwise perturbation, $U_s^+ + u'^+$. The central x-z panel shows the flow at $y/h = 0.65$. The superimposed vectors represent the $(U_s^+ + u'^+, w^+)$ velocities for the x-z panel, $(U_s^+ + u'^+, v^+)$ velocities for the x-y panels and (v^+, w^+) velocities for the y-z panels. The parameters of the simulations are given in Table 1.

scale dominate the flow variance in the outer regions at high Reynolds number [27, 28]. This linear scaling near the wall can be seen in Fig. 8, where the pre-multiplied spectral densities are shown as a function of the distance y and of the spanwise wavelength, k_z, as in [27, 28], for both NL and RNL. In both simulations the spanwise wavelength associated with the spectral density maxima increases linearly with wall distance, and this linear dependence persists up to $y/h \approx 0.5$ (or $y^+ \approx 450$). Beyond $y/h \approx 0.5$ structures assume the widest wavelength allowed in the channel, suggesting that simulations must be performed in wider boxes in future work (cf. discussion in Jiménez & Hoyas [28] and Flores & Jiménez [29]). Corresponding contour plots of spectral energy density as a function of streamwise wavelength and wall-normal distance are shown in Fig. 9. These plots show that the perturbation variance in the inner wall and outer wall region is concentrated in a limited set of streamwise components which is apparent in Fig. 6. The restriction of the streamwise structure is particularly pronounced in the case of RNL, in which the outer layer variance peaks at $hk_x = 4$, which scale the wall-normal velocity inherits. Note that the longest wavelength in these graphs is equal to the streamwise length of the box, not to the infinite wavelength associated with the energy of the streak/roll structure, which, as we will see, is the most energetic structure in the NS but not in the RNL.

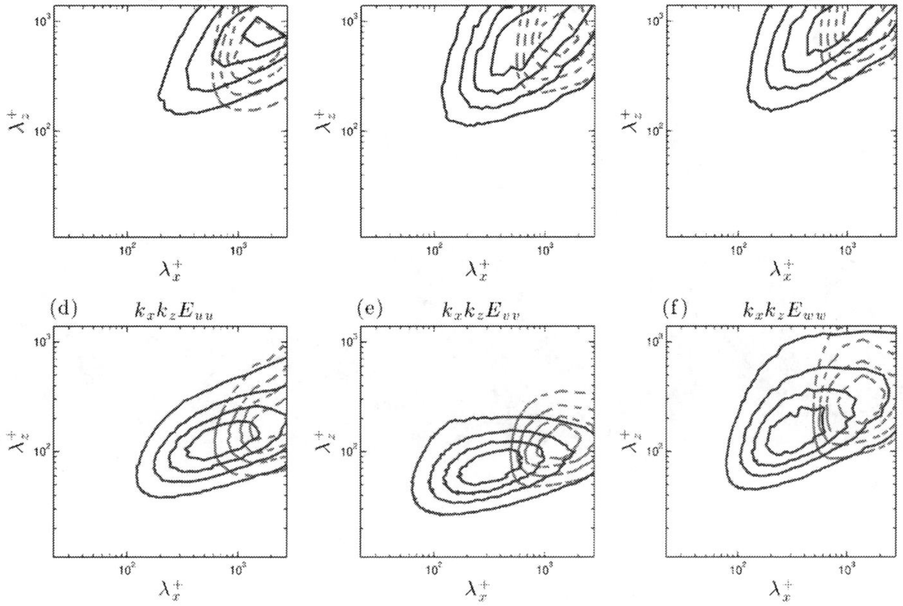

Figure 7. Contours of pre-multiplied power spectra $k_xk_zE_{ff}(k_x,k_z)$ with $f = u, v, w$, as a function of λ_x^+ and λ_z^+ for NS950 (solid) and RNL950 (dashed). (a), (b) and (c) show the spectral energy densities at wall distance $y/h = 0.65$ for the u, v and w respectively, while (d), (e) and (f) show the corresponding spectral energy densities at $y^+ = 20$. Contours are (0.2,0.4,0.6,0.8) times the maximum value of the corresponding spectrum. The maximum λ_x^+ and λ_y^+ are the lengths L_x^+, L_z^+ of the periodic channel.

5. The RNL system as a minimal turbulence model

RNL turbulence has the property that, when initiated with full NS turbulence, it spontaneously transitions to a self-sustaining turbulent state supported by a severely restricted set of streamwise Fourier components. This property is consistent with the structure of the RNL system, which retains only the interaction between the $k_x = 0$ and the $k_x \neq 0$ components, from which it follows that the only energy source for maintaining a perturbation is its interaction with the mean flow [7]. It is remarkable that at these fairly high Reynolds numbers only a small set of streamwise harmonics are maintained by this interaction with the mean flow. Moreover, this is the fundamental interaction maintaining RNL turbulence and, by implication, given the similarity in the structure and dynamics between them, of NS turbulence as well. Even if the RNL dynamics is initialized with an NS flow state with energy in all N_x components, the RNL turbulence eventually reduces to involve only the $n_x \ll N_x$ Fourier components with wavenumber $hk_x = (2\pi/L_x) \times (0, 1, \ldots, n_x - 1)$. We view this transition of NS turbulence to RNL turbulence as a process of distillation by which a small set of structures maintaining the turbulent state is identified, a result that was previously obtained in the case of self-sustained Couette turbulence at Re = 400 and Re = 1000 [7, 21]. In this previous work, a minimal-channel RNL simulation at Re = 400 was shown to self-sustain turbulence by the interaction of only two k_x components: $k_x = 0$ and the first harmonic in the channel. The analogous distillation process for the $\mathrm{Re}_\tau = 950$ simulation is shown in Fig. 10(a). The time evolution of the energy of the first 15 streamwise varying Fourier components in an NS simulation is shown in the left part of the figure ($u_\tau t/h < 100$), while the subsequent evolution of these components is shown in the

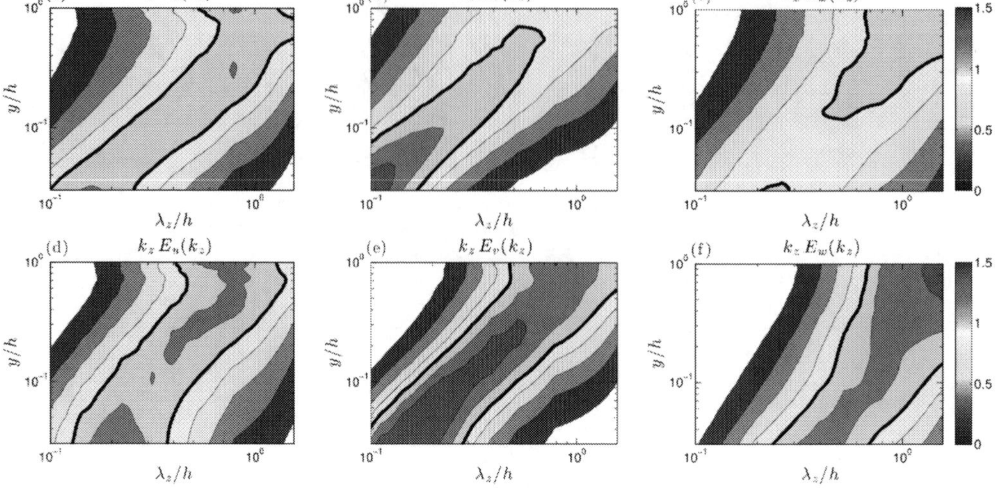

Figure 8. Normalized pre-multiplied spectral densities $k_z E_f(k_z) = k_z \sum_{k_x} E_{ff}(k_x, k_z)$, with $f = u, v, w$, as a function of spanwise wavelength, λ_z/h, and y/h. Spectral densities are normalized so that at each y the total energy, $\sum_{k_z} E_f(k_z)$, is the same. Shown are for NS950 (a) $k_z E_u(k_z)$, (b) $k_z E_v(k_z)$, (c) $k_z E_w(k_z)$ and for RNL950 (d) $k_z E_u(k_z)$, (e) $k_z E_v(k_z)$, (f) $k_z E_w(k_z)$. The isocontours are $0.2, 0.4, \ldots, 1.4$, and the thicker line marks the 1.0 isocontour.

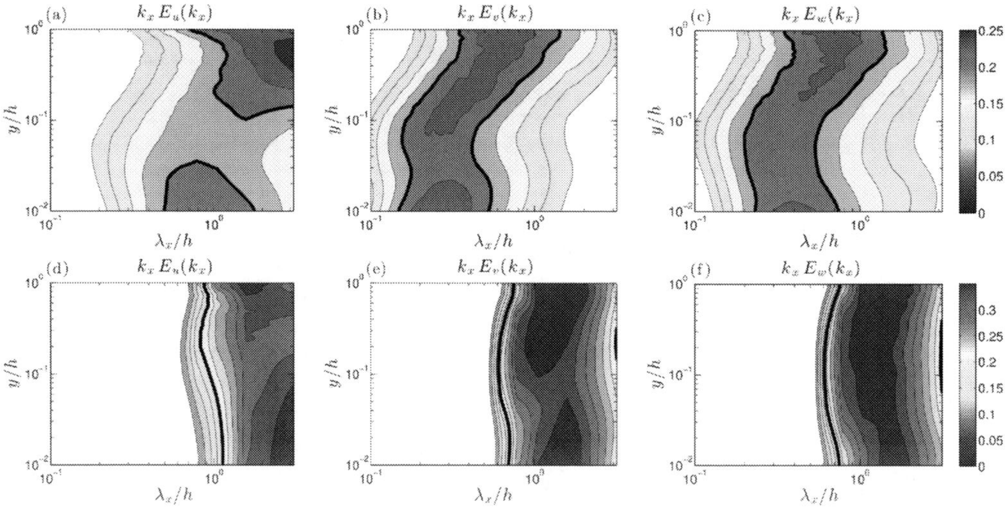

Figure 9. Normalized pre-multiplied spectral densities $k_x E_f(k_x) = k_x \sum_{k_z} E_{ff}(k_x, k_z)$, with $f = u, v, w$, as a function of streamwise wavelength, λ_x/h, and y/h. Spectral densities are normalized so that at each y the total energy, $\sum_{k_x} E_f(k_x)$, is the same. Shown are for NS950 (a) $k_x E_u(k_x)$, (b) $k_x E_v(k_x)$, (c) $k_x E_w(k_x)$ and for RNL950 (d) $k_x E_u(k_x)$, (e) $k_x E_v(k_x)$, (f) $k_x E_w(k_x)$. The isocontours are $0.1, 0.125, \ldots, 0.35$, and the thicker line marks the 0.2 isocontour.

right part, after the perturbation-perturbation interactions are suppressed at the indicated time, so that the turbulence evolves under RNL dynamics. It is evident from Fig. 10(a) that RNL950 turbulence in a channel with $L_x = \pi h$ retains only the six Fourier components $hk_x = 2, 4, \ldots, 12$ out of the $N_x = 127$ streamwise varying components that are present in the NS simulation. All components with $hk_x > 12$ decay exponentially. As a result, a transition occurs in which a reduced-complexity dynamics maintaining turbulence arises, which self-sustains turbulence despite this greatly restricted support in streamwise wavenumber space. That RNL maintains a turbulent state similar to that of NS with nearly the same Re_τ ($\mathrm{Re}_\tau = 882$ vs. $\mathrm{Re}_\tau = 940$) implies that these systems have approximately the same energy production and dissipation and that the n_x components retained in RNL assume the burden of accounting for this energy production and dissipation. Specifically, the components in NS that are not retained in RNL are responsible for approximately one third of the total energy dissipation, which implies that the components that are retained in RNL must increase their dissipation by that much.

6. Streak structure dynamics in NS and RNL

Large scale roll/streak structures are prominent in the inner layer as well as in the outer layer, both in NS and in RNL. The dynamics of this structure can be diagnosed using the time

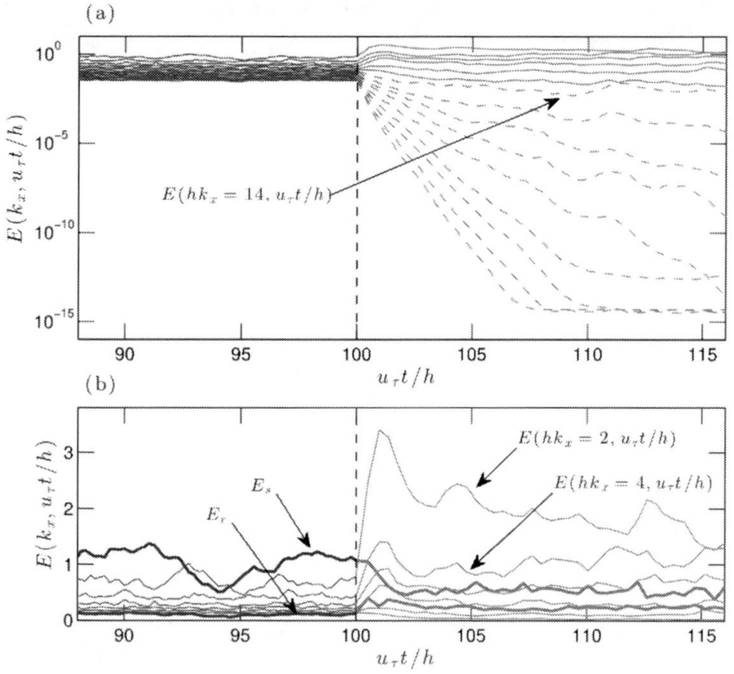

Figure 10. An NS950 simulation up to $u_\tau t/h = 100$ (indicated with the vertical line) is continued subsequently under RNL dynamics. Shown are (a) Energy of the first 15 streamwise varying Fourier components ($hk_x = 2, 4, \ldots, 30$). The energy of the Fourier components decreases monotonically with wavenumber. Decaying Fourier components are indicated with dashed lines. After the transition to RNL dynamics all components with $hk_x \geq 14$ decay ($hk_x = 14$ decays, although it is not shown in this figure). Asymptotically the dynamics of the RNL950 turbulence is maintained by interaction between the set of surviving $hk_x = 2, 4, \ldots, 12$ Fourier components and the mean flow ($k_x = 0$). (b) Detailed view showing the energy of the mean and surviving perturbation components during the transition from NS to RNL dynamics, in which the total energy increased by 10%. For the $k_x = 0$ shown are: the streak energy, $E_s = (hL_z)^{-1} \int \mathrm{d}y\, \mathrm{d}z\, U_s^{+2}/2$, and roll energy, $E_r = (hL_z)^{-1} \int \mathrm{d}y\, \mathrm{d}z\, (V^{+2} + W^{+2})/2$. The energy of the $hk_x = 2, 4, 6, 8$ components increases rapidly during the adjustment after transition to RNL dynamics. Note that the total energy in the perturbation $k_x \neq 0$ components decreases from 0.91 in the NS950 (0.56 being in the components that survive in the RNL) to 0.78 in RNL950. Also the roll/streak energy decreases from 1.1 in NS950 to 0.8 in RNL950, while the energy of the $k_x = k_z = 0$ component increases from 397 to 448.

evolution of the energy of each k_x component during the transition from NS to RNL, shown in Fig. 10(b). It can be seen that in NS the energy associated with the streamwise mean structure with $k_x = 0$ and $k_z \neq 0$ is dominant among the structures that deviate from the mean flow $[U]$. In the inner layer the interaction of roll/streak structures with the $k_x \neq 0$ perturbation field maintains turbulence through an SSP [7, 10, 11]. The RNL system provides an especially simple manifestation of this SSP, as its dynamics comprise only interactions between the mean ($k_x = 0$) and perturbation ($k_x \neq 0$) components. The fact that RNL self-sustains a counterpart turbulent state provides strong evidence that the RNL SSP captures the essential dynamics of turbulence in the inner-wall region.

The structure of the RNL system compels the interpretation that the time dependence of the SSP cycle, which might appear at first to consist of a concatenation of random and essentially unrelated events, is instead an intricate interaction dynamics among streaks, rolls and perturbations that produces the $\mathbf{U}(y, z, t)$ which, when introduced in Eq. (2b), results in the generation of an evolving perturbation Lyapunov structure with exactly zero Lyapunov exponent. S3T identifies this exquisitely contrived SSP cycle, which comprises the generation of the streak through lift-up by the rolls, and the maintenance of the rolls by torques induced by the perturbations which themselves are maintained by an essentially time-dependent parametric non-normal interaction with the streak (rather than e.g. inflectional instability of the streak structure) [7].

Vanishing of the Lyapunov exponent associated with the SSP is indicative of a feedback control process acting between the streaks and the perturbations by which the parametric instability of the perturbations on the time-dependent streak is reduced so that it asymptotically maintains a zero Lyapunov exponent. Examination of the transition from the NS to the RNL, shown by the simulation diagnostics in Fig. 10(b), reveals the action of this controller. When the interaction among the perturbations in Eq. (1b) is switched off so that the simulation is governed by RNL dynamics, we observe a sudden increase of the energy of the surviving $k_x \neq 0$ components, as shown by the rapid increase of the energy in the $hk_x = 2, 4, 6, 8$ components. An increase of the energy of these components is expected because the dissipative effect of the perturbation-perturbation nonlinearity that acts on these components is removed in RNL. As these modes grow, the SSP cycle quickly adjusts to a new turbulent equilibrium state characterized by reduced streak amplitude and increased energy in the largest streamwise scales. This SSP cycle is more efficient in the sense that a self-sustained turbulence with approximately the same Re_τ as that in NS is maintained with smaller mean shear. This turbulent state is dominated by high-amplitude fluctuations of the $hk_x = 2, 4, 6, 8$ components, as well as in the components associated with the wall-normal and spanwise direction. This can be seen in a comparison of the NS and RNL perturbation structure (the velocity field corresponding to $k_x \neq 0$) shown in Fig. 4. The perturbations in RNL simultaneously reduce the shear of the mean flow and maintain a reduced amplitude streak in the outer layer. A comparison of the shear, of the r.m.s. V velocity, and of the r.m.s. streak velocity, U_s, in the outer layer is shown as a function of y in Fig. 11, from which it can be seen that the reduction of the amplitude of the streak in RNL is equal to the reduction in the mean flow shear. It is important to note that these dependencies are integral to the SSP cycle, and specifically of its feedback control that determines the statistical steady state, and must be understood in the context of the cycle.

In the discussion above we have assumed that the presence of roll and streak structure in the outer layer in RNL indicates the existence of an SSP cycle there as well, and by implication also

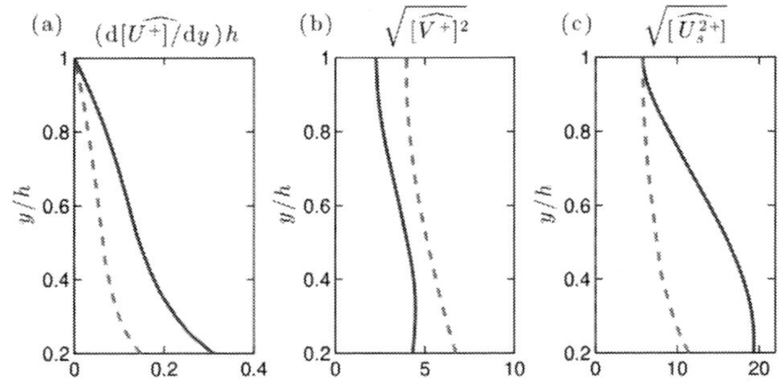

Figure 11. (a) Comparison of the turbulent mean shear, $(\mathrm{d}\widehat{[U^+]}/\mathrm{d}y)h$, (b) the r.m.s. of $[V^+]$, and (c) the r.m.s. of the streak velocity, U_s^+, in the outer layer, $0.2 \le y/h \le 1$. NS950 (solid) and RNL950 (dashed).

in NS. In order to show this consider the momentum equation for the streamwise streak:

$$\partial_t U_s = \underbrace{-\Big(V\,\partial_y U - [V\,\partial_y U]\Big) + \Big(W\,\partial_z U - [W\,\partial_z U]\Big)}_{\mathrm{A}} - \underbrace{-\Big(\overline{v'\,\partial_y u' - [v'\,\partial_y u']}\Big) - \Big(\overline{w'\,\partial_z u' - [w'\,\partial_z u']}\Big)}_{\mathrm{B}} + \nu\,\Delta U_s\ . \qquad (3)$$

Term A in Eq. (3) is the contribution to the streak acceleration by the 'lift-up' mechanism and the 'push-over' mechanism, which represent transfer to streak momentum by the mean wall-normal and spanwise velocities respectively. Term B in Eq. (3) is the contribution to the streak momentum by the perturbation Reynolds stress divergence (structures with $k_x \neq 0$).

In order to identify the mechanism of streak maintenance we examine whether terms A and B accelerate or decelerate the streaks by evaluating from the simulation the contribution of the terms $I_A(t) = h^{-1}\int_R \mathrm{d}y\ \mathcal{I}_A(y,t)$ and $I_B(t) = h^{-1}\int_R \mathrm{d}y\ \mathcal{I}_B(y,t)$, where $\mathcal{I}_A(y,t) = hu_\tau^{-2}L_z^{-1}\int \mathrm{d}z\ \mathrm{sgn}(U_s)\times(\text{Term A})$ and $\mathcal{I}_B(y,t) = hu_\tau^{-2}L_z^{-1}\int \mathrm{d}z\ \mathrm{sgn}(U_s)\times(\text{Term B})$, to streak maintenance as a function of time and region R of the flow (similar results are obtained by multiplying with U_s to obtain an energy rather than a momentum budget for U_s). The results of this calculation are shown in Fig. 12 for R extending over the the inner region, the outer region and the whole channel. In the inner and outer wall regions for both NS and RNL the streak structure is supported primarily by the lift-up mechanism while the Reynolds stress divergences oppose the streak. While the magnitude of the acceleration by the lift-up and the deceleration by the Reynolds stress divergence are nearly the same in both NS and RNL in the inner region, in the outer region the acceleration by the lift-up in the RNL is about half of that in the NS, due to the smaller mean flow shear in RNL. The wall-normal structure of the time-mean $\widehat{\mathcal{I}_A}$ and $\widehat{\mathcal{I}_B}$ are shown in Fig. 13(a,b). We conclude that in NS and RNL the only positive contributions to the outer layer streaks are induced by the lift-up from the roll circulation, despite the small shear in this region. We next consider the dynamics maintaining the roll circulation.

7. Roll dynamics: maintenance of mean streamwise vorticity in NS and RNL

We have established that the roll circulation is not only responsible for streak maintenance in the inner layer but also in the outer layer. We now examine the mechanism of roll maintenance using streamwise averaged vorticity, $\Omega_x = \partial_y W - \partial_z V$, in the outer layer as a diagnostic. In order for roll circulation to be maintained against dissipation there must be a continuous generation of Ω_x. There are two possibilities for the maintenance of Ω_x in the outer layer: either Ω_x is generated locally in the outer layer, or it is advected from the near-wall region.

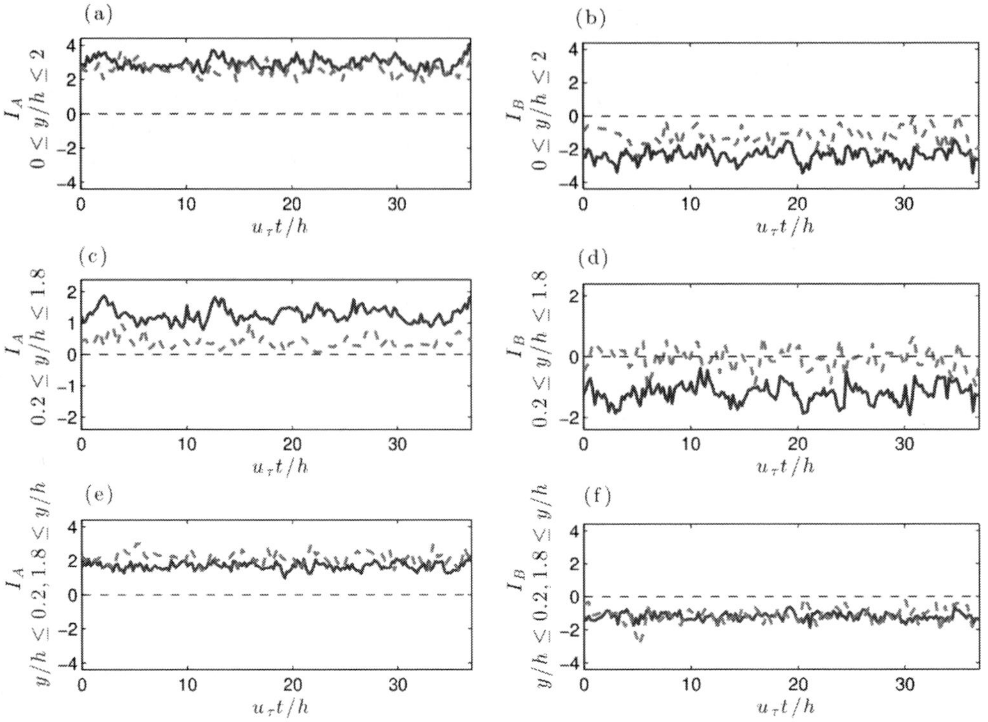

Figure 12. Streak-generation terms. (a) $I_A(t)$ over the whole channel with time-mean value 2.95 for NS950 and 2.59 for RNL950. (c) $I_A(t)$ over the outer region, $0.2 \le y/h \le 1.8$, with time-mean value 1.97 for NS950 and 0.4 for RNL950. (e) $I_A(t)$ over the inner region, $0 \le y/h \le 0.2$ $1.8 \le y/h \le 2$, with time-mean value 1.68 for NS950 and 2.19 for RNL950. (b) $I_B(t)$ over the whole channel with time-mean value -2.4 for NS950 and -1.21 for RNL950. (d) $I_B(t)$ over the outer region, $0.2 \le y/h \le 1.8$, with time-mean value -1.2 for NS950 and -0.08 for RNL950. (f) $I_B(t)$ over the inner region, $0 \le y/h \le 0.2$ $1.8 \le y/h \le 2$, with time-mean value -1.2 for NS950 and -1.13 for RNL950.

From Eq. (1a) we obtain that Ω_x satisfies the equation:

$$\partial_t \Omega_x = \underbrace{-V\,\partial_y \Omega_x - W\,\partial_z \Omega_x}_{\text{C}} + \underbrace{(\partial_{zz} - \partial_{yy})\left(\overline{v'w'}\right) - \partial_{yz}\left(\overline{w'^2} - \overline{v'^2}\right)}_{\text{D}} + \nu\,\Omega_x\,\Delta\Omega_x\ . \tag{4}$$

Term C expresses the streamwise vorticity tendency due to advection of Ω_x by the streamwise mean flow (V, W). Because there is no vortex stretching contribution to Ω_x from the (V, W) velocity field, this term only advects the Ω_x field and cannot sustain it against dissipation. However, this term may be responsible for systematic advection of Ω_x from the inner to the outer layer. Term D is the torque induced by the perturbation field. This is the only term that can maintain Ω_x. The overall budget for square streamwise vorticity in the region R, $y_1 \le y \le y_2$, $0 \le z \le L_z$, is given by:

$$\partial_t \int_{y_1}^{y_2} \mathrm{d}y \left[\frac{1}{2}\Omega_x^2\right] = \underbrace{-\left[\frac{1}{2}\Omega_x^2\,V\right]\Bigg|_{y=y_1}^{y=y_2}}_{h^{-2}u_\tau^3\,I_C} + \underbrace{\int_{y_1}^{y_2} \mathrm{d}y\,\Big[\Omega_x \times \text{Term D}\Big]}_{h^{-2}u_\tau^3\,I_D} + \nu \int_{y_1}^{y_2} \mathrm{d}y\,\Big[\Omega_x\,\Delta\Omega_x\Big]\ . \tag{5}$$

where we have defined $I_C(t) = h^{-1}\int_R \mathrm{d}y\ \mathcal{I}_C(y,t)$ and $I_D(t) = h^{-1}\int_R \mathrm{d}y\ \mathcal{I}_D(y,t)$, with $\mathcal{I}_C(y,t) = h^3 u_\tau^{-3} L_z^{-1} \int \mathrm{d}z\ \Omega_x \times (\text{Term C})$ and $\mathcal{I}_D(y,t) = h^3 u_\tau^{-3} L_z^{-1} \int \mathrm{d}z\ \Omega_x \times (\text{Term D})$. Term

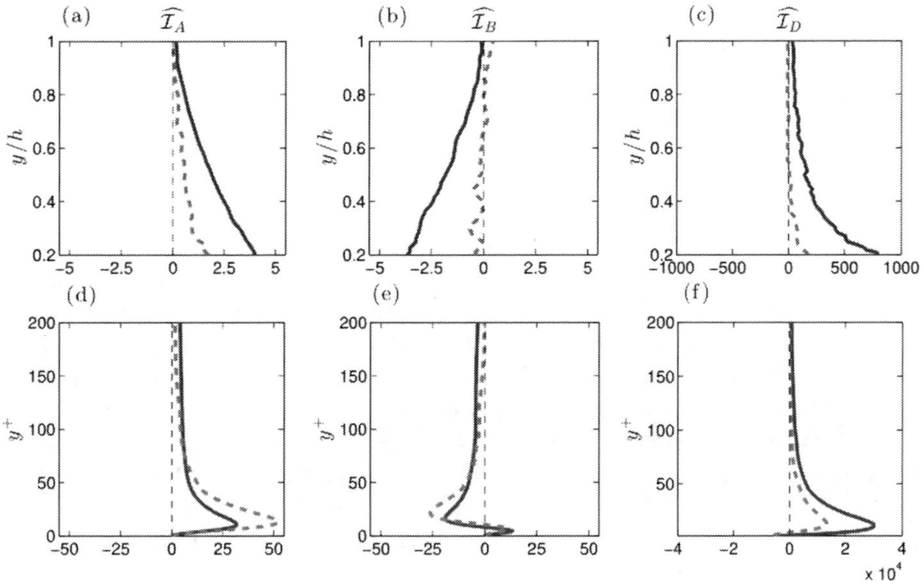

Figure 13. (a) Contribution to streak acceleration from the lift-up mechanism $\widehat{\mathcal{I}_A}$. (b) Contribution to streak acceleration from the perturbation Reynolds stress divergence $\widehat{\mathcal{I}_B}$. (c) contribution to streamwise mean vorticity generation from perturbation Reynolds stress induced torques $\widehat{\mathcal{I}_D}$ (cf. section 7). Results from NS950 (solid) and RNL950 (dashed). Upper panels show structure in the outer layer, $0.2 \leq y/h \leq 1$, lower panels show the structure in the inner layer, $0 \leq y^+ \leq 200$.

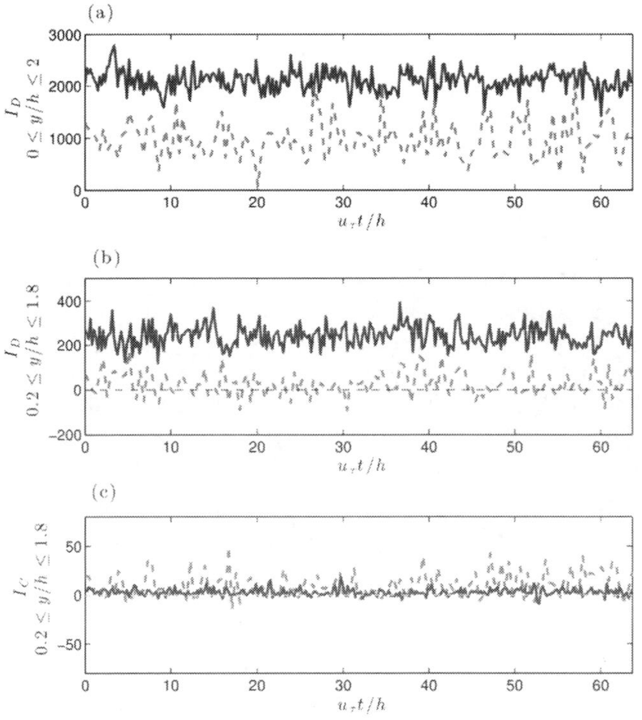

Figure 14. Time series of the contribution to the rate of change of $\int dy\, [\Omega_x^2/2]$ by perturbation torques, $I_D(t)$, and from advection of streamwise mean vorticity by the mean flow, $I_C(t)$, for NS950 (solid) and RNL950 (dashed). (a) I_D for the whole channel, $0 \leq y/h \leq 2$ ($I_C = 0$ in this case). The time-mean $\widehat{I_D}$ is 2103.6 for NS950 and 982.8 for RNL950. (b) I_D over the outer layer, $0.2 \leq y/h \leq 1.8$. The time-mean $\widehat{I_D}$ for this region is 242.5 for NS950 and only 28.7 for RNL950. (c) I_C for the outer layer $0.2 \leq y/h \leq 1.8$. The time-mean $\widehat{I_C}$ is 2.9 for NS950 and 11.2 for RNL950. These figures show that in NS the roll is maintained locally by the perturbation Reynolds stresses and that in RNL the major contribution to the roll maintenance is contributed locally from perturbation induced torques.

I_C represents the flux of vorticity into the region by the streamwise averaged wall-normal velocity, V.

Time series of the contributions from $I_C(t)$ and $I_D(t)$ to the Ω_x production for NS950 and RNL950, shown in Fig. 14, demonstrate that Ω_x is primarily generated in situ by Reynolds stress torques. The corresponding wall-normal structure of the time-mean $\widehat{\mathcal{I}_D}$, representing the local contribution to streamwise mean vorticity generation from perturbation Reynolds stress induced torques is shown in Fig. 13(c). Note that for NS in the outer layer, the streamwise mean vorticity generation by the Reynolds stress is strongly positive at each instant. This is so despite the fact that the r.m.s. V velocity is smaller in the NS than in the RNL, as seen in Fig. 11, implying greater streamwise vorticity dissipation in the NS. It could be argued that the positive value of the generation I_D found in NS and RNL is a consequence of the finite streamwise extent of the channel and as the channel length increases, assuming that there is no systematic correlation between streamwise average torques and streak structure, I_D should decrease as $1/\sqrt{L_x}$ and in the limit of an infinite channel the NS and RNL should sustain no Ω_x. However, S3T theory shows that there is a systematic correlation between streaks and roll generation by the perturbation torque resulting from the deformation of the turbulence by the streak and, as a result, it predicts that in the limit $L_x \to \infty$, I_D should asymptote to a finite non-zero value at least in RNL.

Having established that the streamwise vorticity in the outer layer is generated in situ from local Reynolds stress divergences we conclude that the SSP cycle is operational in the outer layer, just as in the inner one.

8. Discussion and Conclusions

We have established that RNL self-sustains turbulence at moderate Reynolds numbers in pressure-driven channel flow, despite its greatly simplified dynamics when compared to NS. Remarkably, in the RNL system, the turbulent state is maintained by a small set of structures with low streamwise wavenumber Fourier components (at $\mathrm{Re}_\tau = 950$ with the chosen channel the SSP involves only the $k_x = 0$ streamwise mean and the next six streamwise Fourier components). Not only that, but this minimal turbulent dynamics arises spontaneously when the RNL system is initialized by NS turbulence at the same Reynolds number. In this way RNL spontaneously produces a turbulent state of reduced complexity. RNL identifies an exquisitely contrived SSP cycle which has been previously identified to comprise the generation of the streak through lift-up by the rolls, and the maintenance of the rolls by torques induced by the perturbations which themselves are maintained by an essentially time-dependent parametric non-normal interaction with the streak (rather than e.g. inflectional instability of the streak structure) [7]. The vanishing of the Lyapunov exponent associated with the SSP is indicative of a feedback control process acting between the streaks and the perturbations by which the parametric instability that sustains the perturbations on the time dependent streak is reduced to zero Lyapunov exponent, so that the turbulence neither diverges nor decays.

We have established that both NS and RNL produce a roll/streak structure in the outer layer and that an SSP is operating there despite the low shear in this region. It has been shown elsewhere that turbulence self-sustains in the logarithmic layer in the absence of boundaries [13]. This is consistent with our finding that an SSP cycle exists in both the inner-layer and outer-layer.

The turbulence maintained in RNL is closely related to its associated NS turbulence, and both exhibit a logarithmic layer, although with substantially different von Kármán constants. Existence of a logarithmic layer is a fundamental requirement of asymptotic matching between regions with different spatial scaling, as was noted by Millikan [30]. However, the exact value of the von Kármán constant does not have a similar fundamental basis in analysis, and RNL turbulence, which is closely related to NS turbulence but more efficient in producing Reynolds stress, maintains as a consequence a smaller shear and therefore greater von Kármán constant. Specifically, we have determined that the SSP cycle in RNL is characterized by a more energetic and larger-scale perturbation structure, despite having a lower amplitude streak and mean shear.

Formation of roll/streak structures in the logarithmic layer is consistent with the universal mechanism by which turbulence is modified by the presence of a streak in such way as to induce growth of a roll structure configured to lead to continued growth of the original streak. This growth process underlies the non-normal parametric mechanism of maintaining the perturbation variance in the SSP that maintains turbulence if the Reynolds number is large enough [7]. This universal mechanism does not predict nor requires that the roll/streak structures be of finite streamwise extent and, in its simplest form, it has been demonstrated that it supports roll/streak structures with zero streamwise wavenumber. From this point of view, the observed length of roll/streak structures is not a consequence of the primary mechanism of the SSP supporting them, but rather a secondary effect of disruption by the turbulence. In this work we have provided evidence that NS turbulence is persuasively related in its dynamics to RNL turbulence. Moreover, given that the dynamics of RNL turbulence can be understood fundamentally from its direct relation with S3T turbulence, we conclude that the mechanism of turbulence in wall-bounded shear flow is the roll/streak/perturbation SSP that was previously identified to maintain S3T turbulence.

Acknowledgments

This work was funded in part by the Multiflow program of the European Research Council. Navid Constantinou acknowledges the support of the Alexander S. Onassis Public Benefit Foundation. Brian Farrell was supported by NSF AGS-1246929. We thank Dennice Gayme for helpful reviewing comments.

References

[1] Farrell B F and Ioannou P J 1996 Generalized stability. Part I: Autonomous operators *J. Atmos. Sci.* **53** 2025–40
[2] Farrell B F and Ioannou P J 1996 Generalized stability. Part II: Non-autonomous operators *J. Atmos. Sci.* **53** 2041–53
[3] Henningson D S 1996 Comment on "Transition in shear flows. Nonlinear normality versus non-normal linearity" [Phys. Fluids **7**, 3060 (1995)] *Phys. Fluids* **8** 2257–58
[4] Butler K M and Farrell B F 1992 Three-dimensional optimal perturbations in viscous shear flows *Phys. Fluids* **4** 1637–50
[5] Henningson D S and Reddy S C 1994 On the role of linear mechanisms in transition to turbulence *Phys. Fluids* **6** 1396–98
[6] Farrell B F and Ioannou P J 1994 Variance maintained by stochastic forcing of non-normal dynamical systems associated with linearly stable shear flows *Phys. Rev. Lett.* **72** 1118–91
[7] Farrell B F and Ioannou P J 2012 Dynamics of streamwise rolls and streaks in turbulent wall-bounded shear flow *J. Fluid Mech.* **708** 149–96
[8] Jiménez J and Moin P 1991 The minimal flow unit in near-wall turbulence *J. Fluid Mech.* **225** 213–40
[9] Toh S and Itano T 2005 Interaction between a large-scale structure and near-wall structures in channel flow *J. Fluid Mech.* **524** 249–62
[10] Hamilton K, Kim J and Waleffe F 1995 Regeneration Mechanisms of Near-Wall Turbulence Structures *J. Fluid Mech.* **287** 317–48
[11] Jiménez J and Pinelli A 1999 The autonomous cycle of near-wall turbulence *J. Fluid Mech.* **389** 335–59
[12] Tuerke F and Jiménez J 2013 Simulations of turbulent channels with prescribed velocity profiles *J. Fluid Mech.* **723** 587–603
[13] Mizuno Y and Jiménez J 2013 Wall turbulence without walls *J. Fluid Mech.* **723** 429–55
[14] Farrell B F and Ioannou P J 2003 Structural stability of turbulent jets *J. Atmos. Sci.* **60** 2101–18
[15] Farrell B F and Ioannou P J 2007 Structure and spacing of jets in barotropic turbulence *J. Atmos. Sci.* **64** 3652–65
[16] Marston J B, Conover E and Schneider T 2008 Statistics of an unstable barotropic jet from a cumulant expansion *J. Atmos. Sci.* **65** 1955–66

[17] Tobias S M, Dagon K and Marston J B 2011 Astrophysical fluid dynamics via direct statistical simulations *Astrophys. J.* **727** 127
[18] Srinivasan K and Young W R 2012 Zonostrophic instability *J. Atmos. Sci.* **69** 1633–56
[19] Bakas N A and Ioannou P J 2013 Emergence of large scale structure in barotropic β-plane turbulence *Phys. Rev. Lett.* **110**(22) 224501
[20] Constantinou N C, Farrell B F and Ioannou P J 2013 Emergence and equilibration of jets in beta-plane turbulence: applications of Stochastic Structural Stability Theory *J. Atmos. Sci.* Doi:10.1175/JAS-D-13-076.1, in press
[21] Farrell B F, Gayme D F, Ioannou P J, Lieu B K and Jovanović M R 2012 Dynamics of the roll and streak structure in transition and turbulence *Proc. CTR Summer Program* (Stanford University) 43–54
[22] Thomas V, Lieu B K, Jovanović M R, Farrell B F, Ioannou P J and Gayme D F 2014 Self-sustaining turbulence in the RNL system *Phys. Fluids* (submitted)
[23] Gayme D F 2010 *A robust control approach to understanding nonlinear mechanisms in shear flow turbulence* Ph.D. thesis Caltech, Pasadena, CA, USA
[24] Gayme D F, McKeon B J, Papachristodoulou A, Bamieh B and Doyle J C 2010 A streamwise constant model of turbulence in plane Couette flow *J. Fluid Mech.* **665** 99–119
[25] Kim J, Moin P and Moser R 1987 Turbulence statistics in fully developed channel flow at low Reynolds number *J. Fluid Mech.* **177** 133–66
[26] Flores O and Jiménez J 2006 Effect of wall-boundary disturbances on turbulent channel flows *J. Fluid Mech.* **566** 357–76
[27] Jiménez J 1998 The largest scales of turbulent wall flows *CTR Ann. Res. Briefs* (Stanford University) 137–54
[28] Jiménez J and Hoyas S 2008 Turbulent fluctuations above the buffer layer of wall-bounded flows *J. Fluid Mech.* **611** 215–36
[29] Flores O and Jiménez J 2010 Hierarchy of minimal flow units in the logarithmic layer *Phys. Fluids* **22** 071704
[30] Millikan C B 1938 A critical discussion of turbulent flows in channels and circular pipes *Proc. 5th Int. Conf. on Applied Mechanics* (New York: Wiley) pp. 386–92.

The attached reverse and detached forward cascades in wall-turbulent flows

Andrea Cimarelli[1], Elisabetta De Angelis[1], Alessandro Talamelli[1], Carlo Massimo Casciola[2] and Javier Jiménez[3]

[1]Dipartimento di Ingegneria Industriale, Università di Bologna, 47121 Forlì, Italy
[2]Dipartimento di Ingegneria Meccanica e Aerospaziale, Università di Roma La Sapienza, 00185 Roma, Italy
[3]School of Aeronautics, Universidad Politécnica de Madrid, 28040 Madrid, Spain

E-mail: andrea.cimarelli2@unibo.it

Abstract. The present work describes the multidimensional behaviour of wall-bounded turbulence in the space of cross-scales (spanwise and wall-normal) and distances from the wall. This approach allows us to understand the cascade mechanisms by which scale-energy is transmitted scale-by-scale away from the wall, through the overlap layer, and into the bulk flow. Two distinct cascades are identified involving the attached and detached scales of motion, respectively. From the near-wall region, scale-energy is transferred towards the bulk, flowing through the attached scales of motion, while among the detached scales it converges towards small scales, ascending again to the channel centre. It is then argued that the attached scales of wall-bounded turbulence are involved in a reverse cascade process that starts from the wall and ends in the bulk flow. On the other hand, the detached scales belong to a direct forward cascade process towards dissipation. Hence, at a given distance from the wall the attached motion is fed by smaller attached scales located closer to the wall. In turn this attached motion is responsible for creating the scale-energy that sustains larger attached scales farther from the wall and smaller detached scales that are responsible for connecting the scale-energy at large scales to the dissipation at small scales through a forward cascade.

1. Introduction

The presence of a mean velocity gradient makes wall-turbulent flows anisotropic and provides a continuous source of kinetic energy. Shear flows do not need large-scale forcing to maintain turbulence in a statistically steady state. The mean shear makes wall flows also intrinsically inhomogeneous. Thanks to inhomogeneity, wall-bounded flows have been classically studied by dividing the flow domain into well-defined regions. The local production of turbulent fluctuations exceeds the rate of dissipation near the wall, in the so-called buffer layer, while the opposite occurs at the wall and in the bulk of the flow. The turbulent fluctuations in these regions are then fed in part by a spatial flux of energy emerging from the energy excess in the buffer layer. But actually, a third layer exists, forming an intermediate region between the energy source region near the wall and the energy sink region in the bulk flow, the so-called overlap layer. In the classical view, the production of turbulent fluctuations in this region approximately balances dissipation (equilibrium layer) and, hence, the flux of energy from the wall to the bulk is approximately constant. Energy is neither received from nor released to the spatial flux.

Turbulence in this layer is self-similar, and the distance to the wall acts as a similarity scale, leading to a logarithmic velocity profile [1].

The description in physical space alone is, however, insufficient to capture the real dynamics of wall-bounded turbulence, and traditionally it is complemented by a parallel view based on the decomposition of the field into a hierarchy of scales of motion. In fact, every turbulent flow is characterized by interactions of fluctuations of different sizes and intensities and, at sufficiently high Reynolds number, a universal state governed by an energy cascade from large to small scales is expected to take place [2]. Hence, in wall-bounded flows a classical cascade across different eddy sizes is superimposed on a spatial energy transfer.

Then, a fundamental issue in wall-bounded flows is the understanding of the cascade mechanism by which energy is transmitted scale-by-scale away from the wall, through the overlap layer, and into the bulk flow. Recent works dealt with this problem by means of numerical data of turbulent channel flows. In particular, Ref. [3], by using the balance equation for the second-order structure function, the so-called generalized Kolmogorov equation, studied how turbulent energy is transferred from the production to the dissipation regions of the compound space of scales and wall-distances, (r, Y_c). More recently, this work has been extended in [4, 5], accounting also for the anisotropic features of the space of wall-parallel scales (r_x, r_z). In these works, the analysis of the augmented space, (r_x, r_z, Y_c), revealed that the long and wide fluctuating structures are fed by a small-scale region of energy source near the wall through a reverse energy transfer. The need to adopt a multidimensional approach is also shown to be fundamental for the understanding of the interrelated processes governing wall-turbulence. However, there are a number of open issues to be addressed. The first one concerns the behaviour of energy in the complete four-dimensional space, i.e including the wall-normal scales, r_y. Indeed, the analysis of r_y allows us to study directly the behaviour of the so-called attached and detached eddies, i.e. those scales of motion that are of the same order and smaller, respectively, than the wall-distance considered.

The study of the dynamics related to attached and detached scales is fundamental for the comprehension of the high-Reynolds-number asymptotic state of wall turbulence. It is argued that the overlap layer dominates at sufficiently high Reynolds numbers by contributing to the bulk production more than the near-wall region [6]. In turn, the overlap layer is thought to be populated by a hierarchy of attached eddies that are then conjectured to govern the high-Reynolds-number state of wall-turbulence [7, 8].

From the similarity hypothesis and from momentum conservation, which implies that the shear stress, $\tau = -\langle uv \rangle$, is constant and equals u_τ^2 near the wall, the overlap layer should be traversed by a constant spatial flux, $\phi/u_\tau^3 = const$, and, hence, production should balance dissipation. Here, $u_\tau = \sqrt{\tau_w/\rho}$ is the friction velocity, with τ_w the average shear stress at the wall, and ρ the constant density of the fluid. Because the energy-containing scales are small near the wall, while those further away from the wall are large (see figure 1 to have an approximate idea of the topology of the flow field), this transfer of energy is conjectured to be an example of inverse cascade [9, 10]. Energy is transferred to smaller scales at any given location, and to larger ones away from the wall. Since the smallest scales of the flow are thought to be isotropic and to carry no mean shear stress, the larger ones are those responsible for the momentum cascade. Because of the different type of fluxes involved, this cascade process is different from the Kolmogorov one. For scales larger than the wall-distance, the spectrum depends on the momentum flux u_τ^2, and on the wavenumber k, and is independent of the wall-distance. Hence, the k^{-1}-law for the spectrum is obtained, $E(k) \sim u_\tau^2 k^{-1}$ [11]. On the other hand, the smaller scales do not interact with the wall and break down into smaller eddies, following a classical Kolmogorov cascade. This attached/detached eddy picture of energy transfer was first proposed in [1], but still has to be assessed. In this context, the aim of the present work is to verify these theories and to assess the similarities, differences, and interactions between those two cascades.

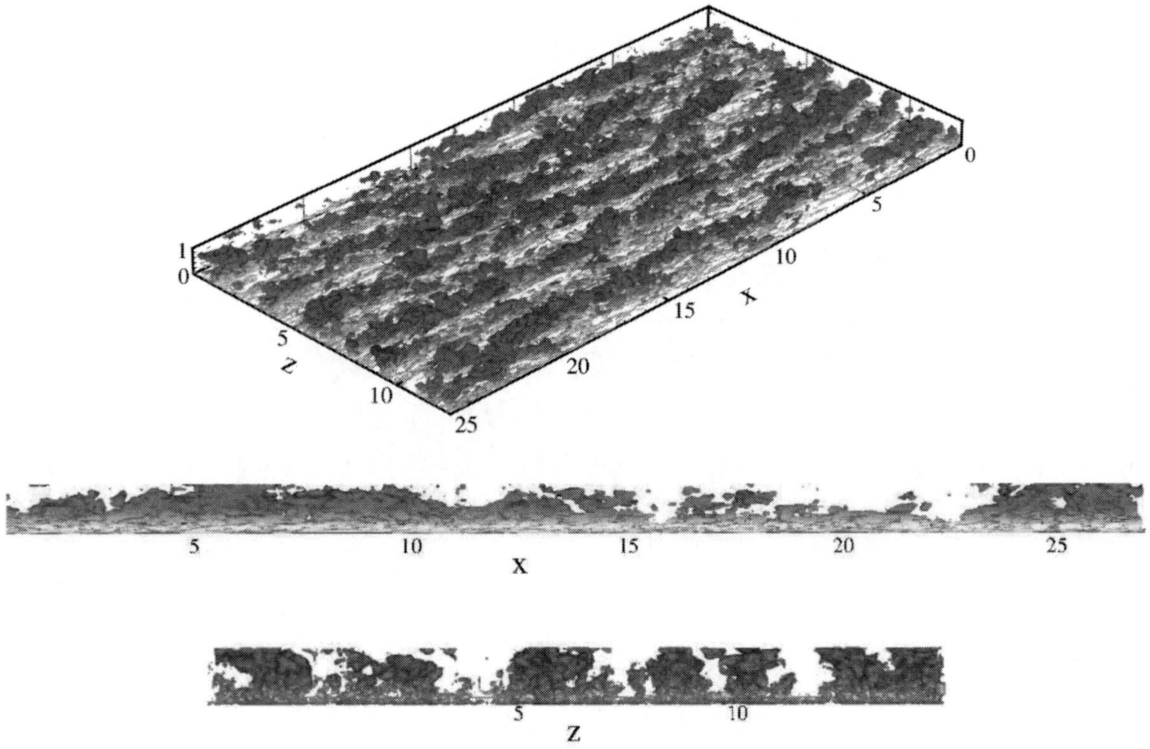

Figure 1. Direct numerical simulation of a turbulent channel flow at $Re_\tau = 550$ [14]. Isosurfaces of a negative value of the streamwise velocity fluctuations, with colours encoding wall distance. Perspective, lateral and frontal views are shown from top to bottom. The streamwise, wall-normal and spanwise directions, (x, y, z), are normalized by the channel half-height, h.

2. The compound (r_y, r_z, Y_c)-space

Energy transfer, production and dissipation are the relevant processes in turbulence. They take place in various ranges of scales and, for inhomogeneous flows, change in different flow regions. As a consequence, a full understanding of these interacting phenomena requires a detailed description of the processes occurring simultaneously in physical and scale space. To this aim, we study the second-order structure function $\langle \delta u^2 \rangle$, where $\delta u^2 = \delta u_i \delta u_i$, $\delta u_i = u_i(X_s + r_s/2) - u_i(X_s - r_s/2)$ is the fluctuating velocity increment, and angle brackets denote ensemble averaging. It is a function of the separation vector r_s and of the mid-point X_s, allowing to describe the scale-dependent mechanisms in the presence of inhomogeneity. Hereafter we will refer to the concept of scale-energy while speaking about the second-order structure function. Although this quantity has the dimensions of kinetic energy, its interpretation in terms of kinetic energy of turbulent structures of a given size is not strict. Indeed, while being strictly related to the energy spectrum, it is not an intensive quantity, i.e. the integral in the space of scales of $\langle \delta u^2 \rangle$ has no meaning. It should be stressed however that, despite some difficulties in its interpretation (see [12] and its appendix), the scale-energy is the natural tool for the statistical analysis of general turbulent flows that lack a classical spectral decomposition due to violation of spatial homogeneity. As it will be shown below, it allows addressing the multidimensional cascade mechanisms and in particular the related attached/detached features.

The governing equation of $\langle \delta u^2 \rangle$ in wall flows is the generalized Kolmogorov equation [13, 3],

Journal of Physics: Conference Series **506** (2014) 012005 doi:10.1088/1742-6596/506/1/012005

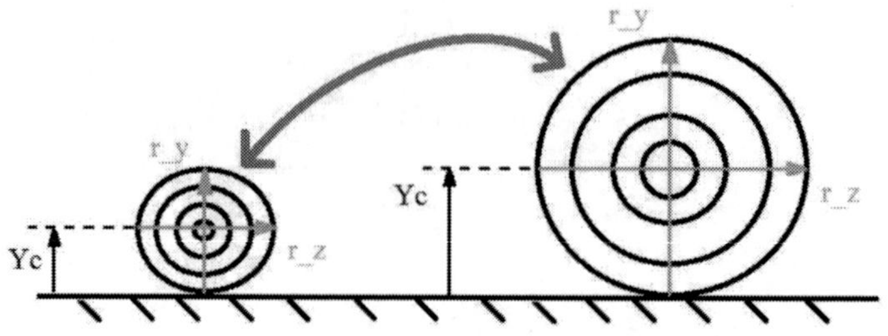

Figure 2. Idealized picture of the (r_y, r_z, Y_c)-space.

which for a turbulent channel flow with longitudinal mean velocity $U(y)$ reads

$$\frac{\partial\langle\delta u^2\delta u_i\rangle}{\partial r_i} + \frac{\partial\langle\delta u^2\delta U\rangle}{\partial r_x} + 2\langle\delta u\delta v\rangle\left(\frac{dU}{dy}\right)^* + 2\langle\delta uv^*\rangle\,\delta\!\left(\frac{dU}{dy}\right) + \frac{\partial\langle v^*\delta u^2\rangle}{\partial Y_c} =$$

$$-4\langle\epsilon^*\rangle + 2\nu\frac{\partial^2\langle\delta u^2\rangle}{\partial r_i\partial r_i} - \frac{2}{\rho}\frac{\partial\langle\delta p\delta v\rangle}{\partial Y_c} + \frac{\nu}{2}\frac{\partial^2\langle\delta u^2\rangle}{\partial Y_c{}^2} \qquad (1)$$

where the asterisk denotes the arithmetic average of a variable at the points $X_s \pm r_s/2$, $Y_c = X_2$ is the wall-normal coordinate of the mid-point, $v = u_2$ is the wall-normal velocity, ν is kinematic viscosity, and $\epsilon = \nu(\partial u_i/\partial x_j)(\partial u_i/\partial x_j)$ is the pseudo-dissipation. As shown in [4, 5], it is useful to recast equation (1) in terms of a four-dimensional vector field, $\mathbf{\Phi} = (\Phi_{r_x}, \Phi_{r_y}, \Phi_{r_z}, \Phi_c)$, hereafter called the scale-energy hyper-flux defined in a four dimensional space (r_x, r_y, r_z, Y_c),

$$\nabla_4 \cdot \mathbf{\Phi}(\mathbf{r}, Y_c) = \xi(\mathbf{r}, Y_c)\,, \qquad (2)$$

where ∇_4 is the four-dimensional gradient and $\xi = -2\langle\delta u\delta v\rangle\,(dU/dy)^* - 2\langle\delta uv^*\rangle\,\delta(dU/dy) - 4\langle\epsilon^*\rangle$ is the scale-energy source/sink. The flux in the three-dimensional space of scales is $\mathbf{\Phi}_r = (\Phi_{r_x}, \Phi_{r_y}, \Phi_{r_z}) = \langle\delta u^2\delta\mathbf{u}\rangle + \langle\delta u^2\delta U\rangle\hat{\mathbf{e}}_x - 2\nu\nabla_r\langle\delta u^2\rangle$, where $\hat{\mathbf{e}}_x$ is the unit vector in the mean flow direction x. In addition to the flux in the space of scales, the generalized Kolmogorov equation features the spatial flux $\Phi_c = \langle v^*\delta u^2\rangle + 2\langle\delta p\delta v\rangle/\rho - \nu/2\partial\langle\delta u^2\rangle/\partial Y_c$. As mentioned above, equation (2) has been analysed in [4, 5] in the hyper-plane $r_y = 0$ of the four-dimensional space. This approach allowed to identify the reverse energy transfer as the crucial mechanisms characterizing wall turbulence, responsible for the formation of the commonly observed very long and wide velocity fluctuations. Here, in order to focus on the conjectured dual cascades involving attached/detached scales, the analysis will be performed in the hyper-plane $r_x = 0$, i.e. in the (r_y, r_z, Y_c)-space. Indeed, the space of the cross-scales most clearly allows to detect the attached/detached feature of the statistics related to scales of motion whose vertical length, r_y, is equal-to/smaller-than twice the wall-distance of their centre, Y_c (see figure 2). The idealized picture of figure 2 sketches also how equation (2) is used to detect the scale-energy transfer in the space of scales, i.e. along the orange (r_y, r_z)-axis, and the transport in physical space, i.e. along the Y_c-axis (red arrow).

Given the definition of velocity increments, the r_y direction is limited by the presence of the wall. In particular, for a given wall-distance Y_c, the space of wall-normal scales extends from zero to twice the distance from the wall, $r_y \in [0, 2Y_c]$. Actually, negative increments, i.e. $r_y \in [-2Y_c, 2Y_c]$, are not considered here since the symmetry of the flow is such that the transformation $\mathbf{r} \to -\mathbf{r}$, $\widetilde{Y_c} = const$ leads to $\mathbf{\Phi}_r \to -\mathbf{\Phi}_r$ and $\Phi_c \to \Phi_c$. This transformation leaves quantities like δu^2 and v^* statistically invariant while reversing the sign of vectors like δu_i

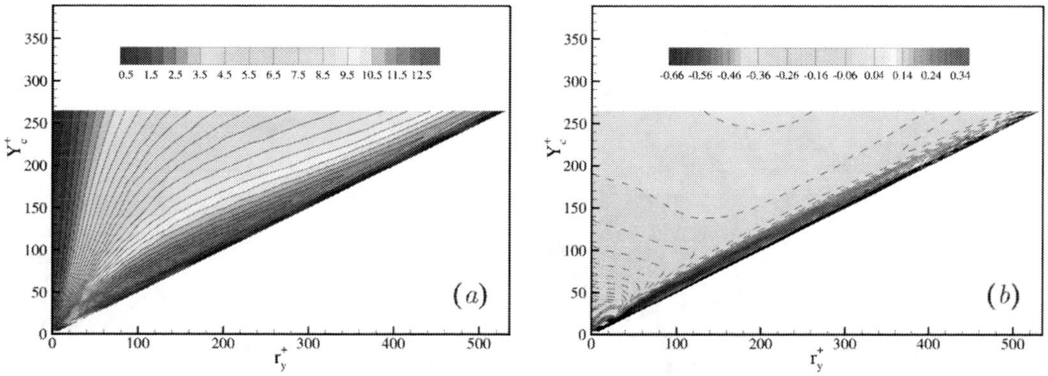

Figure 3. Section of the reduced space (r_y, r_z, Y_c) in the $r_z = 0$-plane. The isocontours show the behaviour of scale-energy, $\langle \delta u^2 \rangle^+$, (left) and of the source/sink term, ξ^+, (right). Solid and dashed lines denote positive and negative values, respectively.

and ∇_r. On the boundary $r_y = 2Y_c$ the scale-energy equals the single-point turbulent kinetic energy,

$$\langle \delta u^2 \rangle = \langle u_{ii}^2 \rangle \Big|_{y=2Y_c} ,$$

where $u_{ii}^2 = u^2 + v^2 + w^2$, while the fluxes with a component normal to the boundary take the form

$$\Phi_{r_y} = \langle u_{ii}^2 v \rangle \Big|_{y=2Y_c} - \nu \frac{\partial \langle u_{ii}^2 \rangle}{\partial y} \Big|_{y=2Y_c}$$

$$\Phi_c = \frac{1}{2} \langle u_{ii}^2 v \rangle \Big|_{y=2Y_c} - \frac{1}{2} \nu \frac{\partial \langle u_{ii}^2 \rangle}{\partial y} \Big|_{y=2Y_c} + \frac{2}{\rho} \langle \delta p (v|_{y=2Y_c}) \rangle .$$

From the above equations we argue that the fluxes could be aligned to the boundary $(r_y = 2Y_c, Y_c)$ only when the pressure term for the spatial flux is negligible, since in this case we would have $\Phi_{r_y} = 2\Phi_c$. As will be shown in what follows, the role of the $(r_y = 2Y_c)$-plane is very important, since the wall-normal scales close to this plane represent the attached features of the flow and, hence, that is the layer where the reverse cascade process and the other paradigms mentioned in the introduction should occur. Smaller wall-normal scales represent instead the detached features of the flow, where an overall forward cascade is expected to take place.

The data used for the present analysis come from a channel-flow DNS at $Re_\tau = u_\tau h / \nu = 550$, where h is the channel half-height. Throughout the paper, inner variables will be used and denoted with $+$ as superscript, implying normalization of lengths with the friction scale ν / u_τ and velocities with u_τ. The simulations have been carried out with a pseudo-spectral code. The computational domain is $8\pi h \times 2h \times 4\pi h$ with $1024 \times 257 \times 1024$ grid points, respectively, corresponding to a resolution in the homogeneous directions of $\Delta x^+ = 13.5$ and $\Delta z^+ = 6.7$ [14].

3. Results

Let us first analyse how the scale-energy and the scale-energy source are distributed as functions of the wall-normal separation, r_y, for different distances from the wall, Y_c. As shown in figure 3(a), the scale-energy attains a clear maximum that, for each wall-distance, is located near the $(r_y = 2Y_c)$-plane. This is clear statistical evidence of fluctuations occurring on a scale extending down to the wall, in the sense of the attached eddies of Townsend [1]. The wall-normal position

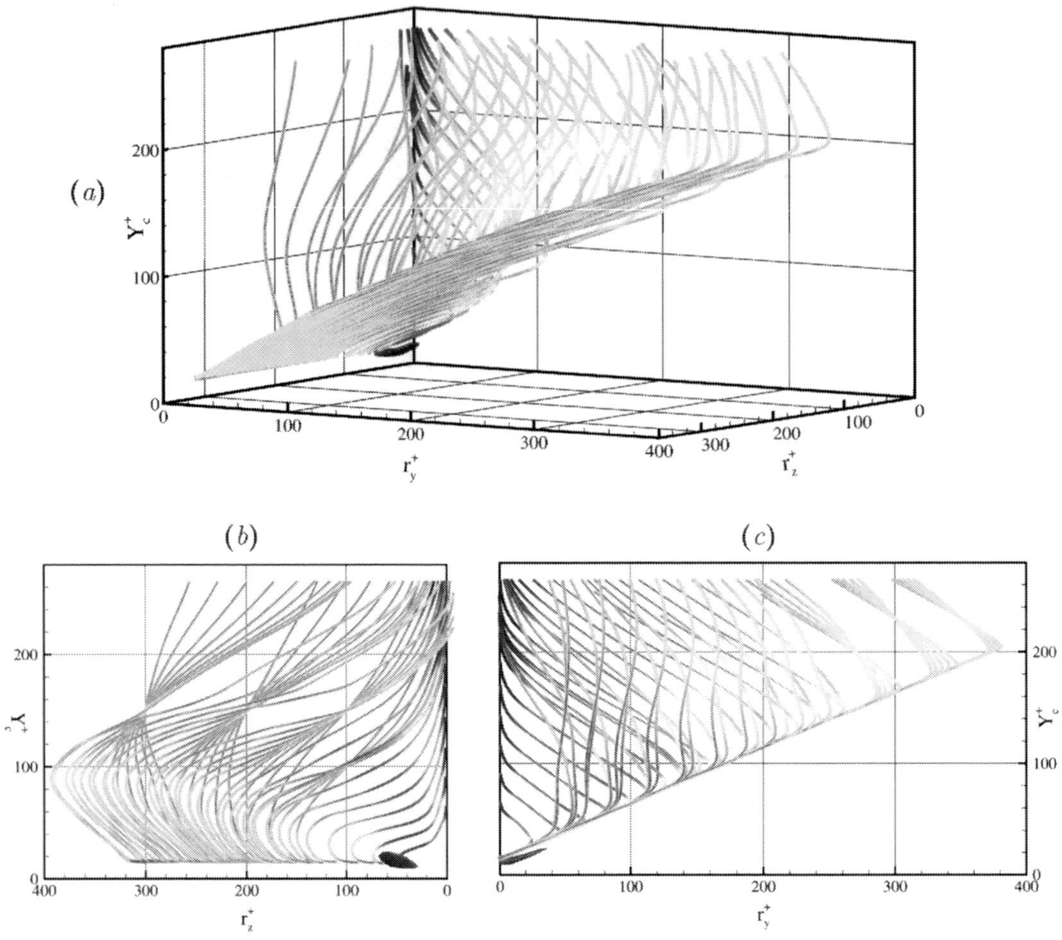

Figure 4. Trajectories of the scale-energy fluxes (colours according to strength) in the (r_y, r_z, Y_c)-space. The red iso-surface shows the maximum for the source term, $\xi^+ = 0.61$.

of the maximum in the scale-energy confirms the relevance of such scales in the putative overlap layer. Namely, these attached scales are not only the most energetic of the $(r_z = 0)$-plane (wall-normal scales, wall-distance) but, as shown in figure 3(b), are also responsible for most of the production of turbulence. Apparently, the source term ξ presents a strong positive peak for scales spanning the distance to the wall, even for Y_c far away from the near-wall production region. This observation is consistent with the conjecture of a second outer turbulent regeneration cycle in the overlap layer [15] involving turbulent fluctuations that scale with the wall-distance [16, 17]. In order to avoid possible misunderstandings, we stress that the source and scale-energy related to the near-wall cycle are not visible in figure 3, which is a section of the (r_y, r_z, Y_c) at $r_z = 0$. Indeed, as it will be shown below and has been discussed already in [4], the near-wall source occurs in a precise range of spanwise scales with $r_z \neq 0$.

Let us now analyse the multidimensional behaviour of scale-energy from production to dissipation in the entire (r_y, r_z, Y_c)-space. As shown by the field lines in figure 4, scale energy emerges from the small scales in the near-wall region and, ascending, feeds larger and taller fluctuations before converging to small dissipative scales. The origin of the fluxes is located in the small scales of the buffer layer, corresponding to the peak of the scale-energy source at $r_y^+ = 7, r_z^+ = 40, Y_c^+ = 13$ (see the red isosurface in figure 4). This is the main source

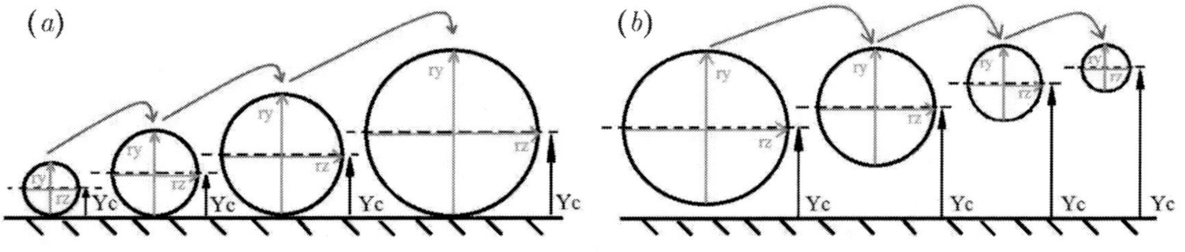

Figure 5. Sketch of the two distinct cascade mechanisms coexisting in wall-flows. (a) The reverse cascade process through attached scales from the wall to the bulk flow. (b) The forward cascade through detached scales up to dissipative scales further away from the wall.

region for the fluxes, while the sinks are distributed continuously with the wall-distance at the smallest scales of motion, close to $r_y = r_z = 0$. From the inspection of the plots it appears that scale-energy flows from the peak of energy source near the wall towards larger fluctuations (larger r_z) following a line of divergence from which the fluxes suddenly depart towards the bulk (see the horizontal envelope at $Y_c^+ \sim 13$ in the bottom-left panel of figure 4). These fluxes become aligned to the ($r_y = 2Y_c$)-plane while ascending (see figure 4c). In turn, this is also a plane of divergence from which the fluxes suddenly depart bending towards smaller scales up to dissipation. Note that the fluxes, while aligned to the ($r_y = 2Y_c$)-plane, move towards larger r_z. In particular, the spanwise scale linearly increases with the wall-distance as half the wall-normal scale, i.e. $r_z \sim r_y/2$. After leaving the ($r_y = 2Y_c$)-plane, the fluxes converge towards smaller r_y and r_z, intercepting narrower and thinner fluctuations. Summarizing, the fluxes leave the line of divergence in the near wall region at $Y_c^+ = 13$ and $r_y^+ = 7$ for different spanwise lengths, denoted as Δz_{nw}^+. The fluxes depart from this line and lie on the ($r_y = 2Y_c$)-plane. In this plane, they intercept wall-normal scales attached to the wall with spanwise scales that linearly increase with wall-distance [18, 19, 20],

$$r_y^+ = 2Y_c^+ \tag{3}$$

$$r_z^+ = \Delta z_{nw}^+ + r_y^+/2 = \Delta z_{nw}^+ + Y_c^+ \sim Y_c^+ \ . \tag{4}$$

It is important to point out that the source term along these paths is always pumping energy, $\xi > 0$. See for example the evident source of scale-energy close to the ($r_y = 2Y_c$)-scales shown in figure 3(b) for $r_z = 0$. As a consequence, the flux is not conserved since a continuous injection of scale-energy occurs along the path. Hence, the attached scales are not simply involved in the reverse cascade process from the wall to the bulk flow. They also play an active role in injecting scale-energy to sustain the reverse cascade mechanism towards attached scales located further away from the wall, and to initialize the forward cascade through detached scales towards dissipation. Summarizing, the above analysis suggests that the turbulent fluctuations attached to the wall govern the overlap layer and provide the scale-energy source contributing to the sustainment of turbulence. Actually, this wall-attached source is the initiator of the Y_c-distributed direct cascade feeding the detached motion down to dissipation and, at the same time, contributes to the attached reverse cascade that feeds turbulence in the bulk of the flow.

It should be noted that in the attached reverse cascade occurring in the ($r_y = 2Y_c$)-plane, all the three components of the flux vector are positive, i.e. scale-energy ascends towards the bulk of the flow moving from small to large scales, as sketched in figure 5(a). In particular, $\Phi_c \sim \Phi_{r_y}/2$, since the fluxes are tangent to the $r_y = 2Y_c$-plane. Furthermore, from figure 4 we also observe that $\Phi_{r_z} \sim \Phi_c$, so that

$$\Phi_{r_z} \sim \Phi_c \sim \Phi_{r_y}/2 > 0 \ ,$$

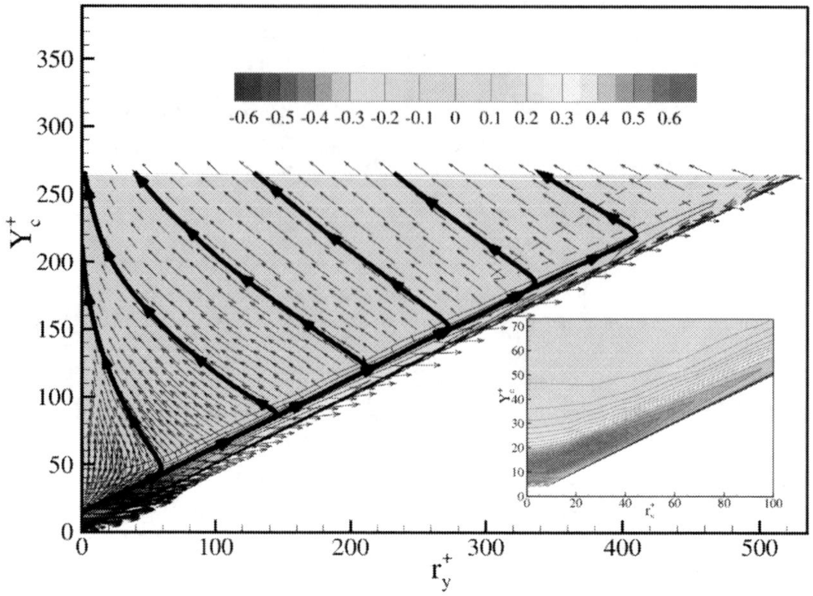

Figure 6. Section of the reduced space (r_y, r_z, Y_c) in the $r_z^+ = 40$-plane. The isocontours show the behaviour of scale-energy source/sink, ξ^+. Solid and dashed lines denote positive and negative values, respectively. Vectors and streamlines show the field of fluxes in this plane. The inset shows a zoom of the source features near the wall.

in agreement with the linear scaling with wall-distance of the spanwise and wall-normal scales in equations (3) and (4). On the contrary, the scale-space fluxes are negative (towards small scales) in the detached direct cascade, while the spatial flux is still positive (towards the bulk of the flow), i.e. a direct cascade ascending to the bulk (see the sketch in figure 5*b*). By looking at the streamlines in figure 6, we observe that, away from the wall and for large wall-normal scales, the fluxes seemingly approach the inclination

$$\Phi_c \sim -\Phi_{r_y}/2 \; .$$

This behaviour is probably related to a putative inertial range and needs to be verified using data at higher Reynolds numbers. The above relation suggests that the top part of the wall-normal scales intercepted by the direct cascade remains at a constant distance from the wall, i.e., if we define $y_{top} = Y_c + r_y/2$ and $y_{bot} = Y_c - r_y/2$, the direct cascade intercepts smaller and smaller wall-normal scales whose y_{top} is constant, while ascending to the bulk, as sketched in figure 5(*b*).

It should finally be stressed that the plane of divergence where the attached reverse cascade process occurs is not exactly the $(r_y = 2Y_c)$-plane. As better shown in figure 6, where a cut of the (r_y, r_z, Y_c)-space is shown at the location, $r_z^+ = 40$, of the peak of energy source near the wall, the streamlines of the fluxes and also the isocontours of the source term are aligned and close to the $(r_y = 2Y_c)$-plane but do not lie exactly on it. Furthermore, let us emphasize the quantitative correspondence of the present results with literature. As also shown in [4], the near-wall source mechanisms show a detached feature with the wall-normal scales of the maximum of energy source independent of the wall-distance. See the inset of figure 6, where the isocontours in the neighbourhood of the source maximum near the wall are apparently horizontal. From this detached feature, the source contours start to be aligned to the $(r_y = 2Y_c)$-plane for wall distances larger than $Y_c^+ \sim 20$. As also pointed out in [7, 8], this wall distance represents the cross-over between the two attached/detached-dominated regions of the flow.

4. Conclusions and final comments

Two cascade mechanisms are found to coexist in wall-turbulence. One involving scales attached to the wall and another characterizing the detached motion. The attached scales are found to be the site of the turbulent production mechanisms driving the overlap layer, and are involved in a reverse cascade process that propagates from the wall to the bulk of the flow. As sketched in figure 5, the attached motion at a given distance Y_c from the wall is fed by a hierarchy of smaller attached scales located closer to the wall. In turn, this attached motion injects scale-energy through a significant scale-energy source, $\xi > 0$, contributing to the propagation of the reverse cascade process towards the bulk, and hence to the sustainment of larger attached scales farther from the wall. The energy source associated to the attached scales is also strong enough to initialize at each distance from the wall a direct forward cascade toward smaller detached scales. This forward cascade represents the Y_c-distributed connection of the production processes in the attached motion with the dissipation at the smallest scales. As sketched in figure 5, this cascade is also superimposed on a spatial flux. At a given distance from the wall a detached scale is intercepted by a flux originating from a hierarchy of larger detached scales located closer to the wall (smaller Y_c), and so on. It is important to point out another peculiar difference between the two cascades here described. The reverse cascade process starts from the wall, is refilled at each wall-distance, and ends up in the bulk of the flow. The forward cascade originates at each wall-distance from the scale-energy source available at attached scales, and ends up further away from the wall at dissipative scales, i.e. the forward cascade feeds the energy dissipation at the smallest scales of each wall-distance.

Let us finally mention the possible implications of the present results for turbulence modelling. In the context of large-eddy simulation (LES), the possibility to analyse the turbulence mechanisms simultaneously in the compound space of scales and positions allows us to understand the overall behaviour of the resolved field as a function of the filter length, and to identify the physics of the small subgrid scales to be modelled [21, 22]. Here, the use of the (r_y, r_z, Y_c)-space directly connects the present results to wall-turbulence LES with near-wall modelling. Indeed, one of the strategies for high-Reynolds-number LES of wall-bounded flow conceives off-wall boundary conditions applied at a Reynolds-number-independent distance from the wall, thereby solving the LES equations only in the bulk flow; see [23] for a review and the references therein. In this context, the understanding of the dynamics in the overlap layer is very important. Indeed, it supports the transfer of momentum from the bulk flow to the wall across a wide range of scales that should be taken into account in the off-wall boundary conditions. Several theories describe the interaction between the inner and outer layer. Here, the attached scales have been shown to be the site of the scale-energy source mechanisms and to provide the transfer to the bulk turbulence, while detached scales have been shown to belong to a direct cascade process simply linking production to dissipation. In this respect, the off-wall boundary condition applied in [15], which mimics the linear dependence with wall-distance of the length scales, seems to be a very promising approach, since it reproduces the attached motion found to play the most important role for the bulk turbulence.

It is worth emphasizing that, from the present results, small attached eddies created near the wall are not implied to migrate to the bulk region to release their energy there. The fact that the overlap layer is traversed by a flux originating from the wall while being refilled by the local scale-energy source does not necessarily mean that the overlap layer dynamics depends on the production features near the wall. Actually, in analogy to the inertial range of turbulence, it appears that the fluxes and other statistics do not depend on how turbulence is generated near the wall [24]. However, in the same way as the inertial range of turbulence needs a large-scale production range to initialize the universal cascade process towards smaller scales, the overlap layer needs the near-wall energy excess to initialize the reverse cascade process. Hence, although the near-wall cycle provides the triggering mechanism, the resulting attached reverse cascade in the overlap layer could still remain universal, i.e. independent of the details of the near-wall turbulence generation. As an example, the reverse cascade in the overlap layer should look the same irrespective of the presence of wall roughness.

Acknowledgments

The authors would like to acknowledge the support of the European Research Council, which funded the Multiflow summer program within which this work was developed. We acknowledge the technical support of Dr. J. Hackl, who helped us to manage the computing resources of the School of Aeronautics, Universidad Politécnica de Madrid, which is also acknowledged.

References

[1] Townsend A A 1976 *The Structure of Turbulent Shear Flows*, (Cambridge: Cambridge University Press)
[2] Kolmogorov A N 1941 Dissipation of energy in the locally isotropic turbulence *Dokl. Akad. SSSR* **32** 19–21
[3] Marati N, Casciola C M and Piva R 2004 Energy cascade and spatial fluxes in wall turbulence *J. Fluid Mech.* **521** 191–215
[4] Cimarelli A, De Angelis E and Casciola C M 2013 Paths of energy in turbulent channel flows *J. Fluid Mech.* **715** 436–451
[5] Cimarelli A, De Angelis E and Casciola C M 2011 Assessment of the turbulent energy paths from the origin to dissipation in wall-turbulence *J. Phys.: Conf. Series* **318** 022007
[6] Smits A J, McKeon B J and Marusic I 2011 High Reynolds-number wall turbulence *Annu. Rev. Fluid Mech.* **43** 353–375
[7] Del Álamo J C, Jiménez J, Zandonade P and Moser R D 2006 Self-similar vortex clusters in the turbulent logarithmic region *J. Fluid Mech.* **561** 329–358
[8] Lozano-Durán A, Flores O and Jiménez J 2012 The three-dimensional structure of momentum transfer in turbulent channels *J. Fluid Mech.* **694** 100–130
[9] Jiménez J 1999 The physics of wall turbulence *Physica A* **263** 252–256
[10] Jiménez J 2012 Cascades in wall-bounded turbulence *Annu. Rev. Fluid Mech.* **44** 27–45
[11] Perry A E, Henbest S and Chong M S 1986 A theoretical and experimental study of wall turbulence *J. Fluid Mech.* **165** 163–199
[12] Cimarelli A 2011 *Statistical Analysis and Simulation Techniques in Wall-Bounded Turbulence* (PhD thesis, University of Bologna)
[13] Hill R J 2002 Exact second-order structure-function relationships *J. Fluid Mech.* **468** 317–326
[14] Segalini A, Cimarelli A, Ruedi J D, De Angelis E and Talamelli A 2011 Effect of the spatial filtering and alignment error of hot-wire probes in a wall-bounded turbulent flow *Meas. Sci. Technol.* **22** 105408
[15] Mizuno Y and Jiménez J 2013 Wall turbulence without walls *J. Fluid Mech.* **723** 429–455
[16] Hutchins N and Marusic I 2007 Large-scale influences in near-wall turbulence *Phil. Trans. R.Soc. A* **365** 647–664
[17] Jiménez J and Hoyas S 2008 Turbulent fluctuations above the buffer layer of wall-bounded flows *J. Fluid Mech.* **611** 215–236
[18] Monty J P, Stewart J A, Williams R C and Chong M S 2007 Large-scale features of turbulent pipe and channel flows *J. Fluid Mech.* **589** 147–156
[19] Bailey S C C and Smits A J 2010 Experimental investigation of the structure of large- and very-large-scale motions in turbulent pipe flow *J. Fluid Mech.* **651** 339–356
[20] Del Álamo J C and Jiménez J 2003 Spectra of the very large anisotropic scales in turbulent channels *Phys. Fluids* **15**, 6 41–44
[21] Cimarelli A and De Angelis E 2011 Analysis of the Kolmogorov equation for filtered wall-turbulent flows *J. Fluid Mech.* **676** 376–395
[22] Cimarelli A and De Angelis E 2012 Anisotropic dynamics and sub-grid energy transfer in wall-turbulence *Phys. Fluids* **24** 015102
[23] Piomelli U and Balaras E 2002 Wall-layer models for large-eddy simulations *Annu. Rev. Fluid Mech.* **34** 349–374
[24] Flores O and Jiménez J 2006 Effect of wall-boundary disturbances on turbulent channel flows *J. Fluid Mech.* **566** 357–376

Journal of Physics: Conference Series **506** (2014) 012006 doi:10.1088/1742-6596/506/1/012006

Granger causality in wall-bounded turbulence

Gilles Tissot[1], Adrian Lozano–Durán[2], Laurent Cordier[1], Javier Jiménez[2], Bernd R Noack[1]

[1]Institut PPRIME, CEAT 43 route de l'aérodrome, 86036 Poitiers, France
[2]School of Aeronautics, Universidad Politécnica, 28040 Madrid, Spain

E-mail: Gilles.Tissot@univ-poitiers.fr, Adrian@torroja.dmt.upm.es, Laurent.Cordier@univ-poitiers.fr, Jimenez@torroja.dmt.upm.es, Bernd.Noack@univ-poitiers.fr

Abstract. Granger causality is based on the idea that if a variable helps to predict another one, then they are probably involved in a causality relationship. This technique is based on the identification of a predictive model for causality detection. The aim of this paper is to use Granger causality to study the dynamics and the energy redistribution between scales and components in wall-bounded turbulent flows. In order to apply it on flows, Granger causality is generalized for snapshot-based observations of large size using linear-model identification methods coming from model reduction. Optimized DMD, a variant of the Dynamic Mode Decomposition, is considered for building a linear model based on snapshots. This method is used to link physical events and extract physical mechanisms associated to the bursting process in the logarithmic layer of a turbulent channel flow.

1. Introduction

Deduction and induction are the two opposite faces of scientific construction. In a deductive approach, accepted principles are used (for instance physical laws) to reach some conclusion by pure logical reasoning. Observations are then only employed for invalidating hypothesis or theory. At the opposite, with inductive reasoning, we try from specific observations to derive broader generalizations and theories. Causality, at the intersection of philosophy and sciences [1], enters directly in this latter category. In physics, different statistical tools have been developed for analysing causes and effects between phenomena [2]. A common approach for inferring information flow from time series is to compute the cross-correlation between two variables for a range of time lags, and determine whether there exists a peak in the correlation at some non-zero lag. However, this approach can be misleading to infer causation since correlation only indicates if two variables are statistically linked. One possible link may be a direct cause-effect relationship between the two variables, but that cause and effect can be also indirect, or due to a confounding variable i.e. an unseen variable that is also correlated with the two variables. Moreover, since correlation is a symmetric measure, it does not give information on the direction of causality. In classical physics, we can exploit the fact that "causes precede the effects" (principle of time-asymmetry of causation). A dynamical model is another way to include a direction in the variation of the time variable. Indeed, the identification of a predictive model is based on the assumption that the system is dictated by a law, and that the system in the same state will evolve in the same way. In other words, the same causes will produce the same effects (physical determinism). In general, this dynamical law is identified by considering statistics of the observations. Based on a regression of observations (often done through the computation of correlations), a model is determined which links at best variables and their temporal evolutions.

In this paper, Granger causality [3, 4] is considered for detecting causation. This method is one of the earliest statistical tests developed to quantify the temporal-causal effect among time series. It has been successfully employed in the past in economy [5], in neuroscience [6] and in climate modelling [7]. There are mainly two manners of using Granger causality for a given set of data: i) to infer the existence and direction of influence from information in the data; ii) to determine which one of several different hypotheses concerning a physical phenomenon of interest is more likely, given the data and assumptions. The first approach is more exploratory, since it does not rely on a priori specification of dominating events and assumptions about the existence and direction of influence between these events. The crucial point is then the selection of the data to be analysed. Once the data are selected, one needs only apply the Granger causality principle to obtain the causations. This approach is typically used in cognitive neuroscience for determining the interactions of activated brain areas from functional brain imaging [6]. In the second approach, one needs to obtain data relevant to the competing hypotheses before using Granger causality tests to provide clues about the direction of causality. This approach was used in [7] to shed light on a debate about two theories proposed for explaining the relationships between hurricanes and climate. In accordance to the first approach, Granger causality is hereafter used to determine causality relationships between different velocity components and scales in a wall-bounded turbulent flow. Formally, x_2 "Granger-causes" x_1 if information in the past of x_2 helps to predict the future of x_1 with better accuracy than it is possible when considering only information in the past of x_1 itself. Mathematically, Granger causality is based on the identification of linear autoregressive (AR) models for the two variables x_1 and x_2. A rigorous definition of Granger causality is postponed to section 2.

In a wall-bounded turbulent flow, elongated structures of low streamwise velocity, denoted as *streaks*, are present in the so-called buffer layer near the wall [8, 9]. These streaks are known to play an important role in the wall-turbulence phenomena (see [10] for an overview). Indeed, it has been observed in [11] with dye visualisations that the streaks tend to lift-up from the wall, oscillate, and then burst. Optimal perturbations [12, 13] showed that streamwise vortices are amplified in wall-bounded flows with a spacing that is consistent with observed streaks. These vortices are responsible for changing the mean flow and for the creation of the streaks. Reference [14] studied in detail the growth of the spanwise instabilities in sufficiently strong streaks, leading to their breakdown. This process is associated with strong ejections of fluid from the wall, called *bursts*, and to interactions with parts of the flow farther from the wall. These phenomena contribute in large part to changes of the Reynolds stresses. Farther from the wall, the logarithmic layer is also organized into streamwise streaks, although they differ from those in the buffer layer by their larger dimensions and their highly multiscale characteristics [15]. These streaks equally burst [16], and the associated strong intermittent wall-normal fluctuations were interpreted by [17] as a linear amplification that can be modelled by the Orr mechanism. The use of the linear Orr-Sommerfeld equation was effectively able to reproduce some of the burst features observed in fully turbulent simulations [17], especially for the wall-normal velocity. Focusing on that kind of events, we propose to adopt the opposite point of view, and to detect Granger causality by the identification of linear models of representative variables in the fully turbulent flow, rather using a linearized model.

Turbulent flows are characterized by a high level of spatio-temporal complexity that leads to dynamical models with a high number of active degrees of freedom. For this type of flows, the traditional approach of Granger causality that relies on identification of autoregressive models through least-squares method is no longer possible. Since the predictive abilities of the linear identified model is essential for the Granger analysis, particular attention should be given to the modelling part. *Reduced-order models* have demonstrated their ability in dynamical modelling for flow control [18, 19] where the constraints in terms of accuracy and representativeness are equally strong. In order to adapt the strategy used by Granger [3] to a snapshot-based point of view, *Optimized Dynamic Mode Decomposition* (DMD) recently introduced by [20] will be used as a linear system identification method.

In section 2, Granger causality is first presented in its original form. The method is adapted in section 3 to the case, typical in fluid mechanics, where the size of the snapshots is very large. For that, optimized DMD is introduced in section 3.1 as a way of identifying a linear system based on snapshots. The use of optimized DMD for flow prediction is demonstrated in section 3.2, and the Granger causality formalism is adapted in section 3.3 to the use of an optimized DMD model. Finally, Granger causalities between velocity components of different scales representative of the streaks are determined in section 4 from numerical data of a channel flow.

2. Granger causality

Let x_1 and x_2 be two discrete-time series of length N_t. The idea of Granger causality [3] formalizes Wiener's intuition that x_2 causes x_1 if knowing x_2 helps to predict the future of x_1. More precisely, we will say that x_2 Granger-causes x_1 if the inclusion of past observations of x_2 reduces the prediction error of x_1 in a linear regression model of x_1 and x_2, as compared to a model which includes only previous observations of x_1. Clearly, the notion of Granger causality does not imply true causality. It only implies forecasting ability.

Assume that x_1 can be described by an univariate autoregressive model of order $p < N_t$, or VAR(p) model, given by

$$x_1(t_k) = \sum_{i=1}^{p} a_i x_1(t_{k-i}) + \eta_1(t_k), \tag{1}$$

and also by a bivariate linear regression model defined as

$$x_1(t_k) = \sum_{i=1}^{p} \widetilde{a}_i x_1(t_{k-i}) + \sum_{i=1}^{p} \widetilde{b}_i x_2(t_{k-i}) + \widetilde{\eta}_1(t_k). \tag{2}$$

In (1) and (2), $\{a_i, \widetilde{a}_i, \widetilde{b}_i\}_{i=1}^{p}$ are called the coefficients of the model, and $\eta_1, \widetilde{\eta}_1$ are white noise processes corresponding to the prediction errors for each time series.

For measuring Granger causality, reference [21] introduced the log ratio of the prediction error variances for the two autoregressive models (1) and (2), i.e.

$$\mathcal{F}_{2\to 1} = \ln\left(\frac{\mathrm{var}(\eta_1)}{\mathrm{var}(\widetilde{\eta}_1)}\right). \tag{3}$$

This definition has several motivations. First, since adding information should statistically improve the prediction, the variance of η_1 is greater than or equal to the variance of $\widetilde{\eta}_1$, and $\mathcal{F}_{2\to 1} \geq 0$, as any measure must be. Second, $\mathcal{F}_{2\to 1}$ is a monotonic function with respect to prediction quality, which can be used to compare the magnitude of Granger causality.

Importantly, the concept of Granger causality is easy to extend to vectorial time series [2]. In that case, the set of N_x time series $\boldsymbol{x}(t_k) = (x_1(t_k), \cdots, x_{N_x}(t_k))^T$ is assumed to be adequately described by the multivariate autoregressive (MVAR) model

$$\boldsymbol{x}(t_k) = \sum_{i=1}^{p} A(i)\boldsymbol{x}(t_{k-i}) + \boldsymbol{\eta}(t_k) \tag{4}$$

where $A(i)$, $i = 1, \cdots, p$ are $N_x \times N_x$ matrices containing the elements $A_{jl}(i)$ that describe the linear interaction at time-lag i from $x_j(t_k)$ to $x_l(t_{k-i})$, and $\boldsymbol{\eta}(t_k) = (\eta_1(t_k), \cdots, \eta_{N_x}(t_k))^T$ is a vector of white noise processes. In terms of MVAR coefficients, x_j Granger-causes x_l if at least one off-diagonal element $A_{jl}(i)$ is significantly different from zero. Indeed, a non-zero value of $A_{jl}(i)$ means that the use of the information of $x_l(t_{k-i})$ statistically improves the estimation of $x_j(t_k)$ leading to a decrease of the corresponding variance.

3. Granger causality for high-dimensional time series

The generalization of Granger causality to MVAR model presented in section 2 seems perfectly well suited to the analysis of turbulent flows. Indeed, considering only one-dimensional time signals for causality detection of turbulent events may not be very conclusive, since we do not exploit the spatio-temporal knowledge coming from highly-resolved numerical simulations [16]. However, high dimensionality poses numerous challenges to statistical methods [22]. Indeed, the classical (ordinary least-squares) or more advanced (e.g., [23]) numerical methods used so far to estimate the coefficients of MVAR models are not appropriate for high-dimensional data. For this reason, we propose in this manuscript an alternative way of estimating the linear model necessary for the Granger analysis. This approach is based on a model reduction algorithm introduced recently in the domain of dynamical systems, and called optimized Dynamical Mode Decomposition (DMD) [20]. This method can be viewed as computing, from flow snapshots, eigenvalues and eigenvectors of a linear model that approximates the underlying dynamics, even if those dynamics are nonlinear. As it is the case for all snapshot-based methods, the modelled dynamics is only the part expressed in the snapshots of the original dynamics. As a consequence, the choice of the snapshots is crucial to obtain a physically reliable model.

The standard optimized DMD algorithm will be first presented in section 3.1. The ability of optimized DMD to perform one-step prediction of snapshots will be detailed in section 3.2 and, finally, this ability of forecasting the dynamics will be exploited in section 3.3 to derive criteria for Granger causality.

3.1. Optimized DMD

Suppose that $\boldsymbol{x}(t_k)$ $(k = 1, \ldots, N_t)$ is a set of N_x-dimensional snapshots. Given $p < N_t$, optimized DMD corresponds to seek complex scalars $\{\lambda_j\}_{j=1}^p$ and vectors $\{\boldsymbol{\Phi}_j\}_{j=1}^p$ such that

$$\Gamma = \sum_{k=1}^{N_t} \|\boldsymbol{e}(t_k)\|_2^2 \tag{5}$$

is minimized, where

$$\boldsymbol{x}(t_k) = \sum_{i=1}^{p} \lambda_i^{k-1} \boldsymbol{\Phi}_i + \boldsymbol{e}(t_k). \tag{6}$$

Formally, this equation is equivalent (see Appendix A.1) to the existence of a linear operator $\mathcal{A}$ that connects the field $\boldsymbol{x}(t_k)$ to the subsequent field $\boldsymbol{x}(t_{k+1})$, that is,

$$\boldsymbol{x}(t_{k+1}) = \mathcal{A}\boldsymbol{x}(t_k) + \boldsymbol{e}(t_{k+1}). \tag{7}$$

The expression of $\mathcal{A}$ in terms of $\{\lambda_j\}_{j=1}^p$ and $\{\boldsymbol{\Phi}_j\}_{j=1}^p$ can be deduced directly from (6). Moreover, it can also be demonstrated (Appendix A.1) that $(\boldsymbol{\Phi}_j, \lambda_j)$ are the eigen-elements of $\mathcal{A}$.

To proceed with the description of the optimized DMD algorithm, we define

$$X = (\boldsymbol{x}(t_1), \cdots, \boldsymbol{x}(t_{N_t})) \in \mathbb{C}^{N_x \times N_t} \tag{8}$$

$$\Phi = (\mathbf{\Phi}_1, \ldots, \mathbf{\Phi}_p) \in \mathbb{C}^{N_x \times p} \tag{9}$$

$$E = (\boldsymbol{e}(t_1), \cdots, \boldsymbol{e}(t_{N_t})), \tag{10}$$

and V the Vandermonde matrix based on λ_j

$$V = \begin{pmatrix} 1 & \lambda_1 & \lambda_1^2 & \cdots & \lambda_1^{N_t-1} \\ 1 & \lambda_2 & \lambda_2^2 & \cdots & \lambda_2^{N_t-1} \\ \vdots & \vdots & \vdots & \ddots & \vdots \\ 1 & \lambda_p & \lambda_p^2 & \cdots & \lambda_p^{N_t-1} \end{pmatrix} \in \mathbb{C}^{p \times N_t}. \tag{11}$$

The optimization problem given by (5) and (6) is then equivalent to find Φ and V such that the Frobenius squared norm $\Gamma = \|E\|_F^2$ is minimized in the model

$$X = \Phi V + E, \tag{12}$$

Of all the choices of Φ, the one which minimizes Γ is

$$\Phi = XV^+ \tag{13}$$

where $V^+ = V^*(VV^*)^{-1}$ is the Moore–Penrose pseudoinverse of V, with V^* the conjugate transpose of V. The residual Γ is therefore given by (see details in Appendix A.3):

$$\Gamma = \|E\|_F^2 = \|X\|_F^2 - \|XV^+V\|_F^2. \tag{14}$$

In [20], a particular global optimization technique was used for computing optimized DMD. The main advantage of global optimization is to seek for a global minimum of the cost function (here Γ). Unfortunately, this class of methods is also very time consuming because of the large number of cost function evaluations that are needed to converge [24]. In this paper, we improved the original algorithm of [20] by using a gradient descent method. For that, the gradient of the cost function Γ with respect to the variations of the eigenvalues λ_j was computed analytically (Appendix A.3). It is formally given by

$$\frac{\partial \Gamma}{\partial \lambda_j} = -\text{tr}\left(X^*X\frac{\partial(V^+V)}{\partial \lambda_j}\right) \tag{15}$$

where tr denotes the trace of a matrix. In the optimized DMD algorithm, we then first determine the eigenvalues $\{\lambda_j\}_{j=1}^p$ with a gradient method, and then employ (13) to determine Φ. The computational cost of the iterative part is then independent of the dimension N_x, which makes it well adapted to high-dimensional snapshots data. Since we know an analytical expression of the gradient, efficient quasi-Newton minimization algorithms can be used leading to an increase of the number of optimization parameters in the minimization problem at constant computational resources. In practice, the Broyden-Fletcher-Goldfarb-Shanno (BFGS) algorithm is employed [25].

3.2. Prediction abilities

By definition of the optimized DMD, the error Γ (see eq. 5) is invariant to permutation of all the snapshots. As a consequence, the residual $\boldsymbol{e}(t)$ determined by optimized DMD cannot be directly used for an analysis of Granger causality for which forecasting ability of the model is necessary. The objective of this section is then to determine a new linear expression for $\boldsymbol{x}$ with an error that can be exploited for the Granger causality. One way of doing that is to first determine an expression of $\boldsymbol{x}$ based on its projection onto the basis Φ as determined by optimized DMD (see section 3.1), and then to propagate the model for $\boldsymbol{x}$ on one time step. We note $\boldsymbol{x}^{\text{proj}}(t_k)$ the orthogonal projection of $\boldsymbol{x}(t_k)$ onto Φ, and $\nu_j(t_k)$ the corresponding projection coefficients (see Fig. 1) i.e.

$$\boldsymbol{x}(t_k) = \boldsymbol{x}^{\text{proj}}(t_k) + \boldsymbol{e}^{\text{proj}}(t_k) = \sum_{i=1}^{p} \nu_i(t_k)\boldsymbol{\Phi}_i + \boldsymbol{e}^{\text{proj}}(t_k), \tag{16}$$

where $\boldsymbol{e}^{\text{proj}}$ is the error introduced in the projection. The projection coefficients $\nu_i(t_k)$ are determined in Appendix A.2. Since $\boldsymbol{x}(t_k)$ is known for $k = 1, \cdots, N_t$, we can also determine $\boldsymbol{x}^{\text{proj}}(t_k)$ for $k = 1, \cdots, N_t$ with (16).

Knowing $\mathcal{A}$, the linear mapping between subsequent snapshots (see Appendix A.1), a one-step prediction of $\boldsymbol{x}(t_k)$ can be written as

$$\mathcal{A}\boldsymbol{x}(t_k) = \sum_{i=1}^{p} \nu_i(t_k)\boldsymbol{\Phi}_i\lambda_i. \tag{17}$$

Finally, since the best prediction which can be done in the projection subspace spanned by Φ is $\boldsymbol{x}^{\text{proj}}(t_{k+1})$, we define the prediction error $\boldsymbol{e}^{\text{pred}}(t_{k+1})$ as

$$\boldsymbol{e}^{\text{pred}}(t_{k+1}) = \boldsymbol{x}^{\text{proj}}(t_{k+1}) - \sum_{i=1}^{p} \nu_i(t_k)\boldsymbol{\Phi}_i\lambda_i. \tag{18}$$

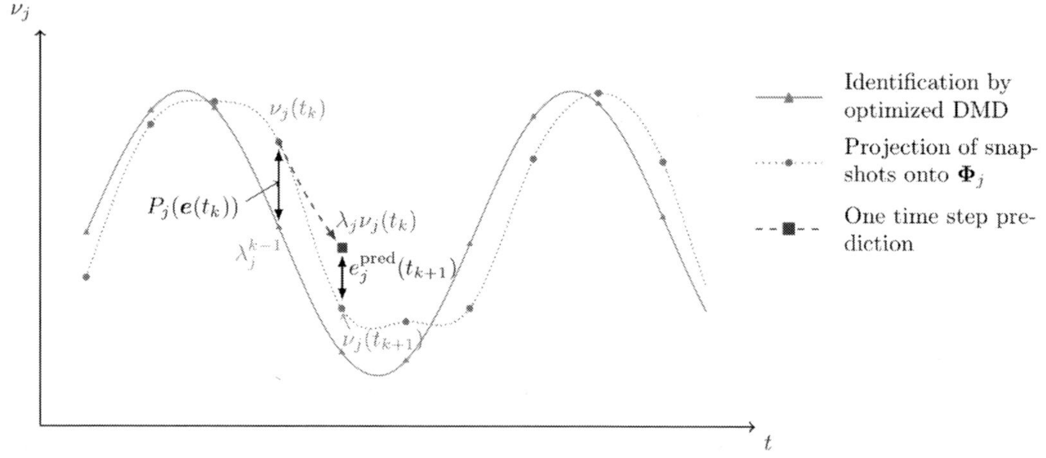

Figure 1. Principle of determination of the prediction error $\boldsymbol{e}^{\text{pred}}$. Linear dynamics identified by optimized DMD (▲), projection of snapshot onto $\boldsymbol{\Phi}_j$ (•) and one time step prediction (■). P_j is the projector onto the vector $\boldsymbol{\Phi}_j$.

The one-step forecast performance is obviously dependent on the model order p. A too low value can lead to a poor representation of the snapshots, whereas a too high value can lead to an "over-fitted" model which can increase the forecast error, as it is demonstrated in [2] for the asymptotic behaviour of estimated MVAR models. The model order may be determined using model selection criteria. The general approach is to choose the value of p which minimizes a criterion that balances the variance accounted for by the model, against the number of coefficients to be estimated. The two most common information criteria are the Akaike (AIC) [26] and the Bayesian Information Criterion (BIC) [27]. These criteria will be used in section 4 for selecting the model order.

3.3. Granger causality using optimized DMD

In sections 3.1 and 3.2, optimized DMD was presented as a method identifying from snapshots a linear model that can be used for one-step prediction. As such, this model can naturally be employed for Granger causality detection.

For the Granger causality, we then consider linear models as given by (18), and assume that λ_i and $\boldsymbol{\Phi}_i$ have already been determined by optimized DMD. For two time series $\boldsymbol{x}_1$ and $\boldsymbol{x}_2$, the Granger causality (see section 2) will then be based on identified linear models given by

$$\begin{cases} \boldsymbol{x}_1^{\text{proj}}(t_{k+1}) = \displaystyle\sum_{i=1}^{p} \nu_{1,i}(t_k)\boldsymbol{\Phi}_{1,i}\lambda_{1,i} + \boldsymbol{e}_1^{\text{pred}}(t_{k+1}) \\ \boldsymbol{x}_2^{\text{proj}}(t_{k+1}) = \displaystyle\sum_{i=1}^{p} \nu_{2,i}(t_k)\boldsymbol{\Phi}_{2,i}\lambda_{2,i} + \boldsymbol{e}_2^{\text{pred}}(t_{k+1}), \end{cases} \tag{19}$$

when the prediction is only based on one time series, and

$$\begin{pmatrix} \boldsymbol{x}_1 \\ \boldsymbol{x}_2 \end{pmatrix}^{\text{proj}} (t_{k+1}) = \sum_{i=1}^{2p} \widetilde{\nu}_i(t_k) \begin{pmatrix} \widetilde{\boldsymbol{\Phi}}_{1,i} \\ \widetilde{\boldsymbol{\Phi}}_{2,i} \end{pmatrix} \widetilde{\lambda}_i + \begin{pmatrix} \widetilde{\boldsymbol{e}}_1^{\text{pred}}(t_{k+1}) \\ \widetilde{\boldsymbol{e}}_2^{\text{pred}}(t_{k+1}) \end{pmatrix} \tag{20}$$

when the prediction is based on both time series. The order of the model (20) is $2p$ to be consistent with the model (2) of the classical formulation of Granger causality (see section 2).

The influence of $\boldsymbol{x}_2$ on $\boldsymbol{x}_1$ ($\mathcal{F}_{2\to1}$), and $\boldsymbol{x}_1$ on $\boldsymbol{x}_2$ ($\mathcal{F}_{1\to2}$) are determined respectively by:

$$\mathcal{F}_{2\to1} = \ln\left(\frac{\displaystyle\int_0^T \|\boldsymbol{e}_1^{\text{pred}}(t)\|^2\,\mathrm{d}t}{\displaystyle\int_0^T \|\widetilde{\boldsymbol{e}}_1^{\text{pred}}(t)\|^2\,\mathrm{d}t}\right) \quad \text{and} \quad \mathcal{F}_{1\to2} = \ln\left(\frac{\displaystyle\int_0^T \|\boldsymbol{e}_2^{\text{pred}}(t)\|^2\,\mathrm{d}t}{\displaystyle\int_0^T \|\widetilde{\boldsymbol{e}}_2^{\text{pred}}(t)\|^2\,\mathrm{d}t}\right). \tag{21}$$

4. Results

In this paper, we propose to analyse the links between different events occurring in a channel flow in the framework of Granger causality. Indeed, although observations seem to link the bursting phenomenon to the evolution of streaks, the mechanisms at the heart of bursts are still not definitely identified, especially in the logarithmic layer. References [11, 17] postulated that the wall-normal velocities are generated by the breakdown of the streamwise-velocity streaks. Here, we suggest to focus on these intermittent events and, for this, we select a set of variables and observations for which we determine their Granger causality following the procedure described in section 3. The objective is to go beyond classical correlations analyses by determining if Granger causality can be useful to understand physical events in the channel flow.

Minimal simulation boxes have demonstrated their ability to generate a single structure in the buffer layer [8] and in the logarithmic layer [15], and also to reproduce statistics of full-size turbulence in the corresponding regions. The data set referred to as W950 in [17] is used in this paper for the Granger analysis. In the following, the streamwise, wall-normal and spanwise directions are denoted by x, y, and z, respectively, and the corresponding velocity components by u, v, and w. The data are obtained from direct numerical simulations of a turbulent channel flow in a minimal box of length $L_x/h = \pi/2$, height $L_y/h = 2$ and width $L_z/h = \pi/4$. These dimensions are considered to be sufficient for isolating typical structures involved in the bursting process in the logarithmic layer [28]. Reference [15] showed that turbulence remains "healthy" roughly below $y \approx 0.3L_z$ corresponding in our computational domain to $y/h \approx 0.25$. The value of the Reynolds number is $Re_\tau = u_\tau h/\nu_{\text{kin}} = 950$, where u_τ is the friction velocity and ν_{kin} is the kinematic viscosity. The incompressible Navier–Stokes equations are integrated in the form of evolution equations for the wall-normal vorticity, and for the Laplacian of the wall-normal velocity, as in [29], and are computed with a constant mass flux. De-aliased Fourier series are used in the spanwise and streamwise coordinate direction, and a Chebychev polynomial representation is employed in the wall-normal direction. A third-order semi-implicit Runge–Kutta method is used for the time integration, as in [30].

The velocity fields u, v and w are first decomposed as a sum of a temporal mean and their fluctuations, that is, $u(x,y,z,t) = U(y) + u'(x,y,z,t)$ (corresponding equations hold for v and w). We then define $\hat{u}(y,t;k_x,k_z)$ the two-dimensional Fourier transform in the streamwise and spanwise directions of the fluctuating velocity field $u'(x,y,z,t)$ where the wave indices k_x and k_z are related to the wavelengths $\lambda_x = \frac{L_x}{k_x}$ and $\lambda_z = \frac{L_z}{k_z}$, respectively. Since the spanwise direction is homogeneous, the optimized DMD algorithm introduced in section 3.1 is applied to snapshots defined as

$$\widetilde{u}_{k_x}(y,t_k) = \sqrt{\frac{1}{L_z}\int_{k_z} |\hat{u}(y,t;k_x,k_z)|^2 \, \mathrm{d}k_z} \quad k = 1,\cdots,N_t. \tag{22}$$

These observations correspond to componentwise fluctuating kinetic energy for one streamwise scale k_x allowing us to track the energy cascade along the components and the scales. In order to represent minimal logarithmic-layer structures and their interactions with the outer part of the flow during the bursts, the full wall-normal profile containing $N_y = 385$ points is kept in the analysis. The low-order Fourier modes in the streamwise direction extract the scales associated to the streaks. We then retain for the Granger analysis the snapshots $\widetilde{u}_{k_x}$, $\widetilde{v}_{k_x}$ and $\widetilde{w}_{k_x}$ for $k_x = 0$ and $k_x = 1$, and also the contribution of all the streamwise wave indices from $k_x = 2$ to the last one that we call the "rest". To simplify the notations, these quantities are referred in the following as $\widetilde{u}_0$, $\widetilde{u}_1$, $\widetilde{u}_r$, $\widetilde{v}_0$, $\widetilde{v}_1$, $\widetilde{v}_r$, $\widetilde{w}_0$, $\widetilde{w}_1$ and $\widetilde{w}_r$, respectively.

Optimized DMD models are then used in accordance to section 3 to determine Granger causality between these nine different variables. First, we characterize the ability of an identified linear model determined by optimized DMD to represent, and predict the snapshots. Since it is difficult to imagine that a low-order linear model can predict the behaviour of a turbulent flow for a statistically long time, the identification is restricted to short-term transient events. In this paper, we consider $N_t = 100$ snapshots taken evenly ($\Delta t^+ \approx 76$) over a time window of $T^+ = T\, u_\tau^2/\nu_{\mathrm{kin}} = 7524$ wall time units, typical of a bursting event time scale for the full channel, at $y^+ \approx 1000$ [15]. In optimized DMD, the order p of the linear model is an intrinsic parameter of the method. A compromise between a good level of representativeness of the snapshots and a low-order representation must be done. Indeed, if choosing $p = N_t$ leads to a perfect reconstruction, considering too many modes in the linear expansion may lead numerically to an "over-fitted" model that is not relevant. Following the Bayesian Information Criterion (BIC) [27] which computes a balance between model accuracy and number of parameters, the order of the model is chosen $p = 8$.

The original history of $\widetilde{v}_0$ is represented in Fig. 2(a). For comparison purposes, the approximation of $\widetilde{v}_0$ according to (6) is shown in Fig. 2(b). In that case, only the essential part of the dynamics is described. The projection of the original data onto the optimized DMD basis as given by (16) is illustrated in Fig. 2(c). The approximation of $\widetilde{v}_0$ is rather good, demonstrating that the description based on optimized DMD is qualitatively accurate. Finally, Eq. (18) is used for doing one-time-step prediction. The result, given in Fig. 2(d), should be compared to Fig. 2(c).

At this point, the ability of optimized DMD to represent and predict the observations is characterized. We can now use these linear models to analyse Granger causalities. The magnitude of the interaction between two variables is measured for the nine quantities considered based on criterion (21). $\mathcal{F}_{i\to j}$ for i, $j = 1,\cdots,9$ determine the level of Granger causality between the variables i and j. These measures are reported in Fig. 3, where i corresponds to

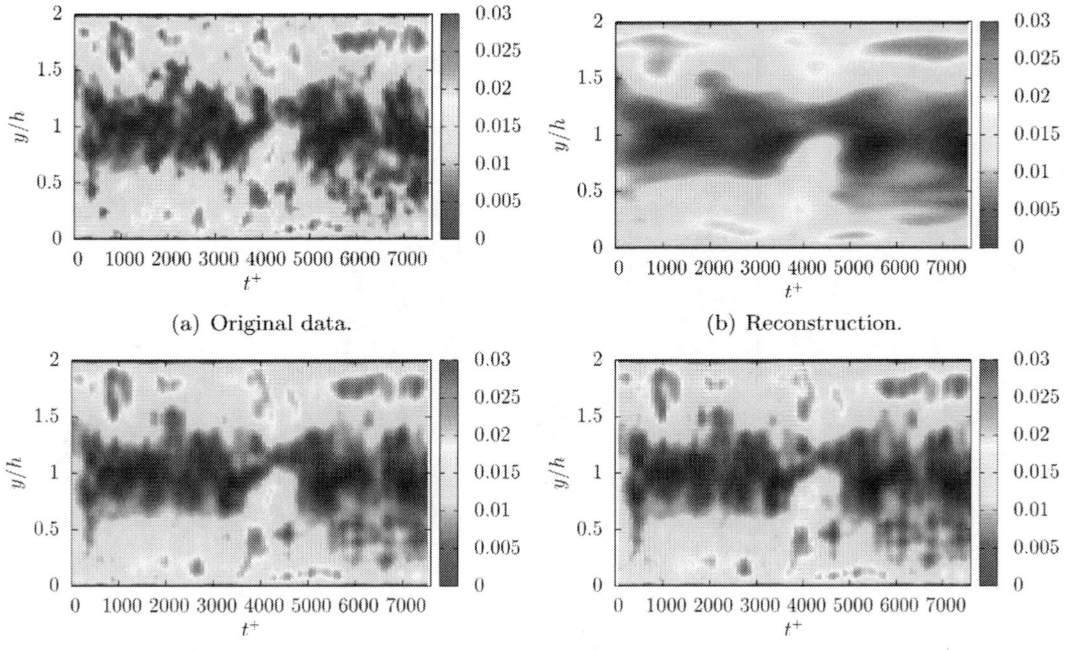

(a) Original data.

(b) Reconstruction.

(c) Projection.

(d) One time step prediction of $\Delta t^+ = 76$.

Figure 2. Linear system identification of $\widetilde{v}_0$.

the ordinates, and j to the abscissas. By convention, the diagonal elements $\mathcal{F}_{i\rightarrow i}$ are equal to 0. We notice immediately that the matrix of Granger causality is not symmetric. The values of the subdiagonal elements are higher than their superdiagonal counterpart, demonstrating globally a direct energy cascade from larger to smaller scales. Moreover, three strong levels of causality are clearly visible, leading us to conclude that: i) $\widetilde{u}_0$ Granger-causes $\widetilde{v}_1$; ii) $\widetilde{u}_0$ Granger-causes $\widetilde{v}_r$; and iii) $\widetilde{u}_1$ Granger-causes $\widetilde{v}_r$. At weaker levels, we also observe the Granger causalities: iv) from $\widetilde{u}_r$ to $\widetilde{v}_r$; v) from $\widetilde{w}_1$ to $\widetilde{v}_1$; and finally, vi) from $\widetilde{w}_r$ to $\widetilde{v}_r$. Remarkably, different scales are involved in the Granger causalities i) to iii), whereas the same scales are participating to the weaker Granger causalities iv) to vi). In terms of wall-turbulence dynamics, the streamwise fluctuating energy for elongated scales, $\widetilde{u}_0$ and $\widetilde{u}_1$, can be interpreted as the presence of meandering streaks, whereas the wall-normal fluctuating energy for the same scales, $\widetilde{v}_0$ and $\widetilde{v}_1$, can be viewed as the sign of ejections and sweeps. Based on these physical interpretations, we can conclude that Granger causalities i) to iii) give strong hints that the dynamical evolution of the streaks causes the ejections and the sweeps. Furthermore, the most likely interpretation of Granger causalities iv) to vi) are the action of oblique vortices.

The variables $\widetilde{u}_0$, $\widetilde{v}_1$, $\widetilde{u}_1$ and $\widetilde{v}_r$ which are involved in the main Granger causalities i) to iii) are represented in Fig. 4. Here, we can see the advantage of a snapshot-based approach that gives an idea of the locations of the involved events. For example, focusing on the $\widetilde{u}_0$–$\widetilde{v}_1$ causality, we can identify at the wall near $t^+ = 4000$ (Fig. 4(a)) an event of strong streamwise fluctuating energy which precedes the wall-normal activity in the outer region of the flow (Fig. 4(b)). Moreover, this event corresponds to streamwise fluctuations that are transformed into wall-normal fluctuations. A plausible interpretation would be that the elongated streamwise fluctuations are transformed in wall-normal fluctuations by the streak disorganization, as observed in [15], creating ejections and sweeps during the bursts. Granger causalities corroborate that the streaks breakdowns cause the bursts, but the way that these streamwise perturbations are transformed into wall-normal fluctuations is not explained by our study.

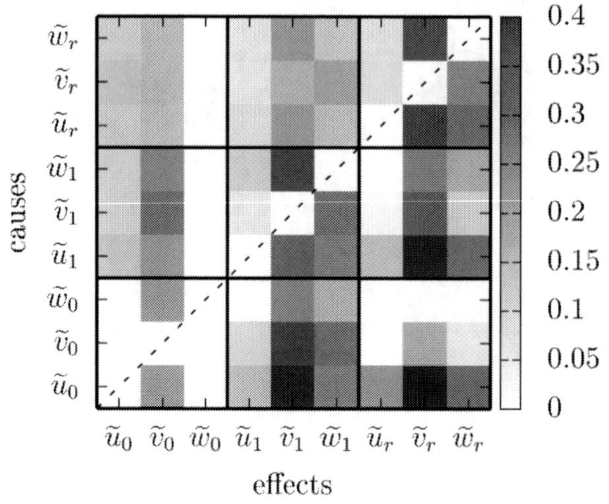

Figure 3. Matrix representation of Granger causalities between $\widetilde{u}_0$, $\widetilde{v}_0$, $\widetilde{w}_0$, $\widetilde{u}_1$, $\widetilde{v}_1$, $\widetilde{w}_1$, $\widetilde{u}_r$, $\widetilde{v}_r$ and $\widetilde{w}_r$ quantified by the prediction improvement $\mathcal{F}_{i\to j}$ as defined in (21). The ordinate i Granger-causes the abscissa j. Heavier lines separate the different scales, whereas the dashed diagonal separates the forward (subdiagonal) and backward (superdiagonal) energy cascade.

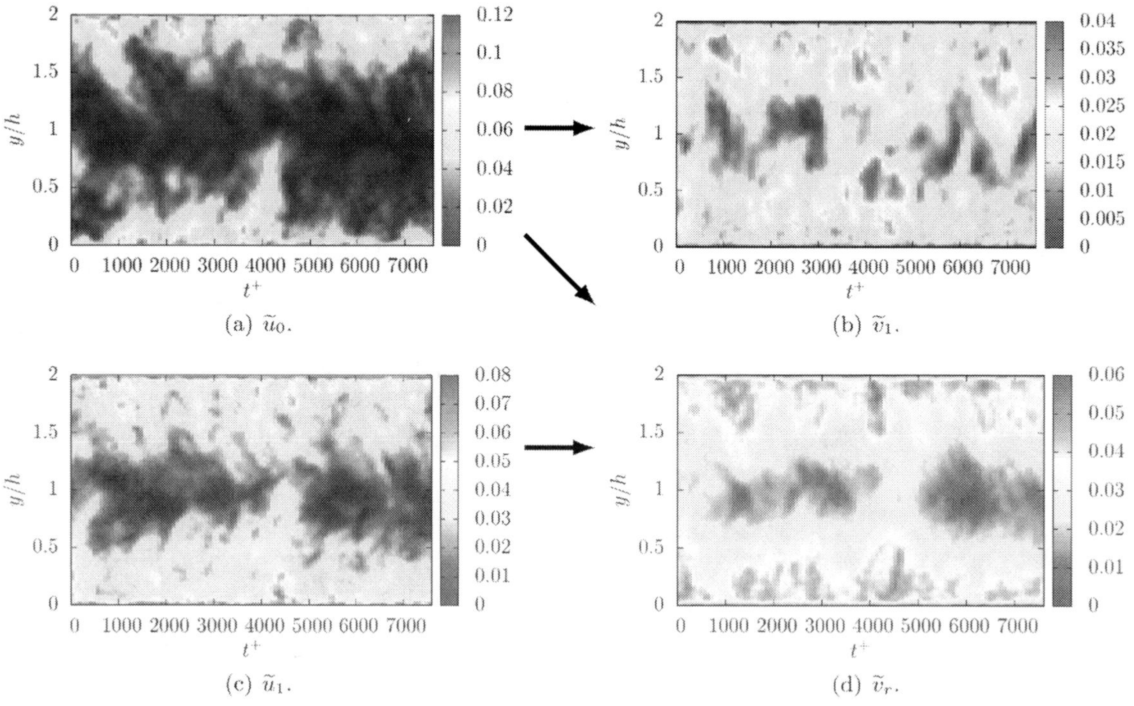

Figure 4. Variables involved in the main Granger causalities. The arrows "→" indicate the direction of causality.

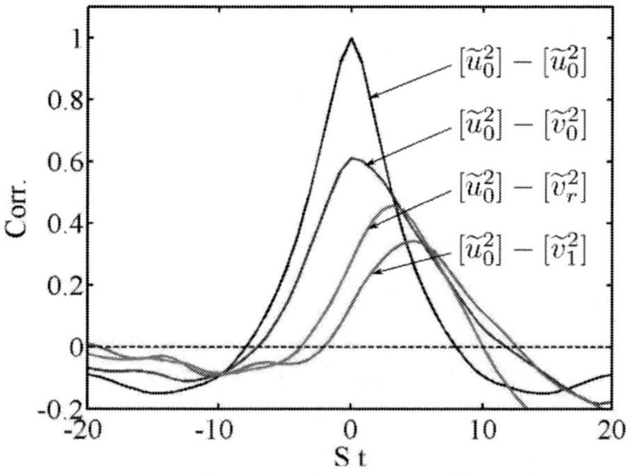

Figure 5. Temporal correlations between $[\widetilde{u}_0^2]$ and $[\widetilde{u}_0^2]$ (black), $[\widetilde{u}_0^2]$ and $[\widetilde{v}_0^2]$ (blue), $[\widetilde{u}_0^2]$ and $[\widetilde{v}_r^2]$ (magenta), and $[\widetilde{u}_0^2]$ and $[\widetilde{v}_1^2]$ (red). Time t is scaled with S, the mean shear averaged over the band from $y_1/h = 0.2$ to $y_2/h = 0.3$.

Those causality behaviours are confirmed by temporal correlations of flow variables integrated over the minimal box corresponding to the dataset W950. For example, $[\widetilde{u}_0^2]$ is computed by averaging the squares of $\widetilde{u}_0$ over a band of y varying from $y/h = 0.2$ to $y/h = 0.3$. Temporal correlations are then calculated for each variable following the method described in [17]. Figure 5 represents the temporal autocorrelation function of $[\widetilde{u}_0^2]$, and the correlations between $[\widetilde{u}_0^2]$ and $[\widetilde{v}_0^2]$, $[\widetilde{v}_1^2]$ and $[\widetilde{v}_r^2]$, respectively. We observe that the level of correlations is strong, and that the correlations of $[\widetilde{u}_0^2]$ with $[\widetilde{v}_r^2]$ and $[\widetilde{v}_1^2]$ are clearly shifted in time suggesting that they are caused by $\widetilde{u}_0$, which is in accordance with the results of Granger causality. We also note that the peak of correlation for $[\widetilde{u}_0^2]-[\widetilde{v}_r^2]$ precedes the corresponding peak for $[\widetilde{u}_0^2]-[\widetilde{v}_1^2]$. In addition, this result is corroborated by a stronger value of Granger causality from $\widetilde{u}_0$ to $\widetilde{v}_r$ than from $\widetilde{u}_0$ to $\widetilde{v}_1$ (Fig. 3), suggesting a more direct influence. All these results indicate that smaller scales are influenced before and more strongly than the larger ones. This observation could be related to Fig. 4, where we notice that $\widetilde{v}_r$, which is associated to smaller scales, is stronger close to the walls than $\widetilde{v}_1$ (see Fig. 4(b)), which is stronger in the outer-region.

In summary, streamwise velocities of low-order streamwise Fourier modes Granger-cause wall-normal velocities of higher-order modes. Alternatively, we can say that "the knowledge of the streaks breakdown helps to predict the bursting process". Moreover, a forward energy redistribution from streaks instabilities to ejections and sweeps was found. This study gives supplementary clues to the fact that strong ejections of fluid are a consequence of the breakdown of the streaks.

5. Conclusion

The capabilities of Granger causality to study physical mechanisms occurring in a turbulent channel flow were illustrated for data obtained from direct numerical simulations. The initial objectives were to improve our understanding of wall-bounded turbulence by studying the energy redistribution between elongated scales in the logarithmic layer. The paper shows that characterizing the causal relationship between significant events is an important but non-trivial aspect of understanding the behaviour of channel flow. As a post-processing approach, the choice of the observations variables is essential in the application of Granger causality. For maximizing our chance of obtaining physically relevant conclusions, we need to choose very carefully the data, especially in turbulence, where the dynamics is complex and the size of the snapshots is large. Another difficulty of wall-bounded flows is the lack of rigorous definitions of streaks and bursts which prevent us so far to re-discuss in terms of Granger causality the Self-Sustaining Process proposed in [31]. On the other hand, we may conclude that, compared to pure correlation studies, the use of a predictive model confers to Granger causality a stronger basis for determining causality between events. This advantage is balanced by a strong modelling effort. Indeed, in turbulence, reduced-order modelling is still a challenge if a predictive behaviour is expected. In the present study, we show that, when the snapshots are high-dimensional, the linear models at the heart of the Granger causality framework can be determined by optimized DMD. Finally, it should be noted that the predictive causality as defined by Granger, though helpful in understanding feedbacks and interactions in complex systems, does not necessarily imply true causality.

The results of our Granger analysis, consistent with correlations studies, highlight causality links between streaks breakdown and wall-normal activity probably transferred by shear deformation. Besides, the role of active oblique vortices in the wall dynamics has been detected by Granger causality. The successful use of linear systems for the identification does not imply that the detected mechanism is linear. However, this fact is in line with the assumption that the linear Orr mechanism is involved in the bursting process.

In future work, the authors will apply the Granger analysis to better selected data and more complex flows. Moreover, sensitivity tests will be also performed by adding small amount of noise to the data to prove that the values of Granger causalities are sufficiently robust.

Acknowledgments
This work was funded in part by the Multiflow program of the European Research Council.

Appendix

Appendix A.1. Identification of the linear mapping $\mathcal{A}$ between subsequent snapshots

Based on the modes $\mathbf{\Phi}_i$ and corresponding amplitudes λ_i determined by optimized DMD (see section 3.1), the best reconstruction of the snapshots $\{\boldsymbol{x}(t_k)\}_{k=1}^{N_t}$ that can be obtained are given by

$$\breve{\boldsymbol{x}}(t_k) = \sum_{i=1}^{p} \mathbf{\Phi}_i \lambda_i^{k-1} = \Phi \boldsymbol{\lambda}_k, \tag{A.1}$$

where $\boldsymbol{\lambda}_k = (\lambda_1^{k-1}, \ldots, \lambda_p^{k-1})^T$. The objective of this section is to determine the expression of the linear mapping $\mathcal{A}$ that connects the field $\breve{\boldsymbol{x}}(t_k)$ to the subsequent field $\breve{\boldsymbol{x}}(t_{k+1})$, that is,

$$\breve{\boldsymbol{x}}(t_{k+1}) = \mathcal{A}\, \breve{\boldsymbol{x}}(t_k). \tag{A.2}$$

The one-step prediction of (A.1) can be written as:

$$\breve{\boldsymbol{x}}(t_{k+1}) = \sum_{i=1}^{p} \mathbf{\Phi}_i \lambda_i^{k} = \sum_{i=1}^{p} \mathbf{\Phi}_i \lambda_i \lambda_i^{k-1} = \Phi \Lambda \boldsymbol{\lambda}_k \tag{A.3}$$

with $\Lambda = \mathrm{diag}\,(\lambda_1, \ldots, \lambda_p)$. To proceed, we would like to introduce $\breve{\boldsymbol{x}}(t_k)$ on the RHS of (A.3). For this, we deduce from (A.1) that $\breve{\boldsymbol{x}}$ is spanned by Φ. We then conclude immediately that $P\breve{\boldsymbol{x}}(t_k) = \breve{\boldsymbol{x}}(t_k)$ where P is the orthogonal projector onto the space spanned by Φ which is introduced in section Appendix A.2, that is,

$$P\breve{\boldsymbol{x}}(t_k) = \Phi \left(\Phi^* \Phi\right)^{-1} \Phi^* \breve{\boldsymbol{x}}(t_k) = \breve{\boldsymbol{x}}(t_k) = \Phi \boldsymbol{\lambda}_k$$

or, by identification, $\boldsymbol{\lambda}_k = \left(\Phi^* \Phi\right)^{-1} \Phi^* \breve{\boldsymbol{x}}(t_k)$. Plugging this expression into (A.3), we obtain that

$$\breve{\boldsymbol{x}}(t_{k+1}) = \Phi \Lambda \left(\Phi^* \Phi\right)^{-1} \Phi^* \breve{\boldsymbol{x}}(t_k)$$

i.e.

$$\mathcal{A} = \Phi\Lambda\left(\Phi^*\Phi\right)^{-1}\Phi^*. \tag{A.4}$$

This result justifies the assertion that optimized DMD can be viewed as a way to identify a linear dynamics representing optimally a set of snapshots. Moreover, we can deduce from (A.4) that

$$\mathcal{A}\Phi = \Phi\Lambda\left(\Phi^*\Phi\right)^{-1}\Phi^*\Phi = \Phi\Lambda, \tag{A.5}$$

demonstrating that $(\boldsymbol{\Phi}_j, \lambda_j)$ are eigen-elements of $\mathcal{A}$.

Finally, consider that the linear mapping $\mathcal{A}$ is applied to any snapshot $\boldsymbol{x}(t_k)$, we have

$$\begin{aligned}\mathcal{A}\boldsymbol{x}(t_k) &= \Phi\Lambda\left(\Phi^*\Phi\right)^{-1}\Phi^*\boldsymbol{x}(t_k) = \Phi\Lambda\boldsymbol{\nu}(t_k)\\ &= \sum_{i=1}^{p}\boldsymbol{\Phi}_i\nu_i(t_k)\lambda_i.\end{aligned} \tag{A.6}$$

The application of $\mathcal{A}$ to $\boldsymbol{x}$ is then equivalent to the multiplication of each component of the optimized DMD expansion by the corresponding optimized DMD eigenvalue.

Appendix A.2. Projection

Let P be the orthogonal projector onto the space spanned by Φ. By definition (see [32]), we have for all $\boldsymbol{x}$

$$P\boldsymbol{x}(t_k) = \sum_{i=1}^{p}\nu_i(t_k)\boldsymbol{\Phi}_i = \Phi\boldsymbol{\nu}(t_k), \tag{A.7}$$

where $P = \Phi(\Phi^*\Phi)^{-1}\Phi^*$. By identification, we then find immediately that

$$\boldsymbol{\nu}(t_k) = (\Phi^*\Phi)^{-1}\Phi^*\boldsymbol{x}(t_k). \tag{A.8}$$

Appendix A.3. Optimized DMD

In this section, we determine the analytical expression of the gradient of $\Gamma = \|E\|_F^2$ with respect to the variation of the DMD eigenvalues λ_j. First, we simplify the expression of Γ using the trace expansion of the Frobenius norm. We find that

$$\begin{aligned}\Gamma &= \|E\|_F^2 = \|X - \Phi V\|_F^2\\ &= \|X - XV^+V\|_F^2 = \|X(\mathbb{I} - V^+V)\|_F^2\\ &= \operatorname{tr}\left(X(\mathbb{I} - V^+V)(\mathbb{I} - V^+V)^*X^*\right)\\ &= \operatorname{tr}\left(X(\mathbb{I} - V^+V - (V^+V)^* + V^+V(V^+V)^*)X^*\right).\end{aligned} \tag{A.9}$$

Since V^+V is a projection and is self-adjoint, we have

$$\begin{aligned}\Gamma &= \operatorname{tr}\left(X(\mathbb{I} - V^+V)X^*\right)\\ &= \operatorname{tr}\left(X(\mathbb{I} - V^+V(V^+V)^*)X^*\right)\\ &= \operatorname{tr}\left(XX^*\right) - \operatorname{tr}\left((XV^+V)(XV^+V)^*\right)\\ &= \|X\|_F^2 - \|XV^+V\|_F^2.\end{aligned} \tag{A.10}$$

The gradient of Γ with respect to λ_j is then given by:

$$\begin{aligned}\frac{\partial \Gamma}{\partial \lambda_j} &= -\frac{\partial}{\partial \lambda_j}\left(\mathrm{tr}\left(X(V^+V)(V^+V)^*X^*\right)\right)\\ &= -\mathrm{tr}\left(X\frac{\partial(V^+V)}{\partial \lambda_j}X^*\right)\\ &= -\mathrm{tr}\left(X^*X\frac{\partial(V^+V)}{\partial \lambda_j}\right).\end{aligned} \tag{A.11}$$

The Gramian matrix X^*X is computed once for all at the beginning of the optimization procedure for reducing the computational cost. Moreover, since $V^+ = V^*(VV^*)^{-1}$, we have to determine the derivative of a product of matrices with respect to a parameter μ, i.e.

$$\frac{\partial(AB)}{\partial \mu} = \frac{\partial A}{\partial \mu}B + A\frac{\partial B}{\partial \mu}, \tag{A.12}$$

and the derivative of the inverse of a matrix A

$$\frac{\partial A^{-1}}{\partial \mu} = -A^{-1}\frac{\partial A}{\partial \mu}A^{-1}. \tag{A.13}$$

Combining all these expressions, we obtain:

$$\frac{\partial(V^+V)}{\partial \lambda_j} = \frac{\partial V^*}{\partial \lambda_j}\left(VV^*\right)^{-1}V - V^*\left(VV^*\right)^{-1}\left(\frac{\partial V}{\partial \lambda_j}V^* + V\frac{\partial V^*}{\partial \lambda_j}\right)\left(VV^*\right)^{-1}V + V^*\left(VV^*\right)^{-1}\frac{\partial V}{\partial \lambda_j} \tag{A.14}$$

where the gradient of V with respect to λ_j is given analytically by

$$\frac{\partial V}{\partial \lambda_j} = \begin{pmatrix} 0 & 0 & 0 & \cdots & 0\\ \vdots & \vdots & \vdots & \vdots & \vdots\\ 0 & 1 & 2\lambda_j & \cdots & (N-1)\lambda_j^{N_t-1}\\ \vdots & \vdots & \vdots & \vdots & \vdots\\ 0 & 0 & 0 & \cdots & 0\end{pmatrix}. \tag{A.15}$$

References

[1] McKay Illari P, Russo F and Williamson J (eds) 2011 *Causality in the Sciences* (Oxford University Press)
[2] Lütkepohl H 2007 *New Introduction to Multiple Time Series Analysis* 2nd ed (Berlin: Springer)
[3] Granger C W J 1969 Investigating causal relations by econometric models and cross-spectral methods *Econometrica* 424–38
[4] Seth A K 2010 A MATLAB toolbox for Granger causal connectivity analysis *J. Neurosci. Meth.* **186** 262–73
[5] Granger C W J, Huang B–N and Yang C–W 2000 A bivariate causality between stock prices and exchange rates: evidence from recent Asian flu *Q. Rev. Econ. Financ.* **40** 337–54
[6] Roebroeck A, Formisano E and Goebel R 2005 Mapping directed influence over the brain using Granger causality and fMRI *Neuroimage* **25** 230–42
[7] Elsner J B 2007 Granger causality and atlantic hurricanes *Tellus A* **59** 476–85
[8] Jiménez J and Moin P 1991 The minimal flow unit in near-wall turbulence. *J. Fluid Mech.* **225** 213–40
[9] Waleffe F 2001 Exact coherent structures in channel flow *J. Fluid Mech.* **435** 93–102
[10] Panton R L 2001 Overview of the self-sustaining mechanisms of wall turbulence *Prog. Aerosp. Sci.* **37** 341–83
[11] Kline S J, Reynolds W C, Schraub F A and Runstadler P W 1967 The structure of turbulent boundary layers *J. Fluid Mech.* **30** 741–73
[12] Butler K M and Farrell B F 1993 Optimal perturbations and streak spacing in wall-bounded turbulent shear flow *Phys. Fluids A* **5** 774–77

[13] Del Álamo J C and Jiménez J 2006 Linear energy amplification in turbulent channels *J. Fluid Mech.* **559** 205–13
[14] Schoppa W and Hussain F 2002 Coherent structure generation in near-wall turbulence *J. Fluid Mech.* **453** 57–108
[15] Flores O and Jiménez J 2010 Hierarchy of minimal flow units in the logarithmic layer *Phys. Fluids* **22** 071704
[16] Jiménez J 2013 Near-wall turbulence *Phys. Fluids* **25** 101302
[17] Jiménez J 2013 How linear is wall-bounded turbulence? *Phys. Fluids* **25** 110814
[18] Ravindran S S 2007 Optimal boundary feedback flow stabilization by model reduction *Comput. Method. Appl. M.* **196** 2555–69
[19] Bergmann M, Cordier L and Brancher J P 2005 Optimal rotary control of the cylinder wake using Proper Orthogonal Decomposition reduced-order model *Phys. Fluids* **17** 097101
[20] Chen K K, Tu J H and Rowley C W 2012 Variants of Dynamic Mode Decomposition: Boundary Condition, Koopman, and Fourier Analyses *J. Nonlinear Sci.* **22** 887–915
[21] Geweke J 1982 Measurement of linear dependence and feedback between multiple time series *J. Am. Stat. Assoc.* **77** 304–13
[22] Fan J, Lv J and Qi L 2011 Sparse high-dimensional models in economics *Annu. Rev. Econ.* **3** 291–317
[23] Schlögl A 2006 A comparison of multivariate autoregressive estimators *Signal Process.* **86** 2426–29
[24] Auger A, Hansen N, Zerpa J M P, Ros R and Schoenauer M 2009 Empirical comparisons of several derivative free optimization algorithms *Acte 9-ième Colloq. Nat. Calcul Structures, CSMA* (Giens, France)
[25] Gilbert J C and Lemaréchal C 2009 The module M1QN3 – Version 3.3 *INRIA Rocquencourt & Rhône-Alpes*
[26] Akaike H 1974 A new look at the statistical model identification *IEEE Trans. Autom. Control* **19** 716–23
[27] Schwarz G 1978 Estimating the dimension of a model *Ann. Stat.* **6** 461–64
[28] Jiménez J 2012 Cascades in wall-bounded turbulence *Annu. Rev. Fluid Mech.* **44** 27–45
[29] Kim J, Moin P and Moser R 1987 Turbulence statistics in fully developed channel flow at low Reynolds number *J. Fluid Mech.* **177** 133–66
[30] Spalart P R, Moser R D and Rogers M M 1991 Spectral method for the Navier–Stokes equations with one infinite and two periodic dimensions *J. Comput. Phys.* **96** 297–324
[31] Waleffe F 1997 On a self-sustaining process in shear flows *Phys. Fluids* **9** 883–900
[32] Meyer C 2000 *Matrix Analysis and Applied Linear Algebra: book and solutions manual* (SIAM)

Analysis of a Turbulent Boundary Layer Subjected to a Strong Adverse Pressure Gradient

Ayse G Gungor[1], Yvan Maciel[2], Mark P Simens[3] and Julio Soria[4]

[1] Faculty of Aeronautics and Astronautics, Istanbul Technical University, 34469 Maslak, Istanbul, Turkey

[2] Department of Mechanical Engineering, Laval University, Quebec City, QC, G1V 0A6 Canada

[3] School of Aeronautics, Universidad Politécnica de Madrid, 28040 Madrid, Spain

[4] Department of Mechanical and Aerospace Engineering, Monash University, Melbourne, 3800 Australia

E-mail: `ayse.gungor@itu.edu.tr`

Abstract. A strongly decelerated turbulent boundary layer is investigated by direct numerical simulation. Transition to turbulence is triggered by a trip wire which is modelled using the immersed boundary method. The Reynolds number close to the exit of the numerical domain is $Re_\theta = 2175$ and the shape-factor is $H = 2.5$. The analysis focuses on the latter portion of the flow with large velocity defect, at higher Reynolds numbers and further from the transition region. Mean velocity profiles do not reveal a logarithmic law. Departure from the law of the wall occurs throughout the inner region. The production and Reynolds stress peaks move to roughly the middle of the boundary layer. The profiles of the uv correlation factor reveal that u and v become less correlated throughout the boundary layer as the mean velocity defect increases, especially near the wall. The structure parameter is low in the present flow, similar to equilibrium APG flows and mixing layers, and decreases as the mean velocity defect increases. The statistics of the upper half of the boundary layer resemble those of a mixing layer. Furthermore, various two-dimensional two-point correlation maps are obtained. The C_{vv} and C_{ww} correlations obtained far from the transition region at $Re_\theta = 2175$ and at $y/\delta = 0.4$ coincide with results obtained for a ZPG boundary layer, implying that the structure of the v, w fluctuations is the same as in ZPG. However, C_{uu} indicates that the structure of the u fluctuation in this APG boundary layer is almost twice as short as the ZPG one. The APG structures are also less correlated with the flow at the wall. The near-wall structures are different from ZPG flow ones in that streaks are much shorter or absent.

1. Introduction

Wall bounded turbulent flows are important to understand in order to be able to reduce the energy losses associated with these turbulent flows, which have a wide range of practical applications. An important sub-group of wall-bounded turbulent flows are adverse-pressure-gradient (APG) boundary layers. These flows are found for instance around surfaces with curvature, as encountered in many aerodynamic applications such as airplane wings, cars and rotors in turbines. Although a significant amount of research has been devoted to understanding channel flows, pipe flows and zero-pressure-gradient (ZPG) boundary layers, the same cannot be said for APG boundary layer flows. A brief overview will be presented of the most significant results which are consequential to the study reported in this article.

The velocity components of the boundary layer flow will be denoted by u, v, w, and the respective spatial coordinates by x, y, z for the streamwise, wall-normal and spanwise directions, respectively. The pressure is denoted by p. Capital letters refer to mean quantities, and primed ones to root-mean-squared fluctuation intensities. The superscript, '+' , denotes viscous-friction scaled quantities, normalized with the x-dependent friction velocity u_τ and the kinematic viscosity ν. The 99% boundary-layer thickness is δ.

Although studied less, there is still a wealth of experimental and numerical data on first- and second-order statistics available for a range of APG boundary layer flows, for example [1–7]. Among other things, these studies clearly demonstrate that the important problem in APG boundary layer flow research is the lack of a canonical set-up and the appropriate design for the corresponding experiments. The lack of such a canonical set-up is to a large extent related to the lack of understanding of which parameters are paramount for the development of the APG boundary layer. This might also explain the reason why APG boundary layer flows have been studied less than other turbulent wall-bounded flows.

One of the more significant results from these previous studies is the observation that, under certain conditions, the near-wall u'-peak is absent and u' reaches a maximum in the middle of the boundary layer. Although the absence of the near-wall u'-peak seems related to a high or prolonged APG [8], one of the outstanding questions that needs to be addressed is: what exactly causes the disappearance of the near-wall u'-peak? Furthermore, it is unknown how these APG boundary layer flows relate to other types of canonical flows, like the ZPG boundary layer flow and free shear flows.

It has long been recognized that turbulent flows contain coherent 3-D structures [9]. Due to the deterministic temporal and spatial behaviour of these structures in a chaotic background environment it has also long been recognized that these structures offer the best opportunity to control the flow. Recently statistical information on these structures in a 3-D framework has been studied [10–12] for the outer part of turbulent channel flow and for APG boundary layer flows.

The purpose of this article is to review results and present new results on strong APG boundary layers that relate the statistical results of these flows with those from other canonical flows, like free shear flows and ZPG boundary layer flows, and to study the velocity correlations and coherent structures for the APG boundary layer flow simulated especially for this study.

The article begins with a discussion of the numerical methodology employed to undertake the DNS of the APG boundary layer flow. This is followed by a discussion of the results, beginning with a general overview of the flow field, followed by a more in-depth presentation of the first- and second-order statistics, as well as a comparison with other results. Then the structure of the boundary layer is examined, by considering the various two-point correlations and uv-structures. The article concludes with a summary of the main results.

2. Numerical methodology

The boundary layer subjected to a strong APG is simulated in a parallelepiped domain over a smooth no-slip wall, with spanwise periodicity and streamwise non-periodic inflow and outflow. The numerical scheme used for the simulation is similar to the one described in Simens *et al* [13]. The Navier-Stokes equations are integrated using a fractional step method on a staggered grid, with third-order Runge-Kutta time integration, fourth-order compact spatial discretisation for the convective and viscous terms, and second-order discretisation for the pressure in the directions perpendicular to the span, which is spectral. The simulation parameters are summarized in table 1.

The desired constant streamwise velocity gradient is controlled by imposing a constant uniform suction at the top simulation boundary. The streamwise and spanwise velocities at the top boundary satisfy free-slip conditions. The laminar Hiemenz profile is prescribed at the

Table 1. Parameters of the turbulent boundary layer simulations. L_x, L_y and L_z are the box dimensions along the three axes. N_x, N_y, and N_z are the corresponding grid sizes, and the Δ values are the resolutions in wall units. The momentum thickness, θ_0, is measured at the inflow. The Kolmogorov length η is computed from the local energy dissipation. The coarsest resolution along x and z in terms of η is found near the wall at reattachment, where $\Delta x/\eta \approx 10$.

Re_θ	$(L_x, L_y, L_z)/\theta_0$	$\Delta x^+, \Delta y^+_{min}, \Delta z^+$	N_x, N_y, N_z	β
150-2200	$2380 \times 450 \times 1100$	$2.2 \times 0.2 \times 2.0$	$1537 \times 201 \times 768$	5-100

inflow, and the velocities at the outflow are estimated by a convective boundary condition, with small corrections to enforce global mass conservation [13]. The turbulent transition is triggered by a disturbance strip with a height of $1.1\theta_0$ located close to the inflow, and modelled using the immersed boundary method [14–18]. The implementation of the immersed boundary method is discussed in detail in Simens and Gungor [19], including a full discussion of the effect of roughness on the development of APG boundary layers.

3. Results and discussion

3.1. Flow description

The flow is initially laminar, separates, transitions within the separation bubble, reattaches and develops into an attached turbulent APG boundary layer, the latter being the flow zone of interest for this study. The presence of a small transitional separation bubble near the inflow helps achieve the fully turbulent regime within a shorter distance than what would be obtained with attached transition through Tollmien-Schlichting waves. The disturbance strip located upstream of the bubble does not cause the flow to transition, but is responsible for generating perturbations that hasten the reattachment of the separation bubble [19]. The flow after the trip element is still laminar, and perturbations due to the trip element hardly grow until the bubble starts to form. The turbulent kinetic energy distribution in the lower half of the domain is shown in figure 1. The time-averaged separated region and the boundary layer growth are also indicated in the same figure with thick solid lines. The flow fluctuations originate at the inflection point of the shear layer formed by the separation bubble. The perturbations generated by the disturbance strip grow in the separated shear layer, and trigger the inviscid Kelvin–Helmholtz instability that is responsible for the breakdown to turbulence. Kelvin–Helmholtz instability is characterized by the formation of large, two-dimensional vortices in the separated shear layer. The large amplitude fluctuations seen in figure 1 are due to the regular shedding of these vortical structures. The growth of these fluctuations in the initial part of the separated region is slow, whereas they grow much faster close to reattachment. This sudden growth of the fluctuations triggers a slowdown of the bubble growth due to the turbulent energy diffusion, and reattachment occurs. The reattachment of the flow is accompanied by an overall maximum in the turbulent intensity. This intensity decreases slowly downstream, while the peak location is shifted outward from the reattachment point to the end of the solution domain.

The effect of the separated region can be seen in figure 2, which displays the two-point cross-correlation coefficient

$$C_{vu}(x, y; x', y') = \frac{\langle v(x, y)u(x', y')\rangle}{(\langle v(x, y)^2\rangle\langle u(x', y')^2\rangle)^{1/2}}, \tag{1}$$

where the averaging $\langle\rangle$ is taken over time and spanwise direction. The correlation is centred at the reattachment point of the separated region in the streamwise direction at $x/\theta_0 = 680$ and at $y/\delta = 0.6$, where δ is the local boundary layer thickness. The influence of the separated

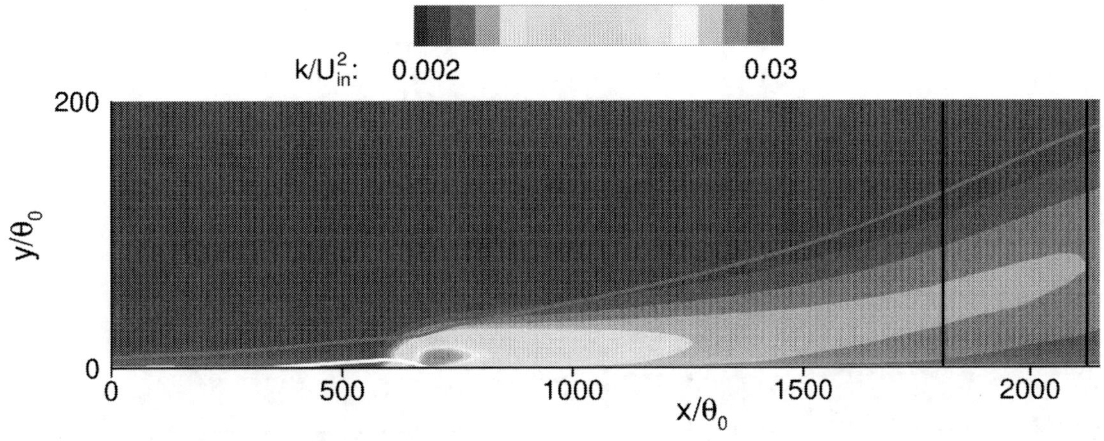

Figure 1. Turbulent kinetic energy normalized with the inflow freestream velocity in the lower half of the domain. The axes are not to scale. The solid white line is the zero contour of the streamwise velocity U, and marks locations of the trip element ($80 \leq x/\theta_0 \leq 130$) and the separation bubble. The solid magenta line shows the boundary layer thickness. The vertical black lines are the selected two stations where $(Re_\theta, H) = (1755, 2.0)$ and $(2175, 2.5)$, respectively.

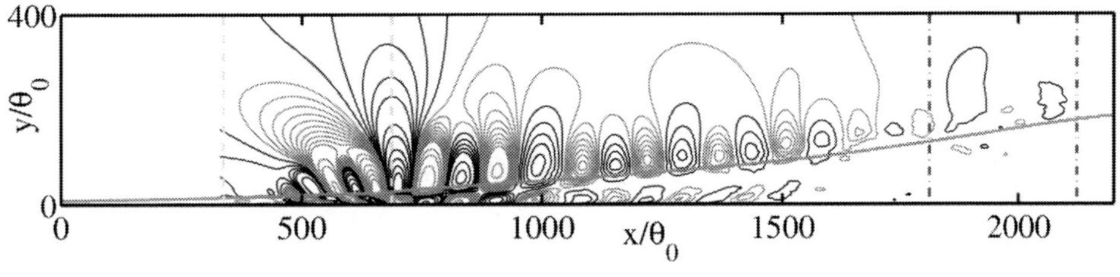

Figure 2. Two-point spatial cross-correlation coefficient of wall-normal and streamwise velocity fluctuations, $C_{vu}(x, y; x', y')$ in xy plane. The correlation is centred at the reattachment location of the time-averaged separation bubble and at $y/\delta = 0.6$. The solid magenta line is the location of the boundary layer thickness. The dashed vertical lines mark the separation and reattachment locations of the averaged separated region. The dotted-dashed blue vertical lines are the selected two stations where $(Re_\theta, H) = (1755, 2.0)$ and $(2175, 2.5)$, respectively.

region survives up to $x/\theta_0 \approx 1800$, and only becomes small beyond that limit. The attached turbulent boundary layer is therefore probably still recovering from the effects of the shear layer instability all the way up to the end of the computational domain. The present study focuses on the region beyond $x/\theta_0 = 1800$, corresponding to $Re_\theta = 1700 - 2200$, where these effects should be relatively small.

In the case of boundary layers with an external flow with streamline curvature, like the present flow, the definition of the external velocity U_e is not straightforward. In order to be as consistent as possible with boundary layer theory, U_e is chosen as the maximum streamwise component of velocity in the wall-normal direction. Figure 3 shows the streamwise evolution of U_e and of the pressure gradient parameter $\beta_{ZS} = -(\delta/U_{zs})(dU_e/dx)$ in the region downstream of the separation bubble, where $U_{zs} = U_e\delta^*/\delta$ is the Zagarola-Smits velocity. Rotta–Clauser's pressure gradient parameter $\beta = -(\Delta/u_\tau)(dU_e/dx)$ is also shown in figure 3 for reference, where

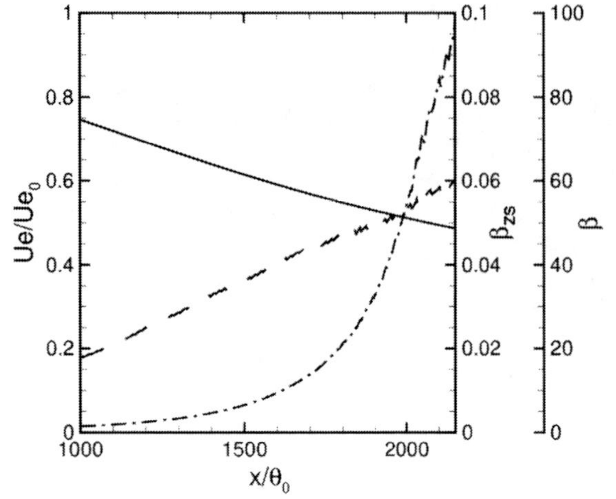

Figure 3. Streamwise evolution of external velocity (——) and pressure gradient parameters downstream of the separation bubble. β_{ZS}, ---- ; β, — · —.

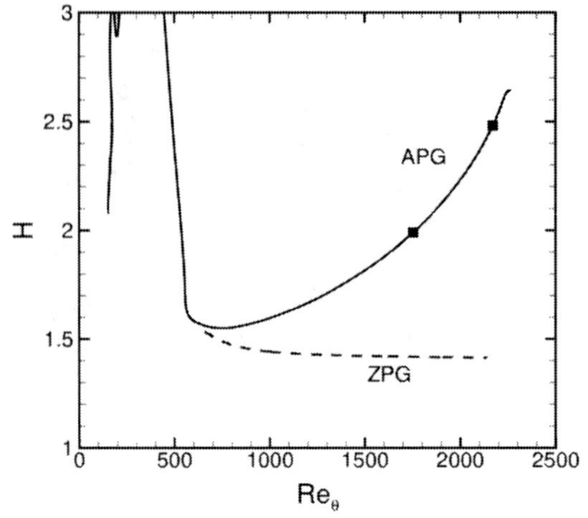

Figure 4. Streamwise evolution of the shape factor.

$\Delta = \delta^* U_e/u_\tau$ is the Rotta–Clauser length scale. β is the traditional pressure gradient parameter that assumes the outer region velocity scale to be u_τ. Because of flow curvature introduced by the outlet and top boundary conditions, U_e does not decrease linearly, but, interestingly, the flow conditions lead to a pressure gradient parameter β_{ZS} that increases in an almost linear fashion. A gradient of β_{ZS} implies that the boundary layer is not in dynamic equilibrium [7]. Moreover, if a positive gradient of β_{ZS} is maintained, the streamwise mean momentum defect of the boundary layer increases. These statements are reflected in the increase of the shape factor H shown in figure 4 for $Re_\theta > 1000$. The increase of H with respect to Re_θ is rapid, indicating the strong non-equilibrium nature of the flow.

3.2. Comparison with other flows
In this section, comparisons with other flows of interest are made at the same time as the velocity statistics of the present flow are discussed. The data for the present flow is taken at two streamwise positions near the end of the domain, where the mean velocity defect is large, $H = 2$ and 2.5 respectively, and the influence of the shear-layer instability of the upstream bubble is expected to be small. The other turbulent boundary layers considered are: the ZPG TBL of Simens et al. [13] at a comparable Reynolds number ($Re_\theta = 1975$); the equilibrium APG TBL of Skare and Krogstad [2] (SK), which has a relatively large velocity defect ($H \cong 2$) and is at high Reynolds numbers ($Re_\theta = 39000$–51000), and the non-equilibrium APG TBL of Maciel et al. [6] (MRL) with increasing velocity defect that includes data at detachment ($H = 1.72$–3.85, $Re_\theta = 3350$–12691). Finally, the single-stream mixing layer of Wygnanski and Fiedler [20] (WF) is also considered because the mean velocity profile of the outer region of a large defect boundary layer resembles that of the high-speed side of a mixing layer.

Outer (or high-speed region) scales common to both types of flows have to be defined in order to be able to compare the mixing layer with boundary layers. For simplicity, we choose scales that are similar to those frequently used for free shear layers. For the boundary layers, the outer length scale remains the boundary layer thickness δ as defined previously, that is the wall-normal distance from the wall to the point where $U = 0.99U_e$. For the single-stream mixing layer, since the data is not accurate in the low-speed region, the outer length scale is defined

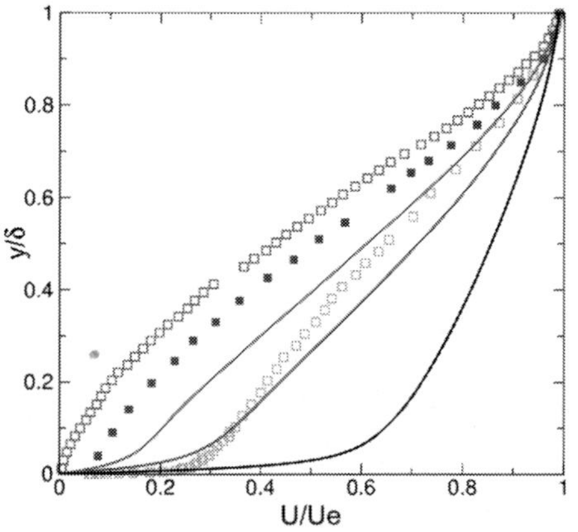

Figure 5. Mean velocity profiles of the present flow for: Blue line, $H = 2$; red line, $H = 2.5$; black line, ZPG TBL; green □, equilibrium APG TBL of SK; red □, profile at detachment of the non-equilibrium APG TBL of MRL; blue ■, mixing layer of WF.

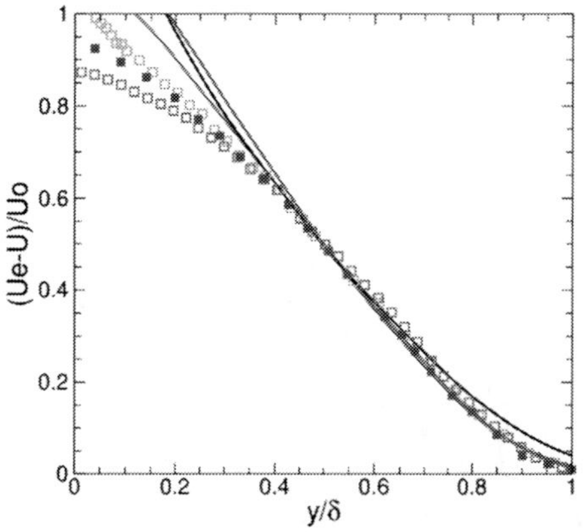

Figure 6. Velocity defect profiles normalized with U_o. Lines and symbols as in figure 5.

only with the high-speed part of the flow:

$$\delta = 2(y_{0.99} - y_{0.5}), \qquad \text{(only for the mixing layer)}$$

where $y_{0.5}$ and $y_{0.99}$ correspond to the positions where $U = 0.5U_e$ and $0.99U_e$, respectively. The outer velocity scale, U_o, is used for both TBLs and mixing layer and is defined as twice the velocity defect at the middle of the shear layer:

$$U_o = 2\left[U_e - U(y = 0.5\delta)\right].$$

It is (only approximately) equivalent to U_{zs}, in the sense that both velocity scales are proportional to the mean velocity deficit of the outer flow.

Selected mean velocity profiles of these various flows are shown in figure 5. The profile shown for the APG TBL of MRL is the one at detachment ($C_f = 0$). Excluding the wall region, the velocity profiles of the large-defect boundary layers clearly resemble the single-stream mixing layer one, with large velocity gradients and the presence of an inflection point. Two of these velocity profiles are at the same shape factor $H = 2$, namely one for the present flow and the equilibrium one of SK, but the shapes of these profiles are very different. Such a shape difference is probably more related to the different streamwise evolution of the flows than to the dissimilarity in Reynolds number. The corresponding velocity defect profiles normalized with U_o are presented in figure 6. With the exception of the ZPG TBL, the velocity defect profiles of all other flows are similar on the high-speed side. It can be seen from figure 6 that the choice of U_o and δ as common outer scales is adequate to compare velocity statistics between these flows.

Figure 7 shows the mean velocity profiles normalized with the viscous-friction scales (u_τ and ν/u_τ) of the present flow and of the ZPG TBL of Simens et al. at a comparable Reynolds number. Even if at a low Reynolds number, the ZPG TBL profile follows fairly closely the law of the wall throughout the inner region. In contrast, the present non-equilibrium TBL deviates from it and the departure increases as the velocity defect increases. It cannot be excluded that a departure from the law of the wall is due or partly due to the presence of the separation bubble upstream, even if the profiles considered are relatively far from it. But because this departure

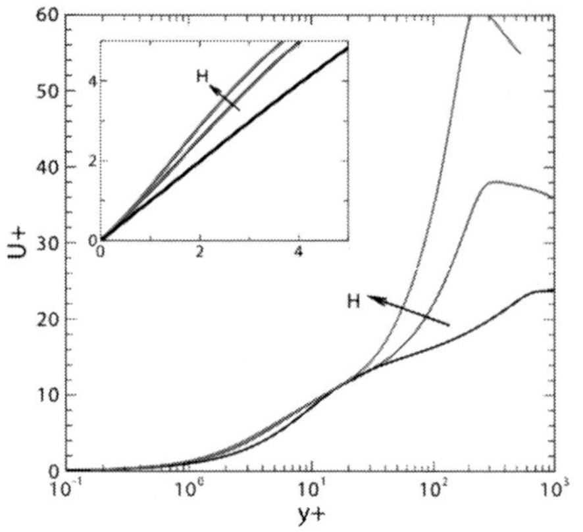

Figure 7. Mean velocity profiles normalized with viscous-friction scales. Near-wall region is shown in the inset. Lines as in figure 5.

exists even very close to the wall, where the impact of the bubble shear layer instability should be small, it might rather be linked to the non-equilibrium state of the boundary layer, while it could also be amplified by the fact that the Reynolds number is low. Such a departure has been observed in many other APG TBLs with strong non-equilibrium conditions [3, 5, 7, 21–26]. Moreover, it is expected near detachment for all TBLs at finite Reynolds number, since in these conditions u_τ and ν/u_τ no longer scale the mean velocity profile near the wall. In contrast, it is not clear if the law of the wall could still hold for equilibrium APG TBLs with large velocity defects. For instance, the equilibrium APG TBL of SK follows the logarithmic law, at least approximately, even if it has a relatively large velocity defect. It is therefore important to stress that it is not so much the strength of the pressure gradient that matters here, but the fact that the pressure gradient parameter is changing. In other words, the fact that the flow is not in dynamic equilibrium.

Selected profiles of the streamwise Reynolds normal stress of the various flows are shown in figure 8. In order to be able to compare the mixing layer and the boundary layers, $\langle u^2 \rangle$ is normalized with the common outer scale U_o and the plot is linear-log in order to clearly reveal the near-wall behaviour. Note that in the case of the TBLs the trends shown in this figure remain the same when $\langle u^2 \rangle$ is normalized with U_{zs}. The five profiles of MRL are included in this figure in order to show the behaviour observed in a non-equilibrium boundary layer with

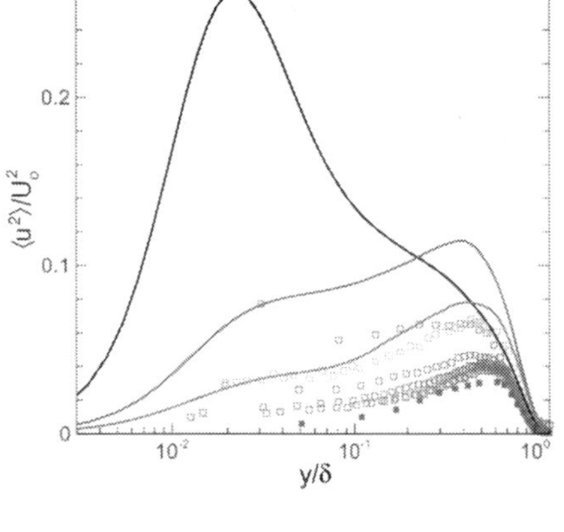

Figure 8. Streamwise Reynolds normal stress normalized with U_o. Lines and symbols as in figure 5.

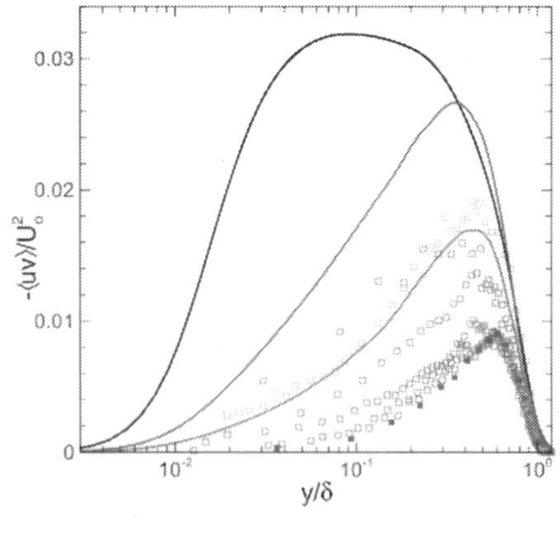

Figure 9. Reynolds shear stress normalized with U_o. Lines and symbols as in figure 5.

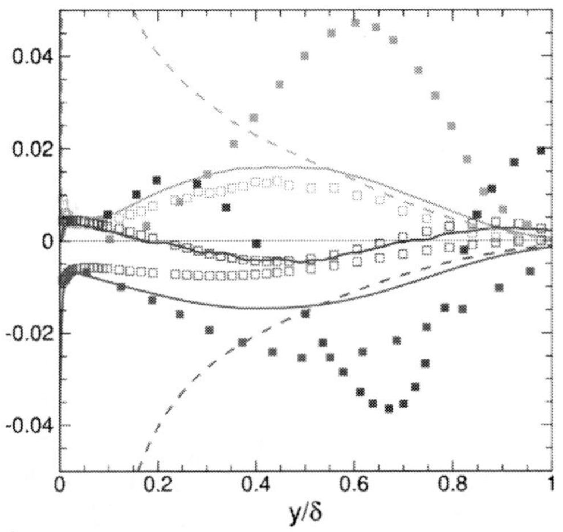

Figure 10. Selected terms of the turbulent kinetic energy budgets normalized with U_o and δ. green, production; red, dissipation; blue, turbulent transport. ——, present flow at $H = 2.5$; ----, ZPG TBL (turbulent transport not shown); □, equilibrium APG TBL of SK; ■, mixing layer of WF.

increasing velocity defect up to detachment. The trend observed by MRL is one of a gradual streamwise decrease of all Reynolds stresses throughout the boundary layer, but especially near the wall. The two most downstream profiles are actually superposed, indicating a TBL that reaches a state of near equilibrium before detachment (the last profile is at detachment). The near-wall peak of $\langle u^2 \rangle$, which is a common feature of canonical wall flows, is not present in the case of all the TBLs subjected to a strong APG, with the possible exception of the first profile of MRL, which is at the beginning of the APG region of the flow ($H = 1.72$). The maximum of $\langle u^2 \rangle$ is in the middle of the boundary layer for all these flows. Another particularity of these strong APG TBLs is that $\langle u^2 \rangle / U_o^2$ is lower than in the ZPG case for $y < 0.5\delta$. In the case of the present flow, the stronger levels of fluctuation in the outer region could be due to the presence of the shear layer instability upstream. This seems to be corroborated by the fairly strong decrease of $\langle u^2 \rangle / U_o^2$ from the first profile to the second one. Interestingly, the profile of $\langle u^2 \rangle / U_o^2$ of the mixing layer is similar to the detachment profile of the MRL flow. In the following, it will be seen that other turbulence statistics are similar between these two flows.

All the observations that have been made above for $\langle u^2 \rangle / U_o^2$ can be repeated for the Reynolds shear stress normalized by U_o, as shown in figure 9. However, the similarity between the mixing layer and the TBL detachment profile is even better, in particular closer to the wall. Another difference is that, in the case of the large-velocity defect boundary layers, the dominance of $-\langle uv \rangle$ in the outer region in comparison to the wall region is even more pronounced than for $\langle u^2 \rangle$. Since U_o is proportional to the mean strain rates present in the outer region, figures 8 and 9 indicate that the strong APG flows are globally less efficient in extracting turbulent energy from the mean flow than the ZPG one.

Figure 10 presents selected terms of the turbulent kinetic energy budgets normalized with U_o and δ for four flows: the present flow at $H = 2.5$, the ZPG TBL of Simens et al., the equilibrium APG TBL of SK and the mixing layer of WF. It is important to note that these budgets are very different from those of canonical wall flows. In the latter, the maximum of the various terms occur in the near-wall region and with values an order of magnitude higher than in the outer region. These facts can be partly appreciated with the ZPG TBL profiles of production and dissipation shown in the figure. In large-velocity-defect TBLs, the maximum of production and dissipation is in the middle of the boundary layer. But this does not necessarily mean that the levels of production and dissipation in the outer region of large-velocity-defect TBLs are higher than those of ZPG TBLs. Indeed, figure 10 shows that these levels are comparable in the upper half of all boundary layers, with the exception of the experimentally deduced dissipation of the equilibrium APG TBL of SK. Note also that a small peak of production is present near the wall in the case of the equilibrium APG TBL. Such a peak also exists further upstream in the present flow, but it gradually decreases and disappears as the outer peak develops.

The budget terms of the mixing layer are seen to be qualitatively similar to those of the large-velocity-defect TBLs. The levels are however higher and the maxima are closer to the edge of the shear layer. Another noticeable difference is the higher level (in relative terms) of turbulent transport of turbulent kinetic energy from the middle of the mixing layer to both edges.

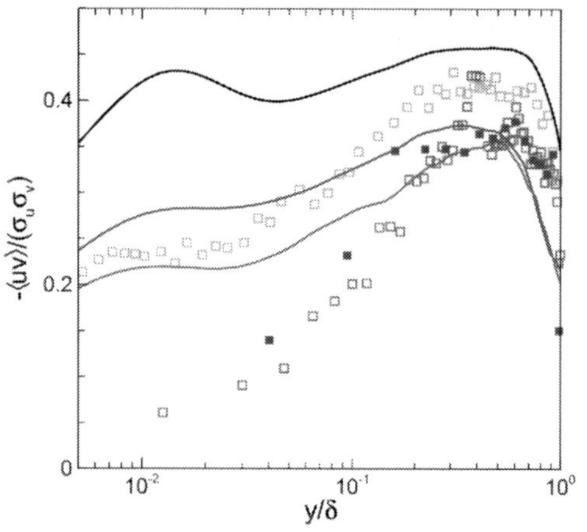

Figure 11. *uv*-correlation factor. Lines and symbols as in figure 5.

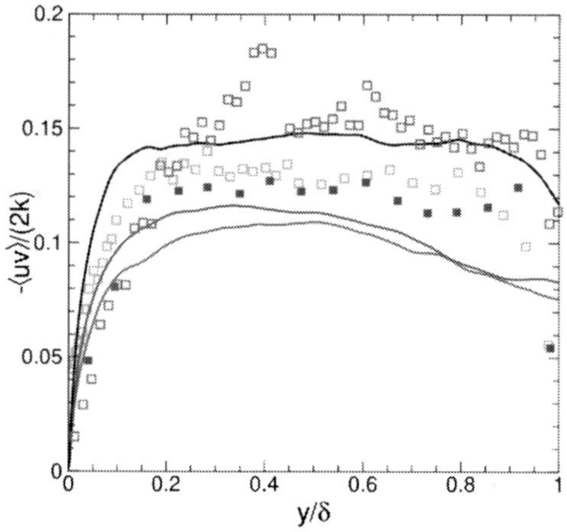

Figure 12. Structure parameter. Lines and symbols as in figure 5.

Figure 11 shows the uv correlation factor for the various flows. The level of correlation between u and v is higher in the ZPG TBL than in all other flows. In the case of the APG TBLs, the correlation of u and v is seen to deteriorate as the mean velocity defect increases, especially near the wall. The correlation profiles of the mixing layer and of the APG TBL of MRL at detachment are strikingly similar. They suggest that turbulence significantly loses its coherence in the low-speed side of these flows.

The structure parameter $-\langle uv \rangle/2k$, where k is the turbulent kinetic energy, is shown in figure 12. With the exception of the detachment profile of MRL in the outer region, all other flows have a lower structure parameter than the ZPG TBL. The structure parameter is particularly low in the present flow. Unlike the Reynolds stresses and the uv correlation factor, the structure parameter of the mixing layer is closer to that of the equilibrium APG TBL of SK than to that of the detachment profile of MRL, except near the wall.

3.3. Flow structure

Figure 3.3 shows the fractional contributions from the four quadrants to the Reynolds shear stress for both ZPG and APG flows: $(-\langle uv \rangle_{Qi})/(-\langle uv \rangle)$ where Qi with $i = 1, 2, 3, 4$ denotes a quadrant in the parameter space of u and v. The pressure gradient is seen to increase the fractional contribution of all quadrants. In the ZPG case, the sweeps ($Q4$ events) contribute the most to the Reynolds shear stress very near the wall while ejections ($Q2$ events) dominate slightly elsewhere. Whereas, for the APG case sweeps contribute more than ejections in the near-wall region and in the lower part of the outer region as well. Correspondingly, the negative contribution of $Q1$ and $Q3$ motions increase near the wall due to the increased energy transfer through the turbulent transport toward the wall [23]. Also as suggested in Krogstad and Skare [27], the strength of $Q1$ motions might be due to reflections of $Q4$ motions at the wall.

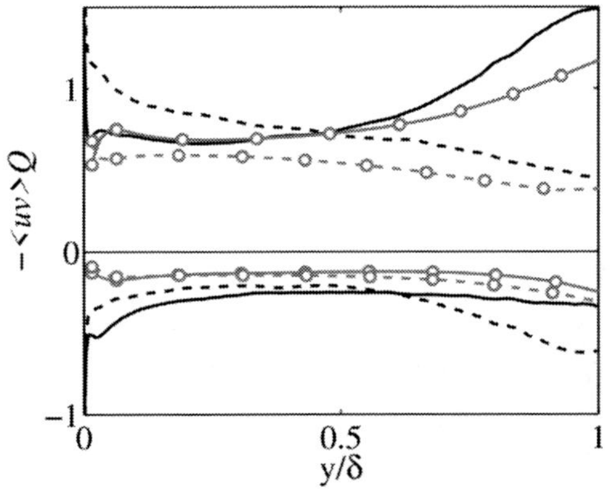

Figure 13. Fractional quadrant contribution to the Reynolds stress $-\langle uv \rangle$. Positive values: —— , Q2; ---- , Q4. Negative values: —— , Q1; ---- , Q3. Lines without symbols are the APG boundary layer at $Re_\theta = 2175$, those with circles are the ZPG boundary layer [28].

The crossing between the sweep-dominated region and the ejection-dominated region takes place much farther from the wall in the APG case. This point is very close to the maximum of the Reynolds stress.

Figure 14 presents the joint p.d.fs of the logarithm of the size in xy and zy of the circumscribing boxes for $Q2$ and $Q4$ three-dimensional structures. The clusters are obtained following the method described in [11]. In particular we use the same threshold $|u(x,y)v(x,y)| > Hu'(x,y)v'(x,y) + 10^{-6}$, with $H = 1.75$ and adding 10^{-6} to prevent anomalies in the free-stream part of the flow. The value for H was chosen in order to directly compare with [11] and was therefore not based on an analysis of the percolation threshold. The clusters are obtained in the region of the flow where $Re_\theta = 1700 - 2200$, which corresponds to the region that is decorrelated from the separation bubble. The size of the boxes was scaled with a δ averaged over this region.

Both the $Q2$ and the $Q4$ structures have very similar sizes, which is in agreement with what was obtained in [11] for turbulent channel flows. However, contrary to channel flow the structures are as long as they are tall and wide, $\Delta_x \approx \Delta_y \approx \Delta_z$. In general the $Q2$ and $Q4$ structures are less elongated than in the channel flow. On the other hand the behaviour for the largest structures is similar to what is found in channel flow, in the sense that the $Q2$ structures are larger than the $Q4$ structures, especially in height. The largest structures are essentially twice as large as the boundary layer thickness.

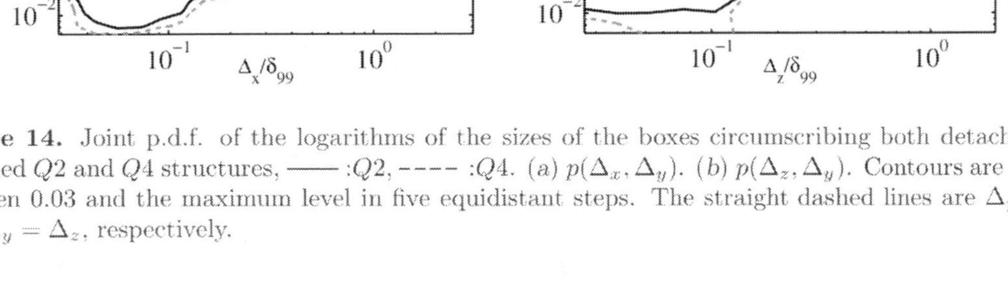

Figure 14. Joint p.d.f. of the logarithms of the sizes of the boxes circumscribing both detached and attached $Q2$ and $Q4$ structures, —— :$Q2$, ---- :$Q4$. (a) $p(\Delta_x, \Delta_y)$. (b) $p(\Delta_z, \Delta_y)$. Contours are plotted between 0.03 and the maximum level in five equidistant steps. The straight dashed lines are $\Delta_y = \Delta_x$ and $\Delta_y = \Delta_z$, respectively.

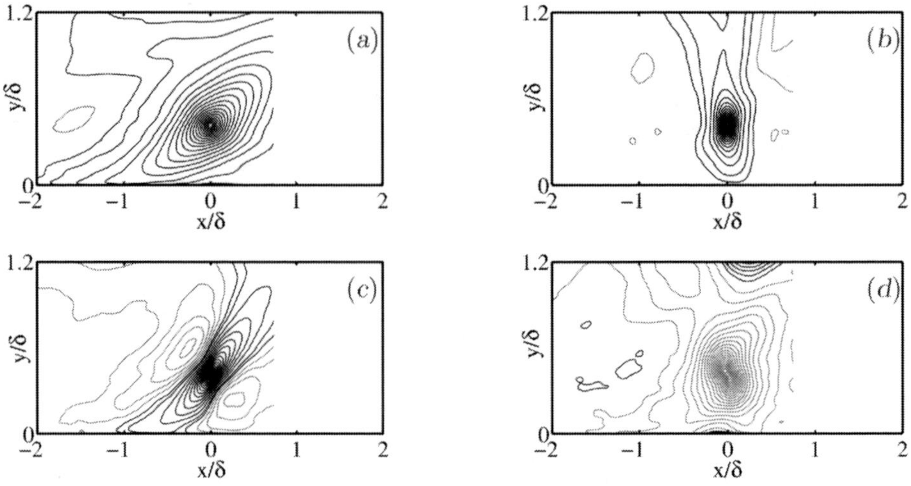

Figure 15. Two-point spatial correlation coefficient centred at $Re_\theta = 2175$ and at $y/\delta = 0.4$. The axes are not to scale. (a) C_{uu}, (b) C_{vv}, (c) C_{ww}. Black: 0.02(0.05)1, Red: -0.02(0.03)1.01. (d) C_{vu}. Black: 0.01(0.01)0.4, Red: −0.01(0.02)0.41.

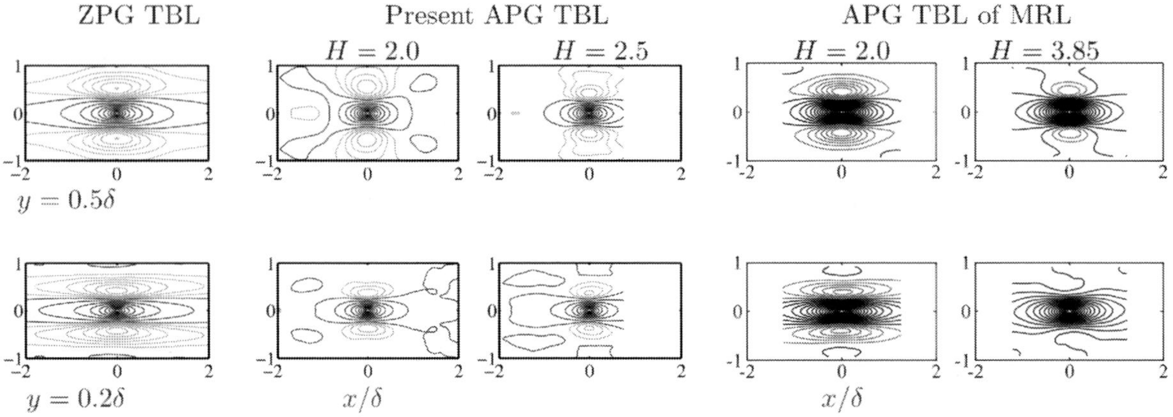

Figure 16. Two-point spatial correlation coefficient of streamwise velocity fluctuations C_{uu} in xz planes for the ZPG TBL, the present APG TBL and the APG TBL of MRL. The axes are not to scale. Black: 0.02(0.05)0.97, Red: −0.02(0.03)0.98.

Figure 15 shows contours of the two-point spatial correlation coefficients of the streamwise, wall-normal and spanwise fluctuating velocities obtained at $Re_\theta = 2175$ and at $y/\delta = 0.4$ in xy planes. Like for ZPG TBLs [28], the contours of C_{uu} are roughly elliptical in shape with the major axis tilted at an angle to the streamwise direction, but they are more inclined than the ZPG TBL ones. The correlation length in the streamwise direction is large. The vertical velocity correlation contours, C_{vv}, are quite different from the streamwise velocity correlations in that the length scales are more limited, and the correlations are slightly elongated in the vertical direction. For ZPG flows, the eddy structures have large correlation lengths in the streamwise direction, indicating that the eddies are more elongated than those in the APG flows. This follows the observation of Krogstad and Skare [27] who show that large-scale structures were shortened and structure inclination angle increased in APG flows. A recent experiment on a strongly decelerated boundary layer by Rahgozar and Maciel [29] has shown that u-structures in the ZPG cases are more elongated in the streamwise direction than those in the APG cases and that, as the turbulent boundary layer continues to decelerate, u-structures become less streamwise elongated.

Figure 16 shows the correlation coefficient of the streamwise velocity fluctuations, C_{uu} in xz planes in the outer layer at $y/\delta = 0.5$, to match the available experimental data [6], and at $y/\delta = 0.2$ at Reynolds numbers of $Re_\theta = 1755$ and 2175 corresponding to $H = 2$ and 2.5, respectively. It can be seen that the correlation contours in the outer part of the boundary layer are shorter and wider compared to the ones closer to the wall. This is consistent with what was found in [27]. It is generally thought that the structures appear less frequently or meander more in an APG flow as compared to the ZPG ones [5, 29, 30]. This is illustrated in the figure, where a ZPG [28] and several strongly decelerated boundary layers with very large velocity defect [6] are shown. The u-fluctuations correlate over much shorter lengths as the velocity defect increases, showing the effect of the APG on the structures.

4. Conclusions

A direct numerical simulation of a non-equilibrium boundary layer subjected to strong adverse pressure gradient is investigated. The DNS is at fairly low Reynolds number ($Re_\theta = 150 - 2200$) and distorted due to the upstream boundary layer which is initially laminar and separates due to the strong pressure gradient. Transition to turbulence is triggered using a trip wire, positioned upstream of the unforced separation point. The two-point cross-correlation coefficient, C_{vu}, with a reference point close to the transition region shows flow periodicity persisting to $Re_\theta \approx 1600$. This periodicity is related to the shear layer instability of the separation bubble created as a result of the APG, and limits the useful range of the attached turbulent boundary layer to the approximate range of $Re_\theta = 1700 - 2200$.

The first- and second-order velocity statistics of the present APG TBL follow the trends generally encountered when the following two conditions are present in an APG TBL: strong non-equilibrium state of the TBL in dynamic terms, and a large mean velocity defect. Note that these two conditions are not necessarily always present at the same time in a given TBL. Like other non-equilibrium TBLs with large velocity defect, the present flow does not follow the law of the wall, even very close to the wall. Similarly to large-velocity-defect TBLs, in this case irrespective of their equilibrium state, the maximum turbulence activity is found to be in the middle of the boundary layer and not near the wall. A near-wall peak of streamwise Reynolds normal stress and of turbulence production does not exist in the present non-equilibrium flow, while the results of Skare and Krogstad [2] suggest that a near-wall production peak might be present in an equilibrium APG TBL with a fairly large velocity defect. The present analysis also shows that u and v are more decorrelated throughout the boundary layer than in ZPG TBLs. Near the wall, u and v become more and more decorrelated as the velocity defect increases.

In connection to the previous observations, an important conclusion that can be drawn from the present analysis is that, in part due to the absence of strong velocity gradients near the wall, large-velocity-defect boundary layers are globally less efficient in extracting turbulent energy from the mean flow than the ZPG TBL. When normalized with an adequate outer region velocity scale (U_o or U_{zs}), the Reynolds stresses and the production of turbulent kinetic energy are weaker in the lower half of the large-velocity-defect boundary layers than in the ZPG TBL.

Due to the resemblance of mean velocity profiles in the outer region of large-velocity-defect TBLs to the high speed part of mixing layers, the present flow and three other TBLs are also compared with a single-stream mixing layer. To be able to do so, an outer region velocity scale analogous to the commonly defined free shear layer velocity scales is introduced. The resemblance between these two types of flows is striking in terms of the distributions of Reynolds stresses, production, dissipation and turbulent transport of turbulent kinetic energy,

uv correlation, and structure parameter $-\langle uv \rangle / 2k$. Although more data is needed to draw a definitive conclusion, the outer-region turbulent statistics of TBLs close to detachment appear to have a similar character to those of the single-stream mixing layer.

The fractional quadrant decomposition of the Reynolds shear stress indicates that sweeps seem to dominate ejections in the near-wall region for the strong APG boundary layers, whilst ejections dominate the outer region. The switch between these two regions is observed to take place close to the maximum of the Reynolds shear stress in the outer layer. The $Q2$ and $Q4$ structures are less elongated than the same structures in channel flow.

The two-point spatial correlation C_{uu} obtained far from the transition region at $Re_\theta = 2175$ and at $y/\delta = 0.4$ indicates that the correlation length of the u fluctuations in an APG boundary layer is almost half that observed in the ZPG TBL. The wall-normal and spanwise correlations, C_{vv} and C_{ww}, are quite different from the C_{uu} correlations in that the length scales are more limited and the correlations are slightly elongated in the vertical direction, as found in ZPG boundary layers, implying that the structure of the v, w fluctuations is the same as in ZPG.

Acknowledgments

Funded in part by the Multiflow program of the European Research Council. AGG, YM and JS were supported in part respectively by ITU-BAP, NSERC of Canada and ARC Discovery Grant of Australia. The computations were made possible by generous grants of computer time from Barcelona supercomputing centre and from the national centre for high performance computing of Turkey. The authors would like to thank Prof. Jiménez for organizing the First Multiflow Summer Workshop and for suggesting the idea of comparing large-defect boundary layers with mixing layers. Finally, they are grateful to Juan Sillero and Adrián Lozano-Durán for providing the data processing programs for the computation of the two-point correlations and for the analysis of the uv-structures, respectively.

References

[1] Simpson RL, Strickland JH and Barr PW 1977 Features of a separating turbulent boundary layer in the vicinity of separation *J. Fluid Mech.* **79** 553–94

[2] Skåre PE and Krogstad P-Å 1994 A turbulent equilibrium boundary layer near separation *J. Fluid Mech.* **272** 319–48

[3] Na Y and Moin P 1998 Direct numerical simulation of separated turbulent boundary layer *J. Fluid Mech.* **374** 379–405

[4] Elsberry K, Loeffler F, Zhou MD and Wygnanski I 2000 An experimental study of a boundary layer that is maintained on the verge of separation *J. Fluid Mech.* **423** 227–61

[5] Skote M and Henningson DS 2002 Direct numerical simulation of separated turbulent boundary layer *J. Fluid Mech.* **471** 107–36

[6] Maciel Y, Rossignol KS and Lemay J 2006 A study of a turbulent boundary layer in stalled-airfoil-type flow conditions *Exp. Fluids* **41** 573–90

[7] Maciel Y, Rossignol KS and Lemay J 2006 Self-similarity in the outer region of adverse-pressure-gradient turbulent boundary layers *AIAA J.* **44** 2450–64

[8] Spalart PR and Coleman GN 1997 Numerical study of a separation bubble with heat transfer *Eur. J. Mech. B: Fluids* **16** 169–89

[9] Kline SJ, Reynolds WC, Schraub FA and Runstadler PW 1967 The structure of turbulent boundary layers *J. Fluid Mech.* **30** 741–73

[10] del Álamo JC, Jiménez J, Zandonade P and Moser RD 2006 Self-similar vortex clusters in the turbulent logarithmic region *J. Fluid Mech.* **561** 329–58

[11] Lozano-Durán A, Flores O and Jiménez J 2012 The three-dimensional structure of momentum transfer in turbulent channels *J. Fluid Mech.* **694** 100–30

[12] Lee JH and Sung HJ 2009 Structures in turbulent boundary layers subjected to adverse pressure gradients *J. Fluid Mech.* **639** 101–31

[13] Simens MP, Jiménez J, Hoyas S and Mizuno Y 2009 A high-resolution code for turbulent boundary layers *J. Comput. Phys.* **228** 4218–31

[14] Fadlun EA, Verzicco R, Orlandi P and Mohd-Yusof J 2000 Combined immersed-boundary finite-difference methods for three-dimensional complex flow simulations *J. Comput. Phys.* **161** 35–60
[15] Mittal R and Iaccarino G 2005 Immersed boundary methods *Ann. Rev. Fluid Mech.* **37** 239–61
[16] Uhlmann M 2005 An immersed boundary method with direct forcing for the simulation of particulate flow *J. Comput. Phys.* **209** 448–76
[17] Pinelli A, Naqavi IZ, Piomelli U and Favier J 2010 Immersed-boundary methods for general finite-difference and finite-volume Navier-Stokes solvers *J. Comput. Phys.* **229** 9073–91
[18] Vanella M and Balaras E 2009 A moving-least-squares reconstruction for embedded-boundary formulations *J. Comput. Phys.* **228** 6617–28
[19] Simens MP and Gungor AG 2013 The effect of surface roughness on laminar separated boundary layers *J. Turbomach.* **136** 031014
[20] Wygnanski I and Fiedler HE 1970 The two-dimensional mixing region *J. Fluid Mech.* **41** 327–61
[21] Simpson RL, Chew Y-T and Shivaprasad BG 1981 The structure of a separating turbulent boundary layer. Part 1: Mean flow and Reynolds stresses *J. Fluid Mech.* **113** 23–51
[22] Nagano Y, Tagawa M and Tsuji T 1993 Effects of adverse pressure gradients on mean flows and turbulence statistics in a boundary layer *Turbulent Shear Flows 8* ed Durst F et al. (Berlin: Springer) pp 7–21
[23] Nagano Y, Tsuji T and Houra T 1998 Structure of turbulent boundary layer subjected to adverse pressure gradient *Intl J. Heat Fluid Flow* **19** 563–72
[24] Spalart PR and Watmuff JH 1993 Experimental and numerical study of a turbulent boundary layer with pressure gradients *J. Fluid Mech.* **249** 337–71
[25] Debisschop JR and Nieuwstadt FTM 1996 Turbulent boundary layer in an adverse pressure gradient: Effectiveness of riblets *AIAA J.* **34** 932–37
[26] Indinger T, Buschmann MT and Gad-el Hak M 2006 Mean-velocity profile of turbulent boundary layers approaching separation *AIAA J.* **44** 2465–74
[27] Krogstad P-Å and Skåre PE 1995 Influence of a strong adverse pressure gradient on the turbulent structure in a boundary layer *Phys. Fluids* **7** 2014–24
[28] Sillero J, Jiménez J and Moser RD 2013 One-point statistics for turbulent wall-bounded flows at Reynolds numbers up to $\delta^+ \approx 2000$ *Phys. Fluids* **25** 105102
[29] Rahgozar S and Maciel Y 2012 Statistical analysis of low- and high-speed large-scale structures in the outer region of an adverse pressure gradient turbulent boundary layer *J. Turbulence* **13** 1–24
[30] Lee JH and Sung HJ 2008 Effects of an adverse pressure gradient on a turbulent boundary layer *Intl J. Heat Fluid Flow* **29** 568–78

Hairpin vortices in turbulent boundary layers

G Eitel-Amor[1], O Flores[2] and P Schlatter[1]
[1]Linné FLOW Centre, KTH Mechanics, Stockholm, Sweden
[2]Universidad Carlos III de Madrid, Madrid, Spain
E-mail: `pschlatt@mech.kth.se`

Abstract. The present work addresses the question whether hairpin vortices are a dominant feature of near-wall turbulence and which role they play during transition. First, the parent-offspring mechanism is investigated in temporal simulations of a single hairpin vortex introduced in a mean shear flow corresponding to turbulent channels and boundary layers up to $Re_\tau = 590$. Using an eddy viscosity computed from resolved simulations, the effect of a turbulent background is also considered. Tracking the vortical structure downstream, it is found that secondary hairpins are created shortly after initialization. Thereafter, all rotational structures decay, whereas this effect is enforced in the presence of an eddy viscosity.

In a second approach, a laminar boundary layer is tripped to transition by insertion of a regular pattern of hairpins by means of defined volumetric forces representing an ejection event. The idea is to create a synthetic turbulent boundary layer dominated by hairpin-like vortices. The flow for $Re_\tau < 250$ is analysed with respect to the lifetime of individual hairpin-like vortices. Both the temporal and spatial simulations demonstrate that the regeneration process is rather short-lived and may not sustain once a turbulent background has formed. From the transitional flow simulations, it is conjectured that the forest of hairpins reported in former DNS studies is an outer layer phenomenon not being connected to the onset of near-wall turbulence.

1. Introduction

Notable progress has been made over the last decades in understanding the physical processes involved in turbulent boundary layers (TBL). We have a relatively complete understanding of the kinematics of the whole TBL, from the near-wall or buffer region to the outer region. The statistics of velocity and vorticity in turbulent flows have been extensively studied, and were reviewed recently in a number of articles, see *e.g.* Refs. [23, 25, 35]. However, our understanding of the dynamics of TBL is more limited [15].

The region of the TBL where we have the most complete understanding of the dynamics of the flow is the buffer region. Since their first visualisation by Kline *et al.* [24] the near-wall streaks with alternating low and high streamwise momentum have been studied thoroughly, confirming their spanwise spacing of about 100 viscous units ℓ_*. Jeong *et al.* [14] applied the λ_2 criterion for vortex eduction to this near-wall region, to conclude that the dominant near-wall structures are staggered quasi-streamwise vortices, surrounding the low-speed streaks. The processes that relate vortices and streaks were studied using a number of different techniques and models [10, 16, 17, 34]. There is now general agreement in that the streaks and the vortices are involved in a self-sustaining non-linear cycle, with a period of about 400 viscous time units. The sinuous instability of the streaks results in the formation of new quasi-streamwise vortices, that interact with the mean shear to generate new streaks. This non-linear cycle is autonomous and takes place even in the absence of an overlying turbulent flow [18], and it is weakly modulated by the intensity of the overlying turbulent structures [12, 13].

On the other hand, the dynamics of the structures populating the logarithmic and outer regions of TBL are less understood. There is some agreement in the predominant role of Townsend's attached eddies [38], although there are several models advocating for different candidates for these eddies. Probably the most widespread model is the hairpin vortex paradigm, which advocates for attached loop-like vortices, the so-called hairpin vortices. Originally proposed by Theodorsen [36], indirect evidence of these specifically shaped vortices has been found by many in experiments [2, 11, 27] and simulations [5, 28]. There exists, however, no unique quantitative definition of a hairpin, rather they describe a category of vortices consisting of a (negatively rotating) head, oriented in the spanwise direction, and two adjacent legs in the upstream and wall-ward direction [11]. From a dynamical point of view, these hairpins grow by merging [37] and by the so-called parent-offspring regeneration mechanism [40]. It is then thought that these conglomerates of similarly-oriented eddies might be responsible for the large-scale structures living in the outer region of the boundary layer [2]. In addition to complete hairpin vortices, non-symmetric incomplete variants (hooks and canes) have been found in the analysis of direct numerical simulation (DNS) data by Robinson [28].

Besides the popularity of hairpin model of TBLs, there are some open questions regarding their relevance in fully developed turbulence. Even if hairpin vortices are very common in various types of transitional flows, for example they occur in classical K and H-type transition [1, 19, 22], their predominance in fully turbulent boundary layers is unclear. In a recent simulation by Wu & Moin [39], a simulated boundary layer was shown to be densely packed with hairpin vortices. The authors classified this as the "forest of hairpins". However, the region investigated was very close to the point of laminar-turbulent transition — in their case via travelling patches of free-stream disturbances — so it can hardly be considered an aspect of fully turbulent flow without further investigation. From stability theory it is known that a hairpin vortex created by transitional mechanisms such as *e.g.* wall-normal shear-layer roll-up or varicose streak instabilities, will advect downstream until it is damped by viscous effects or other (stronger) dynamics [3, 29]. Indeed, recent DNS of TBL reaching up to $Re_\theta = 4300$ [31] show no forest of hairpins downstream of transition, as opposed to Wu & Moin [39].

Besides their predominance, the regeneration process of hairpin vortices is also questionable. Zhou *et al.* [40] showed using DNS that an initial hairpin vortex in a laminar environment, created from a Q2 event, quickly generates a growing hairpin packet which encloses a low-speed region. Later, Kim *et al.* [21] found the packet-generation mechanism to be robust in the presence of background disturbances. They performed simulations where a hairpin was released in a channel flow with a turbulent mean profile disturbed by artificial random noise. Furthermore, the authors observed regeneration of a hairpin introduced in a velocity field extracted from DNS. However, the amplitude of the random noise was only about 15% of the rms known from DNS and the disturbances were not purely dissipative as one would predominantly expect in turbulence. In the case of the DNS flow field, the observation time was very limited, such that only the early stages of regeneration could be observed. The question remains of whether a dissipative background, like the one existing in a real turbulent shear flow, prevents any regeneration processes.

The aim of the present study is to determine and quantify the dynamics of the growth and interaction of hairpin vortices in TBL and to evaluate how these structures contribute to statistical quantities like mean flow and fluctuations. These findings will help answering the question whether hairpin-like structures represent a universal flow structure in TBL, and if they could consequently be used as a valid low-order model for boundary layer turbulence.

This report is split into two main parts. First, a single, artificially generated hairpin vortex is introduced in a mean shear flow and tracked while evolving downstream. A spectral flow solver is employed in perturbation-mode which allows to maintain the base flow for all times and to choose between nonlinear and linear computations. A channel flow as well as a boundary layer with a turbulent mean profile are considered. The hairpin vortex is created by a localized volume forcing which offers more flexibility for follow-up studies. The observed dynamics will provide new details about the lifetime of the streamwise and spanwise vortex constituents in a realistic environment. Additionally, possible regeneration and package-creation processes will be investigated in the context of TBL. To investigate the effects of background noise, an eddy-viscosity approach is followed in this work which has been proposed by Flores [8]. In that regard, the first part of the present work can be viewed as an extension of published work on the evolution of hairpins in laminar flows [1, 9, 21, 40].

In the second part, the transitional region of a boundary layer is simulated using an adapted trip forcing which generates a regular pattern of hairpins with low-amplitude random modulations. The approach is partly similar to the so-called "synthetic turbulence" experimentally investigated by Coles and Barker [7] and Savas and Coles [30]. The obtained data is then analysed to answer a number of questions, *e.g.* how long individual hairpin vortices exist, whether span-/streamwise connection mechanisms can be observed (see [2]), and how the so-called forest of hairpins develops. Furthermore, the proposed set-up may serve as a standard flow case providing a reproducible boundary layer flow for further studies of transitional flow.

2. Hairpin vortices in parallel shear flow

The behaviour of a hairpin vortex in a mean shear flow is studied in the following. First, the numerical set-up and the employed forcing is described. Then, results are presented for a hairpin being tracked in an undisturbed mean flow extending the former study by Zhou *et al.* [40] with respect to higher Reynolds numbers and longer observation times. Furthermore, a temporal boundary layer simulation was performed to examine the effect of a slightly varied mean flow. In a second step, the effect of a turbulent background is included, whereas, in contrast to Kim *et al.* [21], the mean dissipative effect of turbulent fluctuations is accounted for. Both linear and nonlinear simulations are considered to examine the importance of nonlinearity for the regeneration process.

2.1. Numerical setup

In order to reproduce the hairpin regeneration process, a simulation setup similar to the one used by Zhou *et al.* [40] was used. The streamwise, wall-normal, and spanwise directions are x, y, and z respectively. All computational domains were homogeneous in the streamwise and spanwise direction, *i.e.* periodic boundary conditions were imposed and Gauss–Lobatto collocation points were used in the wall-normal direction. The details of the computational domains are given in Tab. 1. Tests with an increased box size clearly showed that this change had no remarkable influence on the results. Numerical integration was performed using the spectral code SIMSON [6] in perturbation formulation (see Appendix), such that a turbulent mean profile could be set as a time-invariant base flow. Mean velocity profiles were taken from DNS data [20, 26], *i.e.* the simulation was started as a undisturbed flow with modified velocity profile. It should be stressed that the mean flow was fixed in the present study in contrast to former simulations by Zhou *et al.* and Kim *et al.* [21, 40].

The disturbance, which is required to develop a hairpin vortex, was introduced by a transient local volume forcing, causing a deceleration and ejection of fluid near the wall. This distinguishes the present study from the one by Zhou *et al.* [40] who employed a conditionally averaged Q2 event extracted from DNS data. However, the velocity amplitudes produced by the present forcing are quite similar to those reported by Zhou *et al.* [40], and the vortex structures possesses all the features described in their work. The exact form of the forcing term was derived in two

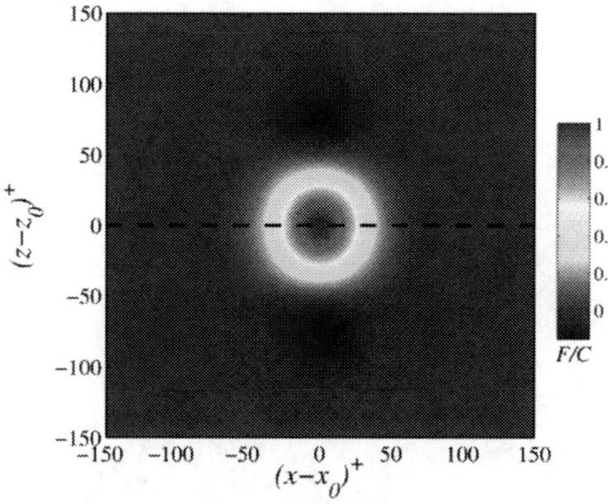

(a)

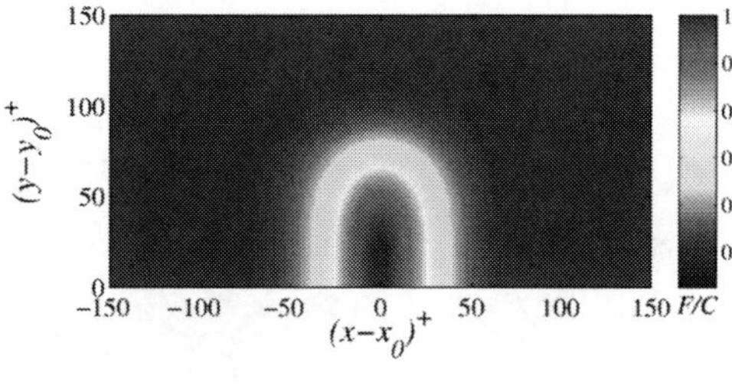

(b)

Figure 1. Distribution of the volume forcing used to produce a single hairpin vortex as described by Eq. 1. (a) top view; (b) lateral view for the plane marked by the dashed line in (a).

Table 1. Parameters for the temporal simulations of a hairpin in a mean shear flow. The channel half-height is h and the boundary layer thickness is δ_{99}. The grid resolutions Δx, Δy, and Δz are obtained from the number of collocation points in physical space.

Re_τ	Box size	Number of spectral modes	Δx^+	Δy^+_{max}	Δz^+
180	$4\pi h \times 2h \times \frac{4}{3}\pi h$	$128 \times 129 \times 128$	11.8	4.4	3.9
590	$2.5\pi h \times 2h \times 2h$	$256 \times 257 \times 256$	12.1	7.2	3.1
501	$2.5\pi\delta_{99} \times 2\delta_{99} \times 2\delta_{99}$	$256 \times 301 \times 256$	10.3	5.2	2.6

steps. First, the amplitudes were chosen in order to resemble the velocity perturbation described by Zhou *et al.*. In a second step, the spatial and temporal dimensions and gradients were tuned until a loop-vortex developed quickly from the initial counter-rotating vortex pair. It was found that a very short and intense pulse with an an increased shear on the top layer gave the best results. The final form of the forcing is given by

$$F_i = C_i \cdot \left[-\left(\frac{z-z_0}{L_z}\right)^2 + 1 \right] \cdot \exp\left[-\left(\frac{x-x_0}{L_x}\right)^2 - \left(\frac{y-y_0}{L_y}\right)^4 - \left(\frac{z-z_0}{L_x}\right)^2 - \left(\frac{t-t_0}{T}\right)^2 \right], \quad (1)$$

with $\boldsymbol{C}^+ = (-4.84,\ 2.42,\ 0)^T$, $L_x^+ = 34$, $L_y^+ = 75.6$, $L_z^+ = 54$, and $T^+ = 1.4$. The values $x_0/h = 0.5$, $y_0 = 0$, and $z_0 = 0$ define the location of maximum forcing, where h is the channel half-height. The factor in the first parenthesis in Eq. 1 causes a weak inverse forcing on the lateral sides of the main peak which supports the onset of a counter-rotating vortex pair (compare Fig. 1). The fourth-order decay in y-direction leads to a stronger shear enhancing the evolution of the first hairpin head. Changing the parameters within a range of $\pm 10\%$ yields very similar structures with minor differences regarding the shape of the hairpin head and the subsequent regeneration process. However, an insufficient amplitude $\boldsymbol{C}$ prevents the initial hairpin generation, whereas the threshold value strongly depends on the time coefficient T. It should be noted that the vortex evolution discussed below is not specific to the exact choice of parameters, but rather very robust.

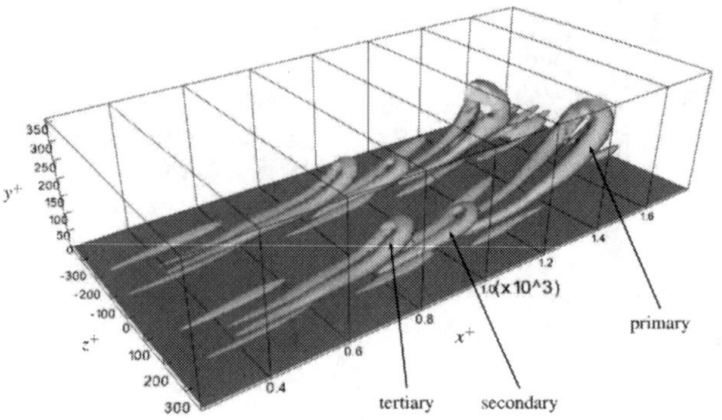

Figure 2. Vortical structures in a mean shear flow at $t^+ = 300$. (background) turbulent channel at $Re_\tau = 180$; (foreground) turbulent channel at $Re_\tau = 590$. The structures are visualised by isocontours of λ_2 for 0.5 % of the minimum λ_2 value. Axes in plus units.

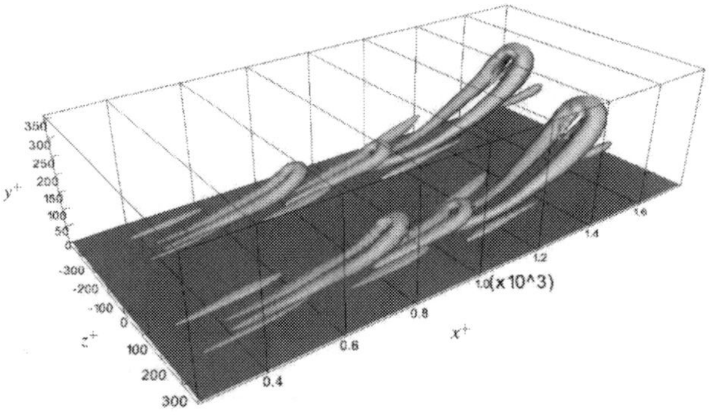

Figure 3. Vortical structures in a mean shear flow at $t^+ = 300$. (background) temporal boundary layer at $Re_\tau = 500$; (foreground) turbulent channel at $Re_\tau = 590$. The structures are visualised by isocontours of λ_2 for 0.5 % of the minimum λ_2 value. Axes in plus units.

2.2. Undisturbed background

In the following, results are presented for a single hairpin vortex in a shear flow using both nonlinear and linear solution methods. First, the nonlinear case is discussed. In Fig. 2 the vortical structures resulting from the above described forcing are shown for Reynolds numbers $Re_\tau = 180$ and 590 at $t^+ = 300$. The origin of the time axis was chosen to be t_0, at which the initial forcing peaks (see Eq. 1). For both Reynolds numbers, the largest loop-like vortex is the one caused by the initial perturbation and is therefore denoted as primary vortex. The primary vortex is stretched by the mean shear and grows away from the wall. At $Re_\tau = 180$, the vortex head extends above the channel's centreline at $y^+ = 180$ and is therefore bent backward. Each of the primary vortices is followed by two smaller upstream vortices induced by the first one. Using the scaled forcing described above, the obtained vortical structures also scale perfectly in inner units as evident from the figure.

Similar simulations were performed using a mean velocity profile for a zero-pressure gradient turbulent boundary layer. Here, the mean velocity was extracted from a DNS of a spatially developing TBL [33] at $Re_\tau = 500$ and implemented as a base flow with with $U(y)$ and $V = 0$. Since the boundary layer is not growing in the y-direction, incompressibility is fulfilled. In contrast to the former simulations, a free-stream boundary condition was imposed at the upper y-boundary. As shown in Fig. 3, the resulting structures look very similar to those from the

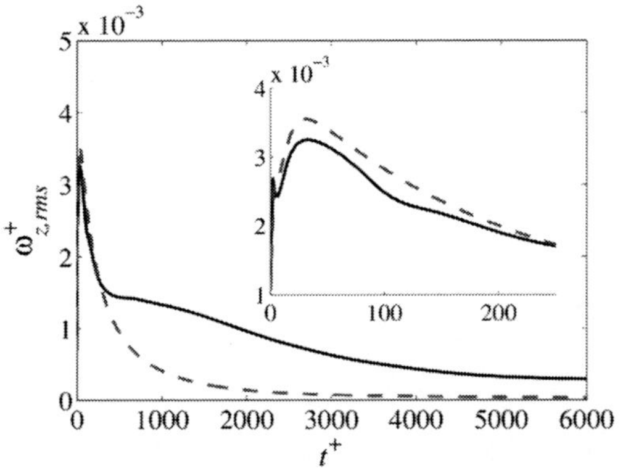

(a)

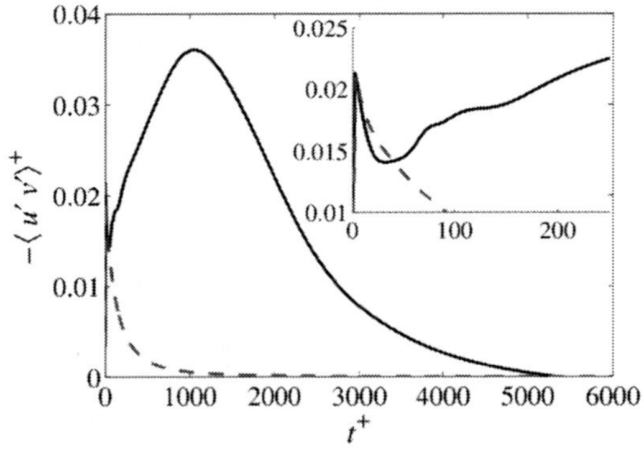

(b)

Figure 4. Hairpin regeneration in channel flow at $Re_\tau = 590$. Temporal development of (a) the rms of the spanwise vorticity $\omega_{z,\mathrm{rms}}$ and (b) the ensemble averaged Reynolds stress $\langle -u'v' \rangle$. Solid lines indicate results from nonlinear simulations, dashed lines from linear simulations. Note that the absolute values depend on the size of the computational domain in which the vortical structure is embedded.

channel case. Slight deviations are due to differences in the convection velocity between channel and boundary layer, with the difference becoming more evident further downstream. To quantify the temporal change of the vortical structures, the rms of the spanwise vorticity ω_z, computed for the whole domain, may serve as an indicator for the strength of the vortex heads. For the channel flow at $Re_\tau = 590$, $\omega_{z,\mathrm{rms}}$ is plotted over time in Fig. 4(a). Directly after the injection of wall-normal momentum, there is a short increase of $\omega_{z,\mathrm{rms}}$, which corresponds to the connection of the initial counter-rotating vortex pair. Later on, $\omega_{z,\mathrm{rms}}$ drops monotonically but the curve flattens $t^+ > 300$, where the hairpin regeneration has come to an end.

The ensemble averaged Reynolds stress $\langle -u'v' \rangle$ passes a minimum at about $t^+ = 30$ and then grows until it reaches a maximum at about 1000 viscous time units (see Fig. 4(b)). The transient hairpin regeneration is notable from a distortion in the growth of $\langle -u'v' \rangle$ around $t^+ = 100$. For times $t^+ > 1000$, $\langle -u'v' \rangle$ slowly decays whereas the lifetime of the vortical structures is large compared to, *e.g.* , the period of the bursting cycle in the buffer layer which is known to be about 400 viscous time units [17]. It is likely that the lifetime of the investigated vortical structures would decrease in a turbulent flow field where interactions with surrounding vortices prohibit an undisturbed growth. It should be remarked that Zhou *et al.* [40] observed a monotonic growth in $\langle -u'v' \rangle$ since their simulations covered a much shorter period of $t^+ = 300$. However, the initial behaviour was very similar to the present findings.

Repeating the above described simulation in linear mode proves that the hairpin regeneration is a nonlinear mechanism (*cf.* Figs. 4(a) and 4(b)). That is, only one initial hairpin is formed in the linear case, which propagates downstream. The vortex is elongated in the streamwise direction and no self-induction takes place. Therefore, after an initial peak $\omega_{z,\mathrm{rms}}$ and $\langle -u'v' \rangle$ decay exponentially.

2.3. Dissipative background

Since in a developed turbulent flow each vortical structure interacts not only with the mean shear and itself but also with all surrounding fluctuations, a more realistic investigation of the regeneration process should invoke those effects. Kim *et al.* [21] extended the study of Zhou *et al.* [40] by including random noise in their channel simulations. Additionally, the initial hairpin was released in a velocity field extracted from DNS. However, the amplitude of the random

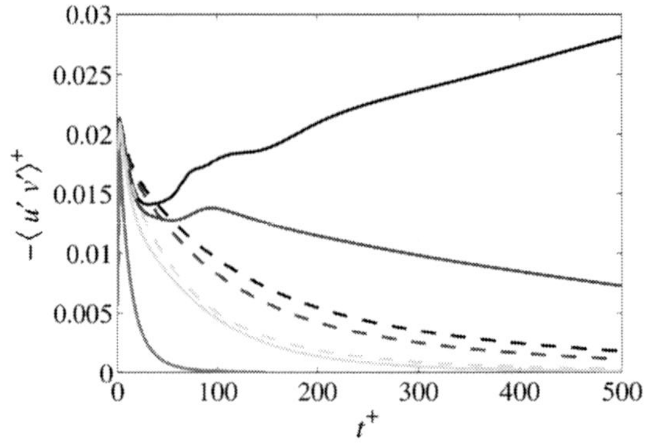

Figure 5. Hairpin regeneration in a mean channel flow at $Re_\tau = 590$ with total viscosity $\nu = \nu_0 + a\nu_t$. Plotted is the temporal development of the ensemble averaged Reynolds stress $\langle -u'v' \rangle$. (black) $a = 0$; (blue) $a = 0.02$; (green) $a = 0.1$; (red) $a = 1$. Solid lines represent nonlinear computations and dashed lines linear ones.

noise was only about 15 % of the rms known from DNS and the disturbances were not purely dissipative. In the case of the DNS flow field, the observation period was strongly limited, such that only the early stages of regeneration could be observed. Nevertheless, the authors concluded that hairpin regeneration is a robust and sustaining process also in fully developed turbulent flow.

In Sec. 2.2 it has been shown in nonlinear simulations that a hairpin packet is formed after a rather short period of time ($t^+ < 200$) with a slow subsequent decay. It appears most likely that a dissipative turbulent background would accelerate the decay and, furthermore, nonlinear effects giving rise to hairpin regeneration should be damped or even inhibited. To investigate which influence a turbulent environment has on the hairpin regeneration process it is feasible to take into account the average dissipative effect of turbulent stresses, namely the turbulent viscosity $\nu_t = -\langle u'v' \rangle / \partial U / \partial y$. This method is limited in the sense that backscatter of turbulent energy from small to large scales is not accounted for.

In the following, the channel flow at $Re_\tau = 590$ from Sec. 2.2 is re-simulated with a total kinematic viscosity of $\nu(y) = \nu_0 + a\nu_t(y)$, where ν_0 is the molecular viscosity, ν_t is extracted from DNS [26], and a is a factor between 0 and 1. The respective trend of $\langle -u'v' \rangle$ is plotted in Fig. 5 for a gradually increasing factor a. For comparison, a linear simulation was performed in each case as well. It is clearly evidenced that an increasing turbulent viscosity strongly damps the hairpin vortices. For $a = 0.02$ one hairpin regeneration takes place, but the structures decay rapidly. Only 10 % of ν_t are sufficient to completely suppress the regeneration process. The nonlinearity of the regeneration process is underlined by the fact that for increasing values of a the curves for nonlinear computations more and more approach the linear solution in which no regeneration was observed. For $a \geq 0.1$ the curves for $\langle -u'v' \rangle$ almost collapse for the nonlinear and linear case (see Fig. 5). Relating back to the results of Zhou *et al.*and Kim *et al.* [21, 40], it appears that the lifetime of hairpin packets and the importance of the parent-offspring process have been overestimated in former studies. It is rather likely that hairpin reproduction occurs only for exceptionally strong loop-vortices on a very short timescale and that these structures are not eligible to sustain turbulent motion.

3. Transitional boundary layer

To investigate how hairpin vortices contribute to transition in a boundary layer and how they evolve downstream, a regular pattern of hairpins is introduced into a laminar boundary layer.

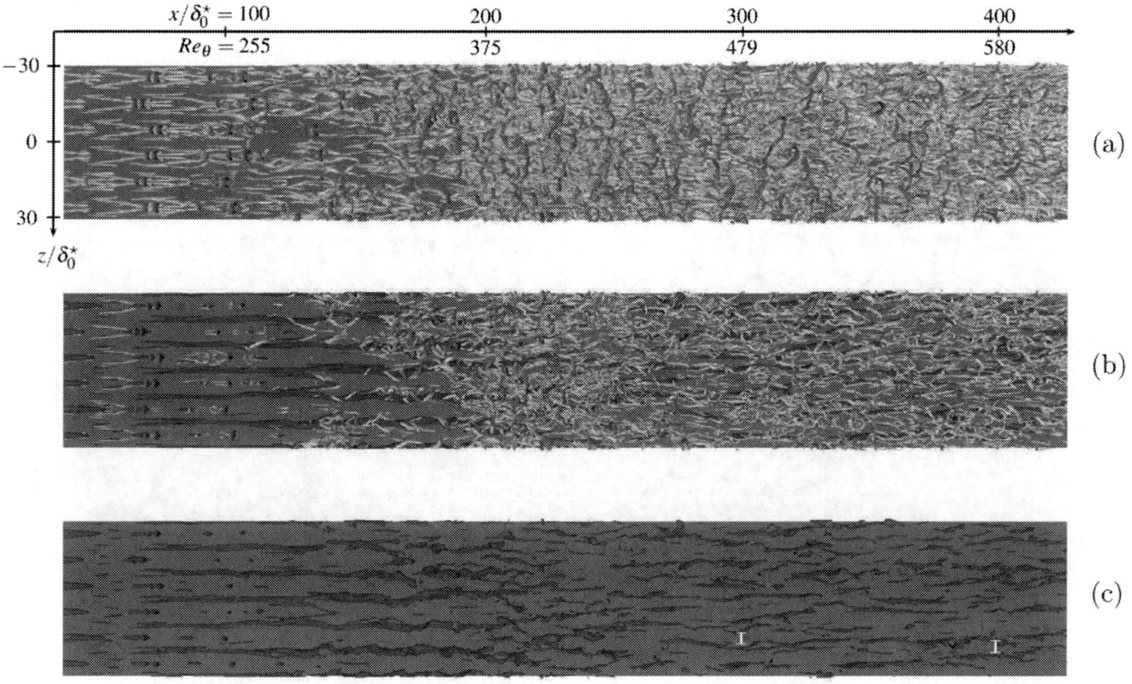

Figure 6. Top view of the synthetic boundary layer going through stages of transition. Flow from left to right with initial hairpin vortices in the laminar region. (a) isocontours of $\lambda_2^+ = -7 \cdot 10^{-3}$; (b) isocontours of $\lambda_2^+ = -26 \cdot 10^{-3}$ and velocity contours for $u'(x,y,z) = u(x,y,z) - U(y) = -0.18\,U_\infty$; (c) velocity contours for $u'(x,y,z) = -0.18\,U_\infty$. The white bars demonstrate a spanwise width of 100 local wall units $\ell_\star$.

To this end, a spatially developing boundary layer is simulated in a computational domain of size $500\,\delta_0^\star \times 40\,\delta_0^\star \times 60\,\delta_0^\star$, with $\delta_0^\star$ being the displacement thickness at the inlet, and $1024 \times 257 \times 256$ spectral collocation points. At the inflow, a Blasius profile is prescribed and a fringe region at the end of the domain ensures streamwise periodicity. The numerical approach is identical to the one used in former DNS studies [31, 33] and recently Schlatter and Örlü reviewed the effect of several tripping techniques and inflow lengths on the development of wall turbulence [32]. In the present study, the trip forcing is modified to place a space-filling spanwise row of six "hairpin emitters" close to the inlet at $x = 25\,\delta_0^\star$, using the ejection scheme described in Sec. 2.1. To be more precise, at each spanwise position an ejection was generated repeatedly with a time period $T_{hp} = 38\,\delta_0^\star/U_\infty$, where U_∞ represents the free stream velocity. The time interval T_{hp} was chosen long enough to produce separated hairpin packets with a slight random variation of the ejection time which breaks the symmetry of the simulation. The time of ejection was allowed to vary by an uncorrelated random shift t_s with $|t_s| < T_{hp}/3$ for each hairpin created. The forcing parameters according to Eq. 1 are taken from the simulation of a temporal boundary layer in Sec. 2.2 and, in units of $\delta_0^\star$ and U_∞, read: $\boldsymbol{C} = (-0.19,\ 0.96,\ 0)^T \cdot (\rho U_\infty^2/\delta_0^\star)$, $L_x = 1.80\,\delta_0^\star$, $L_y = 3.78\,\delta_0^\star$, $L_z = 2.70\,\delta_0^\star$, and $T = 1.56\,\delta_0^\star/U_\infty$.

The developed flow field after approximately three turnover times $500\delta_0^\star/U_\infty$ is shown in Fig. 6. At the inlet, where the flow is still laminar, the six lines of initial hairpins are clearly visible. They undergo a short period of regeneration with secondary and tertiary hairpins while being elongated in the direction of the mean flow. The vortex heads immediately grow towards the edge of the boundary layer and even further. At this stage, which is located around $Re_\theta = 300$, the initial hairpin heads almost disappeared (see Figs 7a and 7b). While the initial

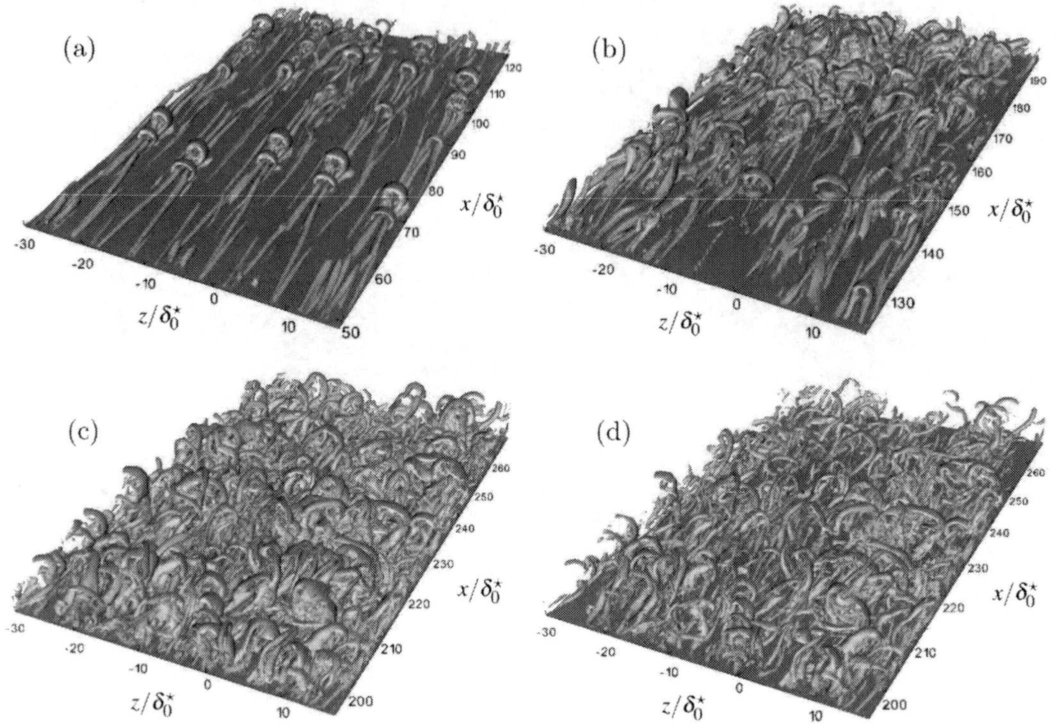

Figure 7. Isoconcontours of λ_2 for different streamwise positions (flow from lower left to upper right). (a) $Re_\theta = 212 - 273$ and $\lambda_2^+ = -7 \cdot 10^{-3}$; (b) $Re_\theta = 273 - 370$ and $\lambda_2^+ = -7 \cdot 10^{-3}$; (c) $Re_\theta = 370 - 445$ and $\lambda_2^+ = -7 \cdot 10^{-3}$; (d) $Re_\theta = 370 - 445$ and $\lambda_2^+ = -26 \cdot 10^{-3}$.

structures are dissipated in the outer layer, there is new dynamics emerging beneath. Namely, secondary low-speed regions, which develop between the original hairpin trails, become sinuously unstable for $Re_\theta > 300$ and burst violently around $Re_\theta = 370$ (see in Figs. 6b and 6c). These unstable streaks shed a large number of vortical structures that grow to the wake region of the boundary layer. At the same time, the near-wall streaks reorganise and establish the pattern well-known for wall turbulence with a streak spacing of about 100 viscous units. From Fig. 6(b), it becomes obvious that the vortical structures near the wall are oriented in a quasi-streamwise manner, while the outer layer structures possess an arch-like shape with a certain amount of spanwise rotation. Depending on the perspective of the observer and on the contour level used for vortex visualisation, these outer structures might give the impression of a forest of hairpins (compare Figs. 7c and 7d). A closer look reveals that the arch-like structures in the outer layer are neither connected to the wall nor to the quasi-streamwise vortices in the near-wall region. Hence, a qualitative analysis of the present data implies that the flow is not dominated by wall-attached hairpins beyond $Re_\theta = 350$, which shall be further investigated in the following.

A snapshot of the fluctuations of the spanwise vorticity $\omega_z' = \omega_z - \langle \omega \rangle$ is presented in Fig. 8. To reveal the different behaviour of the flow in the initial hairpin trails and between those, two xy-planes are shown at the respective positions. The plane at $z/\delta_0^\star = 5$ is centred in one of the hairpin emitters and heads of the corresponding hairpins can be traced up to $x/\delta_0^\star \approx 125$, where a connection to the wall is no longer evident (compare Fig. 7b). At $z = 0$, the secondary low-speed streak, visible as a red area close to the wall, is fully developed at $x/\delta_0^\star = 100$ ($Re_\theta = 255$) and its instability reaches a maximum amplitude at approximately $x = 175\,\delta_0^\star$ ($Re_\theta = 360$). New arch-like vortices are shed from this point into the outer region, which becomes much

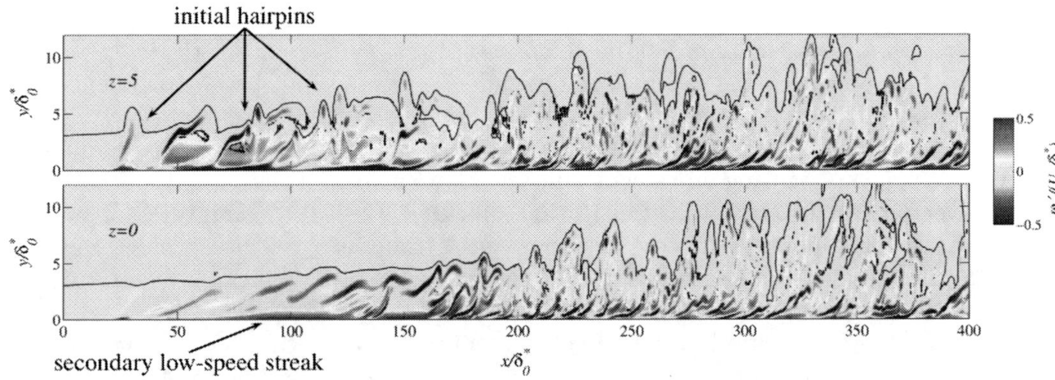

Figure 8. Instantaneous spanwise vorticity fluctuations $\omega_z' = \omega_z - \langle\omega\rangle$ in two xy-planes. (top) in the centre of an initial hairpin trail at $z = 5$; (bottom) between two initial hairpin trails at $z = 0$. Clockwise rotation is defined negative, therefore regions of high shear appear coloured in blue. The black line indicates the enstrophy contour $\omega^2 = 5 \cdot 10^{-4} \, (U_\infty/\delta_0^\star)^2$ with ω being the full vorticity vector.

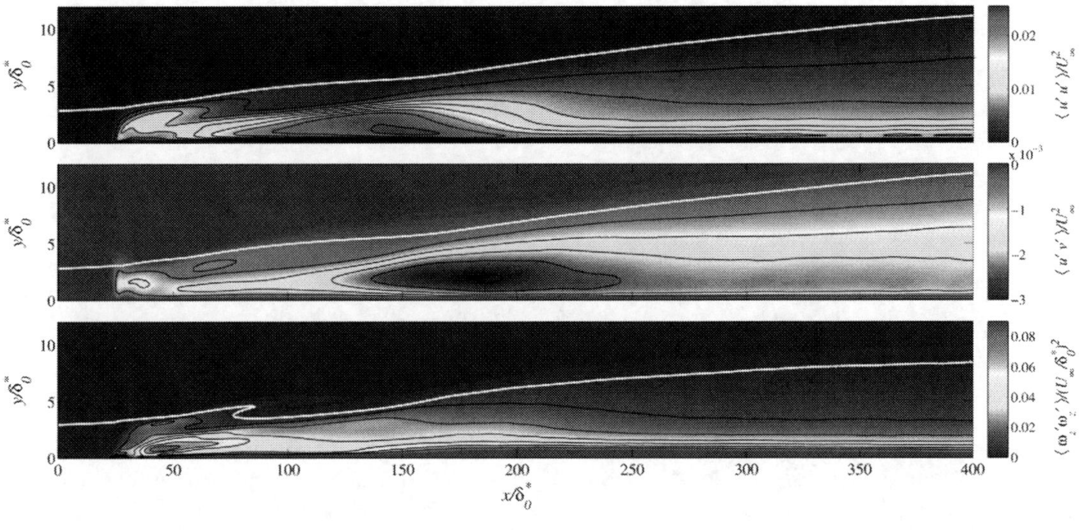

Figure 9. Covariances in the xy-plane for the transitional boundary layer. (top) turbulence intensity $\langle u'u'\rangle$; (middle) Reynolds stress -$\langle -u'v'\rangle$; (bottom) variance of the spanwise vorticity $\langle\omega_z'\omega_z'\rangle$. The white lines mark the boundary layer thickness given by $u = 0.99\,U_\infty$ and $\langle\omega_z\rangle^2 = 5{\cdot}10^{-4} \, (U_\infty/\delta_0^\star)^2$, respectively.

more intermittent. Close to the wall, the newly formed streaks produce numerous wall-attached eddies.

Averaging the flow field in time and spanwise direction yields the distribution of the turbulence intensity $\langle u'u'\rangle$, the Reynolds stress $\langle -u'v'\rangle$, and the variance of the spanwise vorticity $\langle\omega_z'\omega_z'\rangle$, which are summarized in Fig. 9. The upfloating vortex heads at the inflow are visible in all three quantities as a weak outer maximum for $50 < x/\delta_0^\star < 100$. Especially the enstrophy contour indicates that the initial hairpins get lost in the wake region. Crucial for the subsequent transition are the streaks near the wall and their growing instability which manifests itself in a global maximum of $\langle u'u'\rangle$ at $x/\delta_0^\star \approx 150$. In turn, the vortices emitted in this region produce a maximum of the Reynolds stress $\langle -u'v'\rangle$ at $x/\delta_0^\star \approx 180$ and from that point the boundary layer thickness increases rapidly. For $x/\delta_0^\star > 200$, no remarkable activity is found in the wake region and $\langle u'u' \rangle$ and $\langle -u'v' \rangle$ approach the expected y-dependence for developed wall turbulence. It can be concluded from Fig. 9 that the near-wall cycle sets in at $x/\delta_0^\star \approx 200$ and that the arch-like vortical structures observed on top of the boundary layer (*cf.* Fig. 7c) are very weak compared to the near-wall dynamics. Focusing on integral and mean quantities, it becomes apparent that the boundary layer passes through transition rather quickly. Figs. 10(a) and 10(b) show the growth of the momentum thickness and the development of the mean velocity profile. Beyond $x/\delta_0^\star = 200$, the turbulent mean velocity is well developed and $Re_\theta(x)$ increases linearly up to approximately $x/\delta_0^\star = 400$ where the fringe regions starts to take effect.

The shape factor $H_{12} = \delta^\star/\theta$, shown in Fig. 11, confirms that the mean velocity profile possesses the turbulent characteristics for $x/\delta_0^\star > 200$ whereas the friction coefficient $c_f = 2\tau_w/(\rho U_\infty^2)$ approaches the value expected for wall turbulence asymptotically after a brief overshoot.

Interestingly, during the rise of c_f there is an intermediate plateau while the secondary streaks

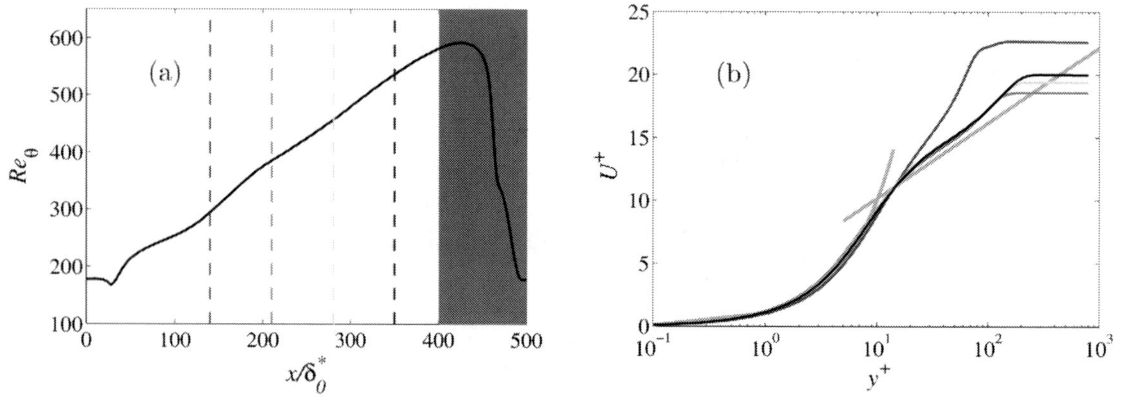

Figure 10. Synthetic boundary layer going through stages of transition. (a) Streamwise development of the momentum thickness Reynolds number Re_θ. The grey area marks the fringe region; (b) Mean velocity profiles at four streamwise positions indicated by dashed lines in (a) with corresponding colours. The grey lines indicate the law of the wall with the constants $\kappa = 0.384$, $B = 4.17$ for the logarithmic region.

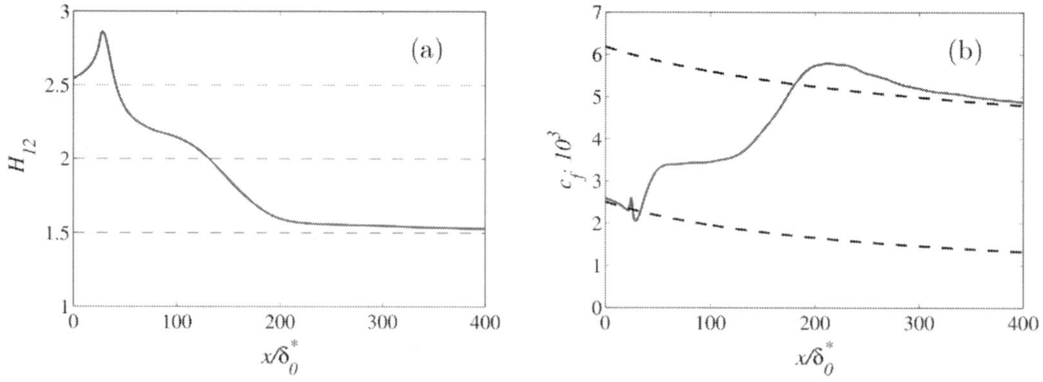

Figure 11. Synthetic boundary layer going through stages of transition. (a) Shape coefficient $H_{12} = \delta^\star/\theta$; (b) Friction coefficient c_f from present data (solid line) compared with the Blasius solution (lower dashed line) and the Coles-Fernholz approximation for the turbulent case [4] (upper dashed line).

are formed ($60 < x/\delta_0^\star < 110$) and c_f grows further as the streaks become unstable.

4. Conclusions

Simulations of a hairpin vortex in a mean shear flow evidenced the regeneration of hairpin vortices and proved the nonlinear nature of the parent-offspring mechanism. In contrast to former studies, the observation time was largely extended, revealing that nonlinear self induction is very short-lived, *i.e.* for convection times greater than $t^+ = 300$, the structures decay constantly. The assumption that hairpin regeneration cannot sustain in turbulent flows is further supported by simulations including a turbulent viscosity. Even a small amount of 10 % background dissipation completely suppresses the generation of new hairpins.

In a second part, a transitional boundary layer with an adapted hairpin trip forcing was set up. Using various flow field visualisations, it was shown that the initial hairpins do regenerate briefly before they are absorbed in the wake region. Therefore, it is very unlikely that hairpin vortices persist in fully developed turbulent boundary layers. Instead, the actual transition is driven by secondary low-momentum streaks which develop a growing instability and finally initiate the near-wall dynamics that sustain turbulence. This behaviour complies with mechanisms known from stability theory, which predicts transient growth of sinuous instabilities while varicose perturbations are less stable [3, 34]. Arch-like vortices were observed in the wake region, which may be associated with the reported phenomenon of a forest of hairpins. However, how these vortices are formed and what causes their absence at higher Reynolds numbers remain open questions.

Since the boundary conditions and the trip forcing used are reproducible, the presented boundary layer simulation could be a valuable tool for further investigations of early stages of transition in the sense of a synthetic boundary layer flow. Variation of the streamwise position of forcing and the number and strength of initial hairpin vortices would allow for a more detailed study of the processes that lead to the onset of wall-turbulence and the formation of large-scale structures.

Acknowledgments

This work was supported in part by the Multiflow program of the European Research Council. The authors would like to thank Adrián Lozano-Durán for his insightful comments on the manuscript. Additional financial support is acknowledged from the Göran Gustafsson Foundation. Computer time was provided by SNIC (Swedish National Infrastructure for Computing).

Appendix: Perturbation mode formulation

In the following, the concept of the perturbation-mode simulations is discussed in more detail. The basic principle is to decompose the velocity field into a constant base flow on the one hand and time dependent deviations from the base flow (perturbations) on the other hand. This approach eases stability analyses since distinct perturbations and their evolution can be studied in an any kind of base flow. In the case of laminar channel flow, the Poiseuille profile would serve as the physically correct base flow and at the same time it represents as a stable (attracting) state of the system. That is, any artificial perturbation will be damped until the laminar profile is restored. If the Reynolds number is high enough and the initial perturbations possess certain characteristics, another stable state of the system, the turbulent regime, can be established and maintained. It is important to note that, in the turbulent state, the base flow is no longer equivalent to the mean flow. As long as, for a given geometry, the base flow is the solution of the (steady) Navier–Stokes equations, the perturbation formulation is identical to the original non-decomposed equations of motion. However, in the present work the turbulent mean profile was used as a base flow, which is not a solution of the steady Navier–Stokes equations since the

Journal of Physics: Conference Series **506** (2014) 012008 doi:10.1088/1742-6596/506/1/012008

Reynolds stress terms are missing. To justify this approach, we will explain the basic concept of the perturbation formulation and discuss the consequences of using a non-physical base flow.

In the incompressible Navier–Stokes equations,

$$\frac{\partial u_i}{\partial t} + u_j \frac{\partial u_i}{\partial x_j} = -\frac{\partial p}{\partial x_i} + \frac{1}{Re}\frac{\partial^2 u_i}{\partial x_j \partial x_j}, \tag{A.1}$$

the velocity can be split into a base flow U_i and a perturbation u_i' component

$$u_i(x,t) = U_i(x) + u_i'(x,t). \tag{A.2}$$

Rewriting Eq. (A.1) for this decomposed velocity field yields

$$\begin{aligned}\frac{\partial U_i}{\partial t} + U_j \frac{\partial U_i}{\partial x_j} + \frac{\partial u_i'}{\partial t} + u_j' \frac{\partial u_i'}{\partial x_j} + U_j \frac{\partial u_i'}{\partial x_j} + u_j' \frac{\partial U_i}{\partial x_j} = \\ -\frac{\partial P}{\partial x_i} + \frac{1}{Re}\frac{\partial^2 U_i}{\partial x_j x_j} - \frac{\partial p'}{\partial x_i} + \frac{1}{Re}\frac{\partial^2 u_i'}{\partial x_j \partial x_j}.\end{aligned} \tag{A.3}$$

Subtracting the pure base flow contributions, we end up with the equations of motion in perturbation form

$$\begin{aligned}\frac{\partial u_i'}{\partial t} + u_j' \frac{\partial u_i'}{\partial x_j} + U_j \frac{\partial u_i'}{\partial x_j} + u_j' \frac{\partial U_i}{\partial x_j} = \\ -\frac{\partial p'}{\partial x_i} + \frac{1}{Re}\frac{\partial^2 u_i'}{\partial x_j \partial x_j}.\end{aligned} \tag{A.4}$$

If the time-invariant base flow $U_i(x)$ fulfills

$$U_j \frac{\partial U_i}{\partial x_j} = -\frac{\partial P}{\partial x_i} + \frac{1}{Re}\frac{\partial^2 U_i}{\partial x_j \partial x_j}, \tag{A.5}$$

then Eq. (A.4) describes the same physics as Eq. (A.1). It should be remarked that nonlinear interactions can easily be cancelled out by neglecting the second term on the left-hand side of Eq. (A.4). If a base flow is employed, that is not a Navier–Stokes solution, one inherently introduces a residual R_i fulfilling

$$U_j \frac{\partial U_i}{\partial x_j} = -\frac{\partial P}{\partial x_i} + \frac{1}{Re}\frac{\partial^2 U_i}{\partial x_j \partial x_j} + R_i \tag{A.6}$$

and therefore one solves for

$$\frac{\partial u_i}{\partial t} + u_j \frac{\partial u_i}{\partial x_j} = -\frac{\partial p}{\partial x_i} + \frac{1}{Re}\frac{\partial^2 u_i}{\partial x_j \partial x_j} + R_i \tag{A.7}$$

instead of Eq. (A.1). In other words, R_i is incorporated as an implicit forcing, which is not explicitly calculated but relaxes the base flow towards the prescribed form. This is similar to the procedure of manually resetting the base flow in every timestep, or explicitly calculating Reynolds stresses and including them in the equations.

References

[1] Acarlar MS and Smith CR 1987 A study of hairpin vortices in a laminar boundary layer. Part 2. Hairpin vortices generated by fluid injection *J. Fluid Mech.* **175** 43–83
[2] Adrian RJ 2007 Hairpin vortex organization in wall turbulence *Phys. Fluids* **19** 041301
[3] Brandt L, Schlatter P and Henningson D 2004 Transition in boundary layers subject to free-stream turbulence *J. Fluid Mech.* **517** 167–198
[4] Chauhan KA, Monkewitz PA and Nagib HM 2009 Criteria for assessing experiments in zero pressure gradient boundary layers *Fluid Dyn. Res.* **41** 021404
[5] Chong MS, Soria J, Perry AE, Chacin J, Cantwell BJ and Na Y 1998 Turbulence structures of wall-bounded flows using DNS data *J. Fluid Mech.* **357** 225–247
[6] Chevalier M, Schlatter P, Lundbladh A and Henningson DS 2007 SIMSON–A Pseudo-Spectral Solver for Incompressible Boundary Layer Flow *Tech. Rep. TRITA-MEK 2007:07, Royal Institute of Technology, Stockholm, Sweden*
[7] Coles D and Barker SJ 1975 Some remarks on a synthetic turbulent boundary layer In *Turbulent Mixing in Nonreactive and Reactive Flows*, pp. 285–292 (New York: Plenum)
[8] Flores O 2007 *The Dynamics of the Outer Region of Wall-Bounded Turbulence* PhD thesis, Universidad Politécnica de Madrid
[9] Haidari AH and Smith CR 1994 The generation and regeneration of single hairpin vortices *J. Fluid Mech.* **277** 135–162
[10] Hamilton JM, Kim J and Waleffe F 1995 Regeneration mechanisms of near-wall turbulence structures *J. Fluid Mech.* **287** 317–348
[11] Head MR and Bandyopadhyay P 1981 New aspects of turbulent boundary-layer structure *J. Fluid Mech.* **107** 297–338
[12] Hoyas S and Jiménez J 2006 Scaling of the velocity fluctuations in turbulent channels up to $Re_\tau = 2003$ *Phys. Fluids* **18** 011702
[13] Hutchins N and Marusic I 2007 Large-scale influences in near-wall turbulence *Phil. Trans. Royal Soc. A* **365** 647–664
[14] Jeong J, Hussain F, Schoppa W and Kim J 1997 Coherent structures near the wall in a turbulent channel flow *J. Fluid Mech.* **332** 185–214
[15] Jiménez J 2013 Near-wall turbulence *Phys. Fluids* **25** 101302
[16] Jiménez J and Moin P 1991 Minimal Flow unit of near-wall turbulence *J. Fluid Mech.* **225** 213–240
[17] Jiménez J, Kawahara G, Simens MP, Nagata M and Shiba M 2005 Characterization of near-wall turbulence in terms of equilibrium and bursting solutions *Phys. Fluids* **17** 015105
[18] Jiménez J and Pinelli A 1999 The autonomous cycle of near-wall turbulence *J. Fluid Mech.* **389** 335–359
[19] Kachanov YS 1994 Physical mechanisms of laminar-boundary-layer transition *Annu. Rev. Fluid Mech.* **26** 411–482
[20] Kim J, Moin P and Moser RD 1987 Turbulence statistics in fully developed channel flow at low Reynolds number *J. Fluid Mech.* **177** 133–166
[21] Kim K, Sung HJ, and Adrian RJ 2008 Effects of background noise on generating coherent packets of hairpin vortices *Phys. Fluids* **20** 105107
[22] Kleiser L and Zang TA 1991 Numerical simulation of transition in wall-bounded shear flows *Annu. Rev. Fluid Mech.* **23** 495–537
[23] Klewicki JC 2010 Reynolds number dependence, scaling and dynamics of turbulent boundary layers *J. Fluid Eng.* **132** 094001
[24] Kline S, Reynolds WC, Schraub F and Runstadler P 1967 The structure of turbulent boundary layers *J. Fluid Mech.* **30** 741–773
[25] Marusic I, McKeon BJ, Monkewitz PA, Nagib HM, Smits AJ and Sreenivasan KR 2010 Wall-bounded turbulent flows at high Reynolds numbers: Recent advances and key issues *Phys. Fluids* **22** 065103
[26] Moser RD, Kim J and Mansour NN 1999 Direct numerical simulation of turbulent channel flow up to $Re_\tau = 590$ *Phys. Fluids* **11** 943–945
[27] Perry AE and Chong MS 1982 On the mechanism of wall turbulence *J. Fluid Mech.* **119** 173–217
[28] Robinson SK 1991 Coherent motions in the turbulent boundary layer *Annu. Rev. Fluid Mech.* **23** 601–639
[29] Sandham ND and Kleiser L 1992 The late stages of transition to turbulence in channel flow *J. Fluid Mech.* **245** 319–348
[30] Savas Ö and Coles D 1985 Coherence measurements in synthetic turbulent boundary layers *J. Fluid Mech.* **160** 421–446
[31] Schlatter P and Örlü R 2010 Assessment of direct numerical simulation data of turbulent boundary layers *J. Fluid Mech.* **659** 116–126
[32] Schlatter P and Örlü R 2012 Turbulent boundary layers at moderate Reynolds numbers. Inflow length and

tripping effects *J. Fluid Mech.* **710** 5–34

[33] Schlatter P, Örlü R, Li Q, Brethouwer G, Fransson JHM, Johansson AV, Alfredsson PH, and Henningson DS 2009 Turbulent boundary layers up to $Re_\theta = 2500$ studied through simulation and experiment *Phys. Fluids* **21** 051702

[34] Schoppa W and Hussain F 2002 Coherent structure generation in near-wall turbulence *J. Fluid Mech.* **453** 57–108

[35] Smits AJ, McKeon BJ and Marusic I 2011. High-Reynolds-number wall turbulence *Annu. Rev. Fluid Mech.* **43** 353–375

[36] Theodorsen T 1952 Mechanism of turbulence *Proc. Second Midwestern Conference on Fluid Mechanics* Ohio State University, Columbus, OH, USA.

[37] Tomkins CD and Adrian RJ 2003 Spanwise structure and scale growth in turbulent boundary layers *J. Fluid Mech.* **490** 37–74

[38] Townsend AA 1961 Equilibrium layers and wall turbulence *J. Fluid Mech.* **11** 97–120

[39] Wu X and Moin P 2009 Forest of hairpins in a low-Reynolds-number zero-pressure-gradient flat-plate boundary layer *Phys. Fluids* **21** 091106

[40] Zhou J, Adrian RJ, Balachandar S and Kendall TM 1999 Mechanisms for generating coherent packets of hairpin vortices in channel flow *J. Fluid Mech.* **387** 353–396

Skin-friction critical points in wall-bounded flows.

José I Cardesa[1], **Jason P Monty**[2], **Julio Soria**[3] **and Min S Chong**[2]

[1]School of Aeronautics, Universidad Politécnica de Madrid, Madrid, SPAIN
[2]Mechanical Engineering Department, University of Melbourne, Victoria 3010, AUSTRALIA
[3]Mechanical Engineering Department, Monash University, Victoria 3800, AUSTRALIA

E-mail: jose@torroja.dmt.upm.es

Abstract. Critical points in the skin friction field of wall-bounded flows were investigated using data from direct numerical simulations of channels at $Re_\tau = 934$ and $Re_\tau = 1834$. A method for their detection and characterisation is outlined, and a statistical description of their properties is reported. Their lifetime, average distance, velocity and area density were computed. Conditionally averaged fields were calculated in order to examine the average flow in the vicinity of a critical point. It was found that the critical points are connected to surprisingly large features in the channel. A mechanism for the generation of the critical points is postulated, based on the conditional averages and analysis of time sequences in the flow above them.

1. Introduction

Using data from full direct numerical simulations (DNS) of wall-bounded flows (pipes, channels and boundary layers), Ref. [1] have shown that critical points, *i.e.* points where the skin friction and surface vorticity are zero, exist on the surface of a no-slip wall. They suggested that, since the invariants of the velocity gradient tensor are zero at the wall, the invariants of the 'no-slip tensor' can be used instead to investigate turbulent structures at this location. The existence of critical points indicates that there are complex three-dimensional separated flow structures at the wall, resulting in negative streamwise skin friction or shear stresses at the wall. Ref. [2] also reported regions of negative skin-friction events in their numerical simulations. Regions at the wall with critical points are rare and there is a need to establish the significance of these regions in the generation of 'turbulence' at the wall. In this paper we will establish accurately the location of the critical points and investigate the statistics and topological properties of these critical points.

2. No-slip critical points

We will adopt a frame of reference where the three orthogonal directions and velocities are labelled as follows:

$$\begin{aligned} \text{streamwise} &\rightarrow u, u_1, x, x_1 \\ \text{spanwise} &\rightarrow v, u_2, y, x_2 \\ \text{wall-normal} &\rightarrow w, u_3, z, x_3. \end{aligned}$$

The unit vectors in the three directions are $\mathbf{e}_1$, $\mathbf{e}_2$, $\mathbf{e}_3$.

Let the velocity gradient tensor be

$$A_{ij} = \frac{\partial u_i}{\partial x_j}. \tag{1}$$

It is easy to show that at a wall, only two components of A_{ij} do not vanish as a result of the no-slip, impermeability and incompressibility conditions: A_{13} and A_{23}. It follows that at such a wall, all three similarity invariants of A_{ij} vanish – P, Q and R as defined by [3]. A local characterisation of the flow field now requires a higher order truncation. Keeping up to second-order terms, the Taylor expansion of the velocity field at a no-slip wall is

$$u(x,y,z) = \dot{x} = \frac{dx}{dt} = z\frac{\partial u}{\partial z} + z\left(x\frac{\partial^2 u}{\partial x \partial z} + y\frac{\partial^2 u}{\partial y \partial z} + \frac{z}{2}\frac{\partial^2 u}{\partial z \partial z}\right) \tag{2}$$

$$v(x,y,z) = \dot{y} = \frac{dy}{dt} = z\frac{\partial v}{\partial z} + z\left(x\frac{\partial^2 v}{\partial x \partial z} + y\frac{\partial^2 v}{\partial y \partial z} + \frac{z}{2}\frac{\partial^2 v}{\partial z \partial z}\right) \tag{3}$$

$$w(x,y,z) = \dot{z} = \frac{dz}{dt} = z\left(x\frac{\partial^2 w}{\partial x \partial z} + y\frac{\partial^2 w}{\partial y \partial z} + \frac{z}{2}\frac{\partial^2 w}{\partial z \partial z}\right), \tag{4}$$

with all derivatives estimated at $z = 0$. We introduce the no-slip tensor $\mathcal{A}_{ij}$

$$\underline{\mathcal{A}} = \begin{bmatrix} \mathcal{A}_{11} & \mathcal{A}_{12} & \mathcal{A}_{13} \\ \mathcal{A}_{21} & \mathcal{A}_{22} & \mathcal{A}_{23} \\ \mathcal{A}_{31} & \mathcal{A}_{32} & \mathcal{A}_{33} \end{bmatrix} = \begin{bmatrix} \frac{\partial^2 u}{\partial x \partial z} & \frac{\partial^2 u}{\partial y \partial z} & \frac{1}{2}\frac{\partial^2 u}{\partial z \partial z} \\ \frac{\partial^2 v}{\partial x \partial z} & \frac{\partial^2 v}{\partial y \partial z} & \frac{1}{2}\frac{\partial^2 v}{\partial z \partial z} \\ \frac{\partial^2 w}{\partial x \partial z} & \frac{\partial^2 w}{\partial y \partial z} & \frac{1}{2}\frac{\partial^2 w}{\partial z \partial z} \end{bmatrix}, \tag{5}$$

and the rescaling

$$d\tau = z dt. \tag{6}$$

Equations (2), (3) and (4) can now be written in matrix form as

$$\begin{bmatrix} \mathring{x} \\ \mathring{y} \\ \mathring{z} \end{bmatrix} = \begin{bmatrix} 0 & 0 & A_{13} \\ 0 & 0 & A_{23} \\ 0 & 0 & 0 \end{bmatrix} + \begin{bmatrix} \mathcal{A}_{11} & \mathcal{A}_{12} & \mathcal{A}_{13} \\ \mathcal{A}_{21} & \mathcal{A}_{22} & \mathcal{A}_{23} \\ \mathcal{A}_{31} & \mathcal{A}_{32} & \mathcal{A}_{33} \end{bmatrix} \begin{bmatrix} x \\ y \\ z \end{bmatrix}, \tag{7}$$

or simply $\mathring{x}_i = A_{ij} + \mathcal{A}_{ij}x_j$. We note in passing that $\mathring{x}_i = dx_i/d\tau \neq \dot{x}_i = dx_i/dt$, and that incompressibility at a no-slip wall implies $\mathcal{A}_{31} = \mathcal{A}_{32} = 0$, $\mathcal{A}_{33} = -(\mathcal{A}_{11} + \mathcal{A}_{22})/2$.

We will refer to the following two-dimensional vector field at the wall as the skin-friction field:

$$\mathbf{s} = A_{13}\mathbf{e}_1 + A_{23}\mathbf{e}_2. \tag{8}$$

The flow's vorticity at the wall reduces to a two-component vector field $\omega = A_{23}\mathbf{e}_1 - A_{13}\mathbf{e}_2$. It is easy to see that at the wall, $\mathbf{s} \cdot \omega = 0$. Hence, the two fields are always mutually orthogonal.

2.1. Definition and characterisation of critical points

A linear approximation to $\mathbf{s}$ around a point $\mathbf{x}_0$ yields $s_i(\mathbf{x}) = A_{i3}(\mathbf{x}_0) + x_j\mathcal{A}_{ij}(\mathbf{x}_0)$, which is valid only in the immediate vicinity of $\mathbf{x}_0$. A critical point (CP) in $\mathbf{s}$ occurs where both its components vanish: $A_{13} = A_{23} = 0$. The topology of the critical point is given by the similarity invariants of the matrix $\mathcal{A}_{ij}$ computed at the critical point. This topology is representative of the CP's immediate vicinity, *i.e.* of the region where the linear approximation holds. Subscripts i and j now span the range $1 \rightarrow 2$ only, so that there are two invariants to characterise the local topology of a CP in $\mathbf{s}$: its trace and determinant. We adopt the convention to label them as $p = -\mathcal{A}_{ii}$ and $q = det(\underline{\mathcal{A}})$. A detailed description of the possible local topologies based on p and q can be found in [4]. Here, we will only make the distinction between two types of CPs: saddles ($q < 0$) and nodes ($q > 0$). Using this characterisation, it can be shown that a CP in $\mathbf{s}$ is also a CP of the same type in ω because their corresponding q is equal.

3. Data sets

The data used for the present study comes from DNS of turbulent channel flow. The numerical scheme integrates the fluid flow equations following the decomposition described in [5]. On wall-parallel planes, space is discretised using Fourier expansions while Chebychev polynomials are used in the wall-normal direction. Two data sets are used, M950 and S1900. The latter is the same as case W1900 from [6], whereas the former is a smaller version of case L950 in [7] – but time-resolved. The details of the simulations are summarised in table 1.

The set S1900 contains 700 independent fields. They were used for standard statistics, *i.e.* statistics that do not require time resolution. The data set M950 was run at constant time step δt, instead of constant CFL as is usually done. The chosen δt of 0.02^+ corresponded to a CFL of approximately 0.3. One in every fifteen fields was stored, thus creating a time series where consecutive fields are separated by 0.3^+. The full time series stored spans 889.5^+, which is almost one large-eddy turnover time. It was used for standard statistics as well as for statistical analysis of time-resolved events. For the two data sets, we stored data from both walls. This meant 1400 independent wall realisations from S1900 and two independent time series from M950, each spanning $t^+ = 889.5$.

Table 1. Properties of the two DNS data sets used. Δx and Δy are the streamwise and spanwise resolutions, respectively, after dealiasing. L_x and L_y are the domain extents in those two directions. N_x and N_y are the number of coefficients in the Fourier expansions in those two directions, while N_z is the number of Chebychev polynomials of the wall-normal expansion. If h is the channel half-height, h/Re_τ yields the wall length unit with which all lengths with superscript "+" are normalised. u_τ is the usual friction velocity.

	Re_τ	L_x^+	L_y^+	Δx^+	Δy^+	N_x	N_y	N_z	$u_\tau \times 10^2$
M950	934	5870	2935	11.5	5.7	512	511	385	4.539
S1900	1834	2880	1440	11.3	5.6	256	255	769	4.079

3.1. Finding critical points

In a numerical data base, the components of $\mathbf{s}$ are only available at discrete locations on a regular grid and stored with finite precision. Naively looking for a zero in both A_{13} and A_{23} yields no CPs. Instead, methods based on thresholding and local minima can be thought of, but we used a different approach which we will now describe in some detail.

We start by assuming the linear approximation $s_i(\mathbf{x}) = A_{i3}(\mathbf{x}_0) + x_j \mathcal{A}_{ij}(\mathbf{x}_0)$ to be valid. In particular, we ask the following question at each discrete point $\mathbf{x}_0$ of the data sets A_{i3} and $\mathcal{A}_{ij}$: at which distance $\mathbf{x}$ from $\mathbf{x}_0$ does the linear approximation predict a zero in $\mathbf{s}$? In matrix form, this question is equivalent to solving the system

$$\begin{bmatrix} s_1 \\ s_2 \end{bmatrix} = \begin{bmatrix} A_{13} \\ A_{23} \end{bmatrix} + \begin{bmatrix} \mathcal{A}_{11} & \mathcal{A}_{12} \\ \mathcal{A}_{21} & \mathcal{A}_{22} \end{bmatrix} \begin{bmatrix} x_1 \\ x_2 \end{bmatrix} = \begin{bmatrix} 0 \\ 0 \end{bmatrix}. \tag{9}$$

A solution for x_1 and x_2 is found at every grid point in terms of A_{i3} and $\mathcal{A}_{ij}$ at those points. However, we only keep those solutions found which comply to the requirement that $\sqrt{x_1^2 + x_2^2} < l$, where l is the data's grid spacing on a wall plane. Any CP found further away than this from the grid point is considered outside of the region where the linear approximation applies. The latter is assumed to be a circle of radius l centred at the interrogated grid point – see figure 1. Each CP thus found should be detected from at least two contiguous grid points – see figure 2. Since each grid point predicts the location of the CP according to the linear approximation valid at the interrogated grid point, two slightly different CP locations will be predicted from two different contiguous grid points for the same CP. We use this redundancy in our favour, and discard any CP found which does not possess at least one neighbour within a circle of radius $|\mathbf{r}| = l$ centred at the CP – see figure 2. Most of the time, CPs are found in sets of four: one from each grid point within a distance l. The CP finding algorithm then screens each "cloud" of such CPs which fit within a circle of radius l and discards all but one – chosen randomly. In the case where a CP was found to have no neighbours within this circle, it is discarded as spurious.

Once the position of the CP is known, the values of $\mathcal{A}_{ij}$ need to be estimated at the CP location. Since its position does not correspond to a grid point, we carry out a subgrid interpolation. This involves the use of a square of 11×11 neighbouring grid points which has the CP location at its centre. The square patch of variables $\mathcal{A}_{ij}$ is given to Matlab®'s `interp2` function which applies cubic spline interpolation to infer their value at the CP location.

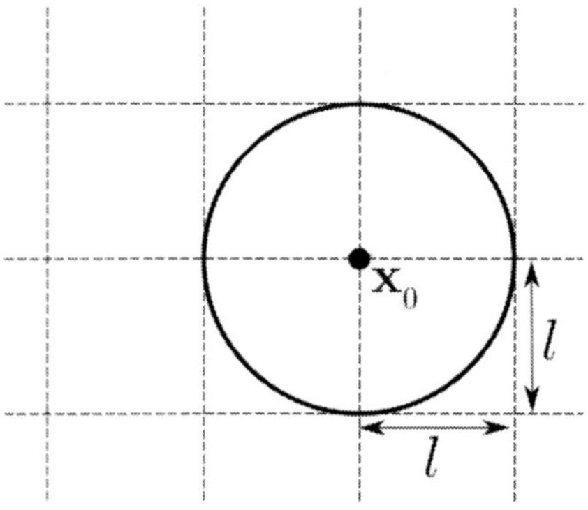

Figure 1. Region of assumed validity for a CP location predicted by the linear expansion at grid point $\mathbf{x}_0$.

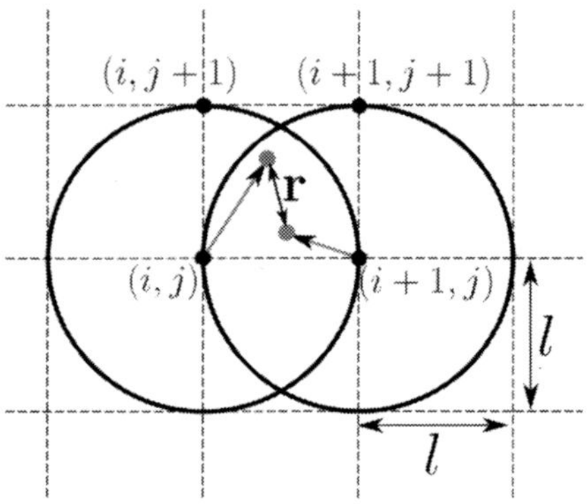

Figure 2. Neighbouring grid points predicting two different CP locations. CPs within $|\mathbf{r}| < 2l$ from each other are considered to be the same CP.

The method that has been described relies heavily on the assumption that the grid spacing, l, is small. This aspect will now be discussed.

3.2. Data interpolation

Section 3.1 concluded by emphasising the strong dependence of the CP-finding routine on l, the streamwise and spanwise spacing between data points. We artificially decreased this spacing by taking advantage of the periodicity in the streamwise and spanwise directions. The data sets of A_{i3} and $\mathcal{A}_{ij}$ were Fourier transformed, padded with zeros in Fourier space, and then transformed back into real space. The fields obtained in this manner differed from the original ones in several ways which are perhaps best explained with a specific example as follows.

Any flow variable can be obtained from the Fourier and Chebychev coefficients of the wall-normal vorticity (vor) and Laplacian of wall-normal velocity (ϕ), which are the two variables the DNS code solves for [5]. A single field of A_{13} from the series M950 would usually yield a real-space representation of $768 \times 768 \times 385$ points for a single quantity ($x \times y \times z$) when using the settings outlined in table 1. However, we obtain a single field representation of A_{13} on a $1536 \times 1536 \times 769$ grid after zero-padding the coefficients of vor and ϕ in the post-processing routine which generates A_{i3} and $\mathcal{A}_{ij}$. This has doubled the number of points in each direction. Only the bottom ($z = 0$) and top ($z = 2$) planes of A_{i3} and $\mathcal{A}_{ij}$ are stored to disk. A patch of field $\mathbf{s}$ extending $12^+ \times 12^+$ in the x and y directions is shown on figure 3(a), as it is obtained directly from the 1536×1536 slice that is stored to disk. It can be seen that the streamwise grid spacing is twice the one in the spanwise direction, which is a consequence of the DNS settings – see table 1. In the next step, the planes of fields A_{i3} and $\mathcal{A}_{ij}$ are then Fourier-transformed. Zero padding is applied to the transforms of A_{i3} and $\mathcal{A}_{ij}$ in order to add $n-1$ data points in y between two contiguous grid points of the original grid, and $2n-1$ additional points in x. The resulting grid now features a constant spacing in both directions. Figure 3(b) shows the same patch of $\mathbf{s}$ as in figure 3(a) after applying this interpolation procedure with $n=5$. Two CPs have become more evident in the interpolated data set. The Fourier-based interpolation was done prior to running the CP finding routine, and hence before applying the final stage cubic spline interpolation on the patch surrounding a detected CP.

As explained in [1], the number of saddles has to be equal to the number of nodes by a topological constraint which follows from the Poincaré-Hopf index theorem. Using the CP finding routine outlined in this section, it was found that the data set M950 obeyed this one-to-one ratio between saddles and nodes in 85.9% of the fields analysed. In 13.41% there was one more CP of one type than the other, in 0.72% there were 2 more CPs of one type than the other and in 0.02% there was an imbalance of three. Within the data set S1900, 95.14% of the fields had the same number of saddles as nodes, 4.72% had an imbalance of one CP, 0.07% had an imbalance of two CPs and 0.07% of three CPs. No trend was observed of one type of CP being found more than the other.

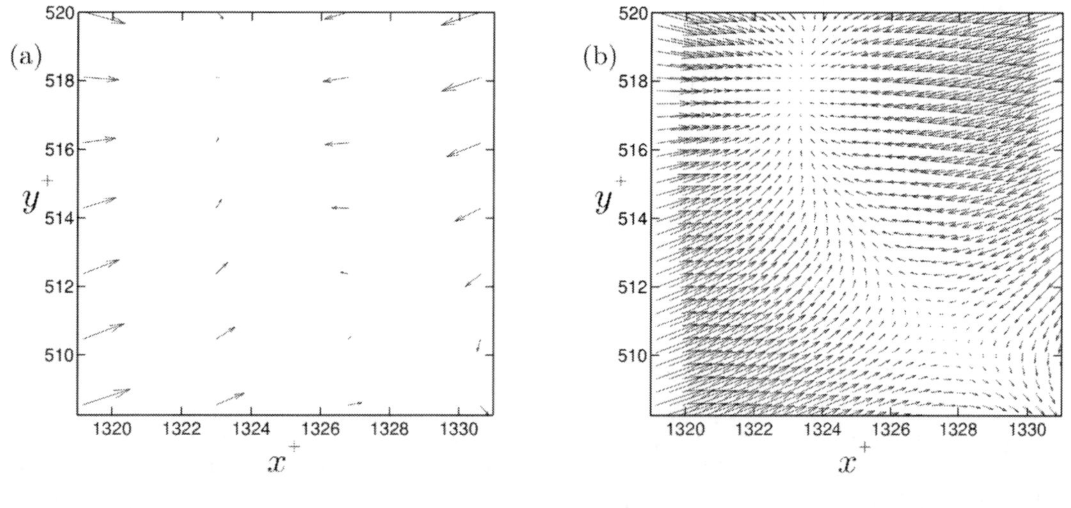

Figure 3. Vector field $\mathbf{s} = A_{13}\mathbf{e}_1 + A_{23}\mathbf{e}_2$ over a wall. Selected region contains two CPs. (a) Original field from rectangular DNS grid; (b) Field after interpolation using zero-padding in Fourier space. 4 and 9 additional points have been added by interpolation between 2 original points in y and x, respectively.

4. Statistics at the wall

We proceed to present various CP statistics measured at the walls of both flows: S1900 and M950. Before doing so, however, it is perhaps best to illustrate at this point what kind of data is being used in a statistical manner.

The data set S1900 is made of instantaneous snapshots of independent fields. The CPs found in M1900 can be used for statistics like average number of CPs per wall plane, average distance between CPs, *etc.*... M950 contains a time-resolved data series, which extends for long enough in time so as to extract the same type of statistics as from S1900. In addition, a look at figure 4 shows what other possibilities this data set offers. Figure 4(a) displays the typical life of 2 CPs. A saddle and a node are born simultaneously and near each other. During their life, they travel predominantly in the streamwise direction before coming close again and disappearing together. The topological constraint that there be an equal number of saddles and nodes is thus satisfied. However, this condition is also compatible with more elaborate life patterns. Figure 4(b) depicts such a scenario extracted from M950. The lives of three CP pairs overlap, without any of them disappearing next to the CP they were born with. From this type of data, we can extract dynamical information like average life times and average convection velocity of the CPs, which we will report in the next sections.

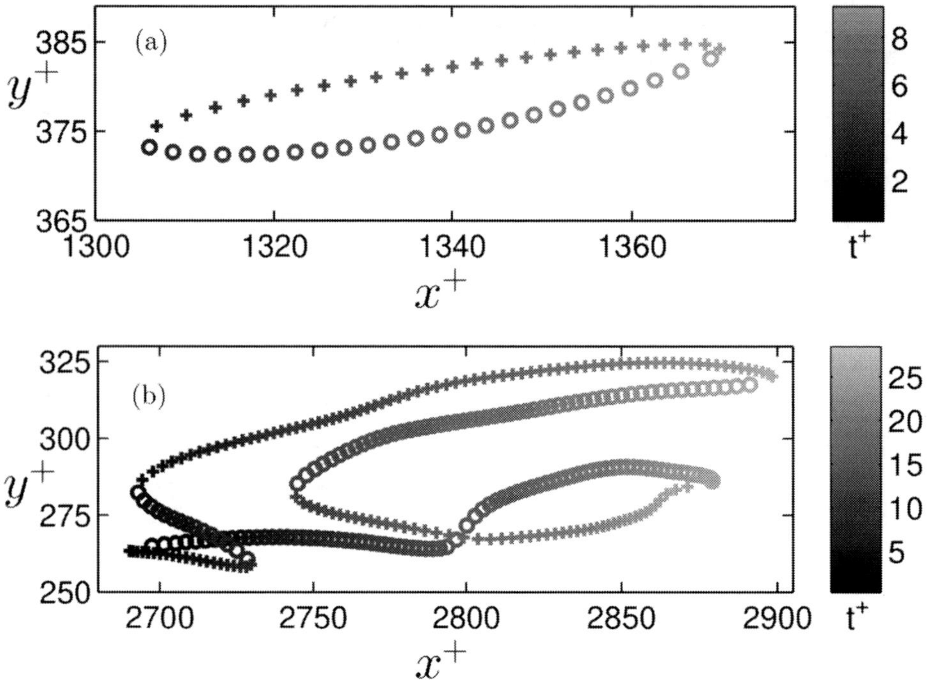

Figure 4. Examples of patterns in CP pair lives. Symbols represent nodes (○) and saddles (+). The time gap between two consecutive points in the series is 0.3 wall time units. (a) A single saddle/node pair. (b) Three saddle/node pairs with overlapping lives.

4.1. Lifetimes

Once we have computed the CP positions at each snapshot of the series M950, we still need to obtain for each CP the list of locations it occupied during its life. This involves inferring if a CP at time t is the same as a CP already detected at time $t - \Delta t$ earlier on. To carry out this task, a post-processing routine made several assumptions in order to establish if two CPs at different instants are actually the same CP. These assumptions are listed below, where $\Delta t = 0.3^+$:

- At time t, a CP is of the same type (saddle or node) as at $t + \Delta t$.
- At $t + \Delta t$, a CP's new position has to be within a radius of up to 15^+ from its position at time t.
- At $t + \Delta t$, a CP's new streamwise position can be at most 3^+ upstream from its position at time t.

- If more than one CP at $t + \Delta t$ meets the above conditions to be the successor of a CP detected at time t, then the closest CP at $t + \Delta t$ to the CP at time t is chosen to be the valid successor.

The data sampling rate is not sufficient to infer any lifetimes shorter than 0.3^+. With all these constraints in mind, the probability density function (PDF) of the CP lifetimes looks as that shown on figure 5. The mean lifetime of 6.6^+ appears to be representative of events similar to those on figure 4(a), while lifetimes as long as 25^+ are, although uncommon, occurring every now and then – see for example figure 4(b). In computing the PDF on figure 5, we took care not to include the CPs born before the data series' first 60^+ wall time units or during the last 60^+. These edges in time could have introduced CPs which were short lived as a consequence of being already born when the series started, or which disappeared not because of natural death but because of the data series' end.

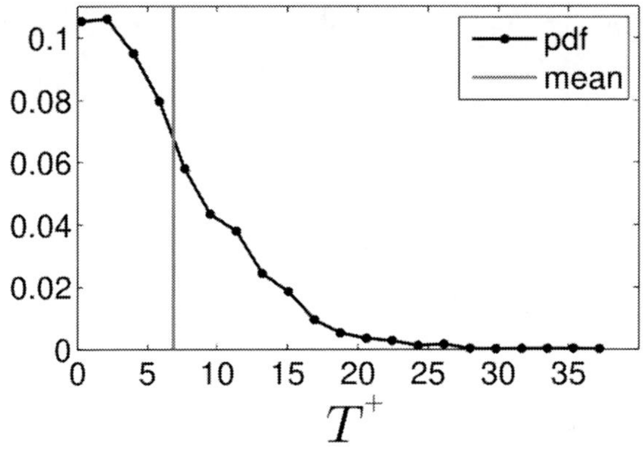

Figure 5. PDF of the CP lifetimes. The computed mean is 6.6^+.

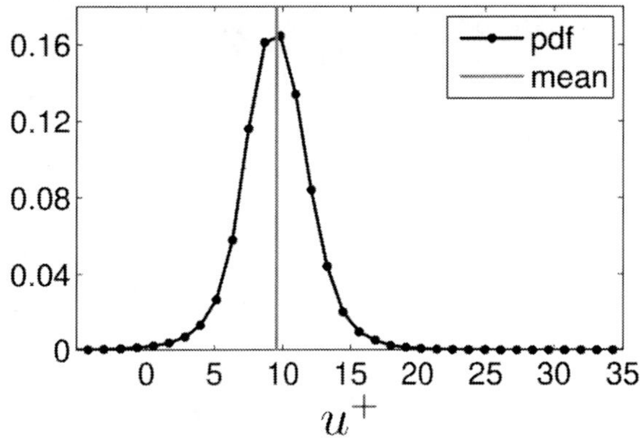

Figure 6. PDF of the streamwise velocity of CPs. The computed mean is 9.5^+.

4.2. Convection velocity

The velocity at which the CPs move along the wall is a question we analysed based on the post-processed data of M950. The displacements measured from snapshot to snapshot lead to velocity estimates which are an average over the time lapse of $\Delta t = 0.3^+$. This way of measuring CP velocities is not ideal, but it yields nevertheless a range of velocities as that seen on figure 6. As an *a posteriori* justification of the maximum displacements assumed in section 4.1, we argue that the PDF does not exhibit abrupt cuts at its edges. On the contrary, the PDF's tails tend smoothly towards zero, hinting at the fact that no particle displacement reached the maxima allowed by the post-processing routine – see section 4.1. Moreover, the likelihood of a slightly negative streamwise velocity appears to be small but non-vanishing, as confirmed by some CP trajectories of the kind displayed on figure 4 but not shown here. The mean streamwise velocity of 9.5^+ corresponds to the flow's mean velocity at $z^+ \approx 9$. The CPs are therefore advecting – in the mean – at the same velocity as the structures in the buffer layer. The mean spanwise velocity measured was 0.017^+ – which will be considered as zero – and the mean total velocity was 9.8^+.

4.3. Average distance between CPs and CP density

We now turn our attention to the PDF of the distance between a CP and all other CPs present at the wall. In computing this, we tried to avoid obtaining distance histograms which are biased. Particular problems arise when, for example, a CP is located at the corner of the channel's rectangular wall plane. Short distances are only available within a quarter of a circle around such a point, as opposed to within a full circle if the point was located in the middle of the wall. On the other hand, a point located in the middle of the wall can be at a maximum distance from another CP of half a diagonal length, whereas a point in a corner can be spaced by a full diagonal from another one. In order to remove the disparity in possibilities depending on where the point is located, we proceeded as follows. Taking the wall dimensions to be $L_x \times L_y$ in the streamwise and spanwise directions, respectively, we constructed a rectangle of length $L_x - L_y/2$ and width $L_y/2$ with its lower-left corner located at $(L_y/4; L_y/4)$ – see figure 7. Only the CPs located inside the inner rectangle were used to compute the distances *from*, while all points within a radius of $L_y/4$ from such a CP were used to compute the distances *to* – this includes distances to points outside of the inner rectangle. In this manner, any distance away from the reference CP is equally likely within that radius. The price paid is that no distance above $L_y/4$ is measured. We discarded the repeated distances which inevitably arise when following this procedure.

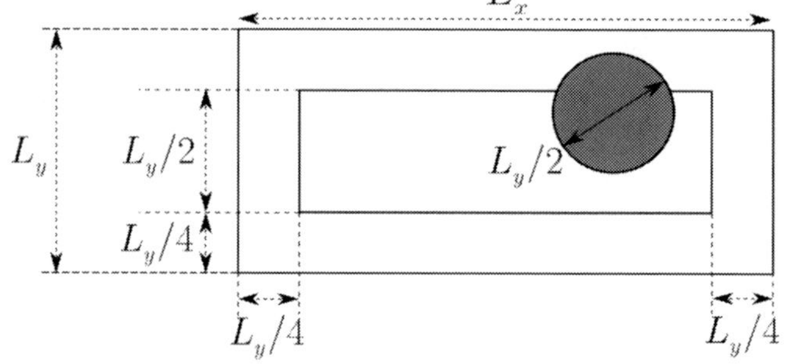

Figure 7. Sketch illustrating which distances were measured within a wall plane. At each CP inside the inner rectangle, a circle of radius $L_y/4$ centred at the CP gave the region within which to measure distances to other CPs. The CPs inside both rectangles were used to measure distances *to*, but only the CPs inside the inner rectangle were used to measure distances *from*.

The PDF of the distance between a CP and all other CPs within the allowed radius is shown on figure 8. The fact that the S1900 curve is higher than that for M950 is entirely due to the difference in the domain size. For S1900, $L_y^+/4 = 360$, while for M950 $L_y^+/4 = 734$. This means the range of distances in M950 extends for longer. Hence, the normalisation in order to yield a total area of 1 under each curve varies, leaving the S1900 curve slightly higher than that of M950.

Two separate regimes are distinguished in the curves shown on figure 8. At lower spacing values Δ^+, the CP separation is governed by the typical distance between CPs which are somehow related, or which interact. In other words, the CPs that appear and/or disappear together, like those shown on figure 4. There is a peak in the distance likelihood at 20^+, which is consistent with what is observed on figure 4. If we call a cloud of interacting CPs a *cluster* of CPs, then the CP spacing within a cluster is given by the left-hand hump of the PDF in figure 8. The right-hand side of the curves, or the apparently never-ending increase in likelihood of distance, is governed by the inter-cluster spacing. This part can increase as much as the domain allows it to, and is perhaps less informative as a result.

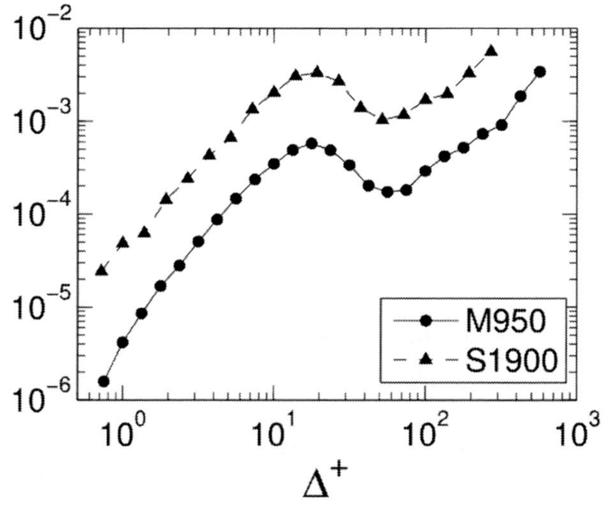

Figure 8. PDF of the distance between CPs at the two Reynolds numbers – see table 1.

Finally, the average CP density estimated using all CPs over the entire domain available yields 2 CPs per million wall square units. The same density was found at both Reynolds numbers. This leads to the estimate that within a wall plane of M950 and S1900, there are on average 34 and 8 CPs, respectively.

5. The flow field around critical points

Although critical points are spatially compact and only rarely found in the flow field, it will be shown in this section that they are associated with extreme events in the flow and a relatively large-scale surrounding flow feature.

5.1. Conditional shear stresses

Firstly, PDFs of the wall shear stress are shown in figure 9 normalised by their peak value. The unconditional PDFs show the well-known characteristics of viscous-scaled wall shear stress: the streamwise component has a mean value of 1 and is positively skewed while the spanwise component has a mean of zero and a lower standard deviation. The PDFs conditioned on the existence of a critical point were determined by analysing the wall shear stress in a circle of radius 20 wall units around a critical point. The PDF of the streamwise component shows that most of the shear stress fluctuations that occur around a critical point are at the extreme ends of the unconditional PDFs. In other words, the wall shear stress is at its extreme lowest around critical points. The variance is also much higher in the conditional sense, illustrating the extreme fluctuations about a critical point. The spanwise conditional average confirms this with a much larger variance (longer tails in the conditional PDF) than the overall flow.

The average skin friction field conditioned on the existence of a critical point at (0,0) is shown in figure 10. The contours are scaled with viscous-scaling so that the mean streamwise wall shear stress is unity. Near the critical point, the wall shear stress is considerably lower than the mean on average. Surrounding this low shear stress field, there are lobes of streamwise wall shear stress greater than the mean value. The pattern thus left is one of sign-alternating large spanwise wall shear stresses. This points to a converging-diverging behaviour in the flow near a critical point. Upstream and downstream of the critical point, the wall shear stress is also lower than the mean. Somewhat surprisingly, there is a signature in the flow downstream of the critical point persisting hundreds of wall units downstream of the event. The lower-than-average wall shear stress is accompanied by opposite-sign spanwise wall shear stress, suggesting

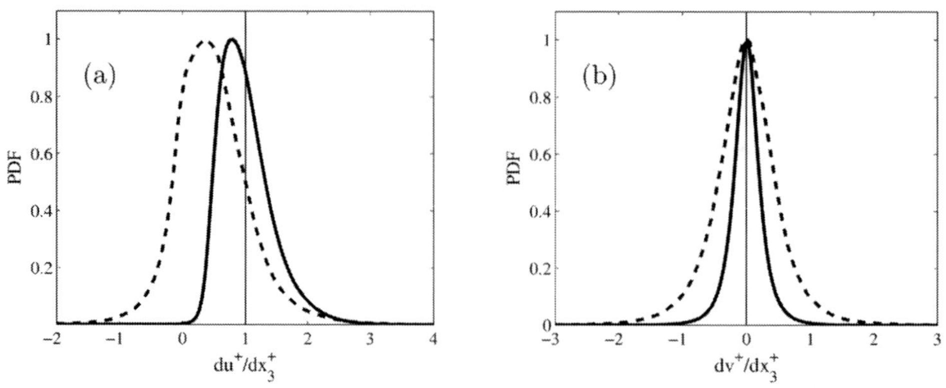

Figure 9. PDFs (solid) and conditional PDFs (dashed) of: (a) streamwise wall shear stress and (b) spanwise wall shear stress.

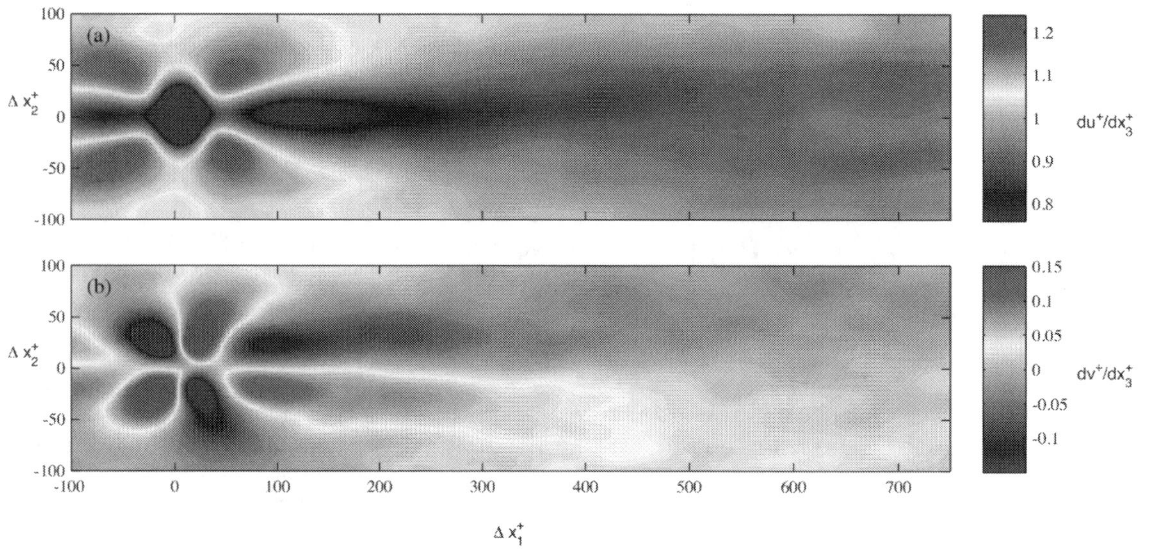

Figure 10. Conditionally averaged fields of wall shear stress: (a) Streamwise; (b) Spanwise wall shear stress.

a convergence of the flow downstream of the critical point. The extended coherence of the flow downstream of the critical point encourages an analysis of the flow above as discussed in the following sections.

5.2. Conditional velocity fields

In figure 11 a fluctuating streamwise velocity field is shown. The velocity fluctuations, u'_c are calculated by averaging velocity fluctuations (about the unconditional local mean streamwise velocity) above a critical point. Conditional averages for both Reynolds numbers are shown and there is little difference between them. Since the Reynolds number effect appears minimal, no further analysis of the high Reynolds number data will be included in this paper (the low Reynolds number simulation was selected for further analysis owing to the larger box size and availability of time-resolved data).

As seen in the conditionally averaged wall shear stress plots, the streamwise coherence

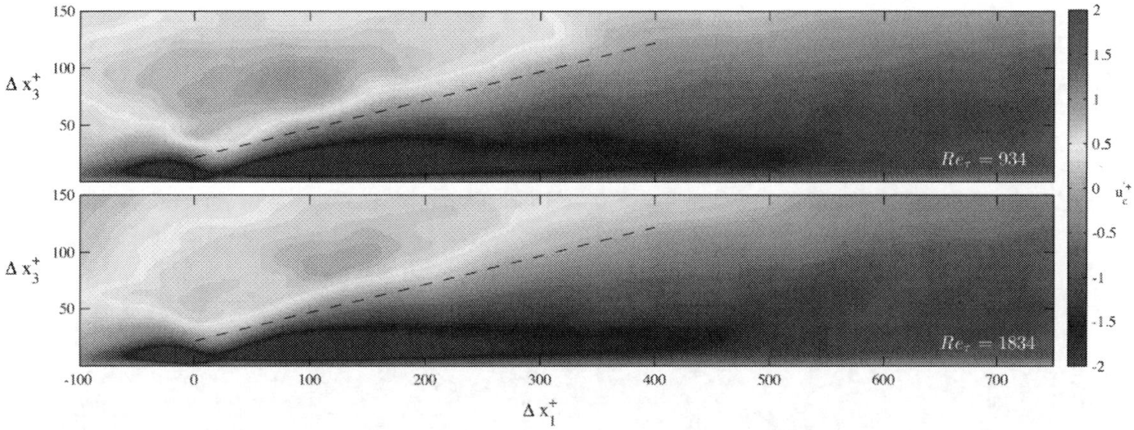

Figure 11. Conditionally averaged streamwise velocity fields at low and high Reynolds numbers. The dashed line illustrates an angle of 14°.

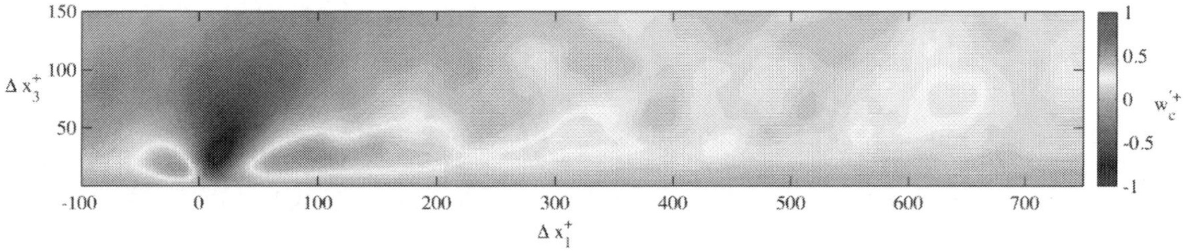

Figure 12. Conditionally averaged wall-normal velocity field at $Re_\tau = 934$.

persists hundreds of wall units downstream of the critical point. Above the wall, there is a large, low-speed region as might be expected above the large, low-shear-stress producing event. Interestingly, the wall-normal coherence also extends well into the logarithmic region. A high-speed region centred at the base of the logarithmic region was found above and downstream of a critical point. The demarcation between the high-speed and low-speed regions is approximated by a line of angle 14° to the wall. This angle is similar to the well-document 'structure angle' of high Reynolds number wall-turbulence [8], providing the first evidence that critical points are somehow associated with large-scale structures.

Conditionally averaged wall-normal velocity fluctuations are shown in figure 12. Although the coherence is somewhat weaker, it is clear that there is an upward flow upstream of a critical point, followed by a downward flow downstream. The downward flow begins in the logarithmic region and intensifies as the critical point is approached. In the far field, there is a general, albeit weak, upward flow in the low streamwise velocity region downstream of the critical point.

The combination of the streamwise and wall-normal fluctuation behaviour near a critical point points to a spanwise vortical structure above a critical point. In figure 13 the fractional difference of spanwise vorticity, $\Delta\Omega_{yc}^+$ is shown. The fractional difference is defined as

$$\Delta\Omega_{yc}^+ = \frac{\langle\Omega_y\rangle_c^+ - \langle\Omega_y\rangle^+}{\langle\Omega_y\rangle^+}, \tag{10}$$

where $\langle\rangle$ represents the unconditional average and $\langle\rangle_c$, the conditional average. The increase in spanwise vorticity is both large (>100% of the unconditional mean spanwise vorticity) and extended over a long distance in both streamwise and wall-normal directions. Clearly there is enhanced vorticity directly above the critical point. Ref. [2] also found a vortex directly above regions of negative streamwise wall shear stress. This is perhaps not too surprising since there *must* be a region of negative wall shear stress in the near vicinity of a critical point, however, the converse is not true (there need not be a critical point near all regions of negative streamwise wall shear stress). As might also have been expected from the conditional averages of streamwise velocity, there exists a shear layer of intense vorticity downstream of the critical point, extending into the logarithmic region and in excess of 700 wall units downstream.

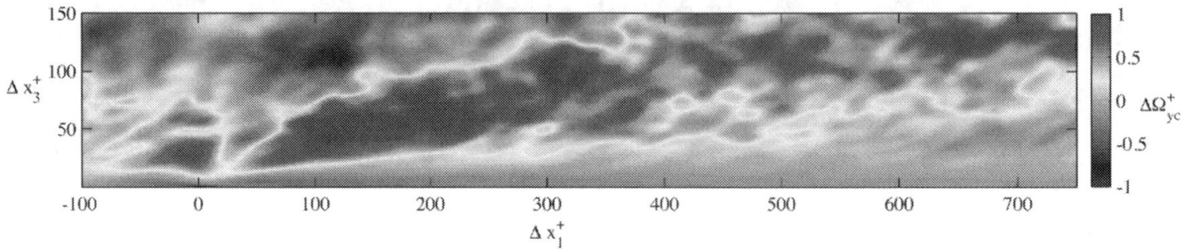

Figure 13. Conditionally averaged spanwise vorticity field at $Re_\tau = 934$.

Finally, the decomposition of the Reynolds shear stress into conditional Q2 ($\overline{uw} < 0, w > 0$) and Q4 ($\overline{uw} < 0, w < 0$) is shown in figure 14. Note that the contours shown display the Reynolds shear stress component as a fraction of the unconditional Reynolds shear stress (in the manner of equation 10). There are clearly strong ejection events immediately upstream and far downstream of the critical point, even hundreds of wall units downstream. Sweep events are less pronounced, except immediately downstream and above the critical point. In fact, the sweep events can occur high in the logarithmic region above the critical point (corresponding with the high-streamwise-velocity region and the negative-wall-normal-velocity region).

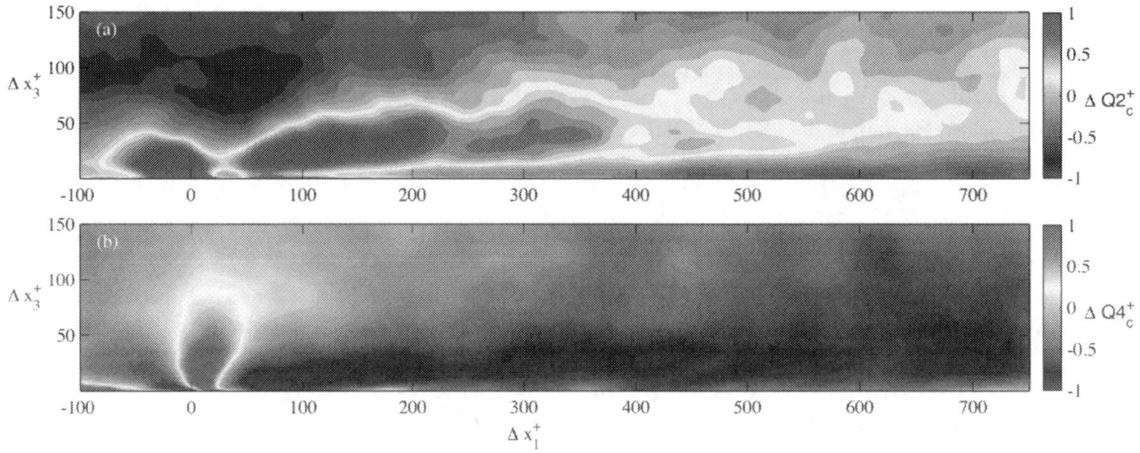

Figure 14. Conditionally averaged (a) Q2 and (b) Q4 events at $Re_\tau = 934$.

5.3. Vorticity transport

For turbulent channel flow, the Reynolds-averaged mean momentum balance results in

$$\frac{\partial^2 \overline{U}^+}{\partial x_3^{+2}} + \frac{1}{Re_\tau} - \frac{\partial \overline{uw}^+}{\partial x_3^+} = 0. \tag{11}$$

It can be shown that the third term is simply the streamwise component of the Lamb vector:

$$-\frac{\partial \overline{uw}}{\partial x_3} = \overline{w\Omega_y} - \overline{v\Omega_z}. \tag{12}$$

The right-hand side of this equation can be interpreted as the difference between velocity-vorticity correlations. The first term represents the wall-normal transport of spanwise vorticity. A negative value leads to an increased momentum source, as can be seen from equation 11. Figure 15(a) shows the transport of spanwise vorticity in the vicinity of a critical point. Also shown in figure 15(b) is the spanwise vorticity fluctuation about the unconditional mean, with the blue region highlighting the vortex above a critical point and inclined shear layer downstream. The sign of vorticity associated with this vortex is negative such that the vortex is 'prograde' ($\Omega_y < 0$). The vortex above the critical point is associated with regions of positive *and* negative transport of vorticity. However, the average transport, $\langle \overline{w\Omega_y} \rangle_c$, over the whole vortex is positive, indicating a tendency for the vortex to be transported toward the wall ($w < 0, \Omega_y < 0$) thus acting as a momentum source.

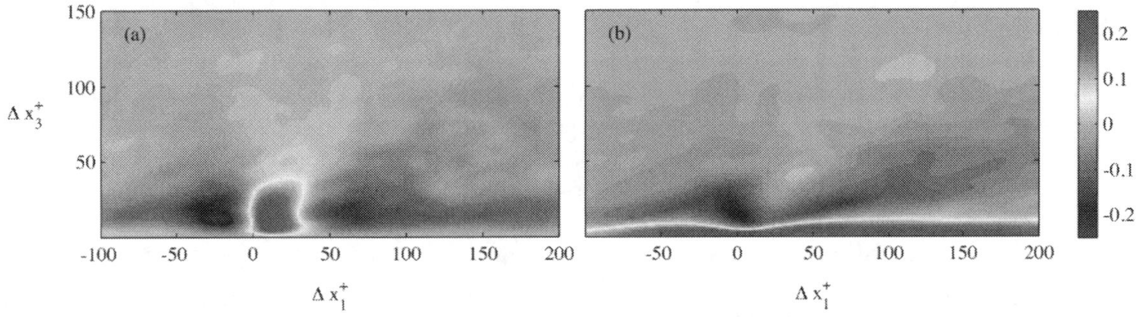

Figure 15. (a) Conditionally averaged Velocity-Vorticity correlation, $\langle w\Omega_y \rangle_c^+$ at $Re_\tau = 934$. (b) Conditional spanwise vorticity, Ω_{yc}^+ relative to the unconditional local mean spanwise vorticity. Color key applies to both subfigures.

6. Time sequence

The vorticity transport deduced in the previous section was also observed by viewing movies of the flow above a critical point. Although this information is difficult to convey in static images, an attempt is made in figure 16, where contours of spanwise vorticity are plotted in the $x_1 - x_3$ plane around a critical point. The time sequence is separated by $t^+ = 5$ and the abscissa is relative to an observer moving with the flow at a velocity of $U_c/u_\tau = 10$. A shear layer exists initially, even before the critical point pair is created, with a notable vortex centred at approximately $x_3^+ = 20$. Shortly after (but before $\Delta t^+ = 5$) the critical point pair is born as the noted vortex travels toward the wall. At $\Delta t^+ = 10$ the vortex has moved even closer to the wall and intensified significantly. Although not shown for brevity, the wall shear stress around the critical points similarly intensifies as the vortex moves closer to the wall.

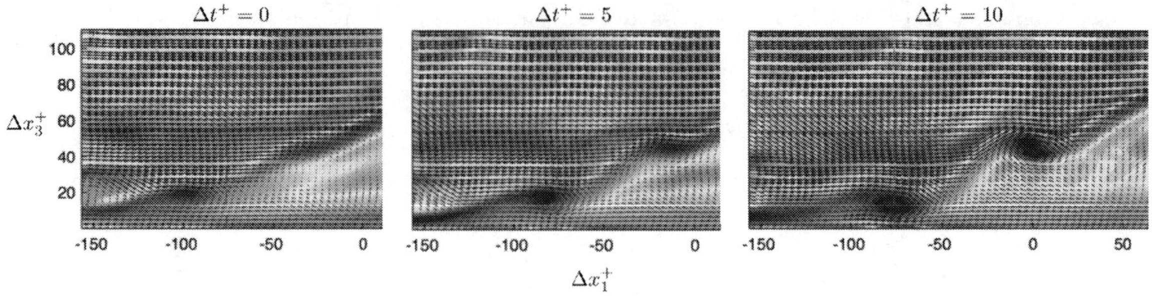

Figure 16. Three snapshots illustrating the evolution of the shear layer above a critical point. A critical point pair was created between $\Delta t^+ = 0$ and 5. The location of the critical point is roughly indicated by the red dashed line. Contours show spanwise vorticity and vectors are instantaneous velocity vectors in the streamwise-wall-normal plane.

The overall picture, then, is that a critical point marks the tail end of some large-scale structure (not necessarily all large-scale structures). The critical point is formed when a negatively signed (prograde) vortex at the tail end of the large-scale structure is transported toward the wall. As the vortex moves closer to the wall, it also intensifies, causing larger fluctuations in all flow variables on the wall and above it. Eventually the vortex decays until the critical points below it come together and die.

7. Summary and conclusions

There are locations on the wall of wall-bounded turbulent flows where the skin friction vector has zero magnitude. These locations are called critical points. This study employed the time-resolved data from two relatively high Reynolds number channel flow simulations to answer some important questions regarding the statistical nature and the life of critical points.

It was found that critical points are always born in saddle-node pairs at the same spatial location. During their life they may encounter other saddle-node pairs, possibly swapping partners, before the saddles and nodes come together and die collocated in space (as they were born). The average lifetime of the critical points is short at $t^+ = 6.6$, however, some critical points can last more than 30 viscous time units. The critical points are typically spaced by 20 viscous wall units and convect at around $U_c/u_\tau = 9.5$. The number of critical points per unit area appears to be fixed with Reynolds number such that there is, on average, one pair of critical points per million square wall units.

Conditional averages confirm the extreme nature of critical points. Moreover, the critical points appear to be at the tail of a large-scale structure that extends from the wall to the outer edge of the logarithmic region. In the streamwise direction, a low-streamwise-velocity region extends over 800 wall units downstream of a critical point. This low-speed region is bounded by a shear layer and is the site of ejection events. There is also a strong spanwise vortex directly above the critical point which is transported toward the wall on average. It is postulated that critical points are created when such a vortex, which is at the tail of a larger collection of vortices, gets pushed sufficiently close to the wall.

Somewhat surprisingly, no Reynolds number dependence was found in the large-scale coherence of the flow around a critical point. This despite the fixed concentration of critical points with Reynolds number. This suggests that only some of the large-scale structures in the flow are tailed by critical points. Understanding the connection between large-scales and the critical points generated will be the focus of ongoing study.

Acknowledgments
Funded in part by the Multiflow program of the European Research Council. Dr Monty acknowledges the support of the Future Fellowship program of the Australian Research Council.

References

[1] Chong MS, Monty JP, Chin C and Marusic I 2012 The topology of skin friction and surface vorticity fields in wall-bounded flows *J. Turb.* **13** 6, 1-10
[2] Lenaers P, Li Q, Brethouwer G, Schlatter P and Örlü R 2012 Rare backflow and extreme wall-normal velocity fluctuations in near-wall turbulence *Phys. Fluids* **24** 035110
[3] Chong MS, Perry AE and Cantwell BJ 1990 A general classification of three-dimensional flow fields *Phys. Fluids* **2** 765-77
[4] Perry AE and Chong MS 1987 A description of eddying motions and flow patterns using critical-point concepts *Annu. Rev. Fluid Mech.* **19** 125-55
[5] Kim J, Moin P and Moser R 1987 Turbulence statistics in fully developed channel flow at low Reynolds number *J. Fluid Mech.* **177** 133-66
[6] Flores O and Jiménez J 2010 Hierarchy of minimal flow units in the logarithmic layer *Phys. Fluids* **22** 071704
[7] del Álamo JC, Jiménez J, Zandonade P and Moser RD 2006 Self-similar vortex clusters in the turbulent logarithmic region *J. Fluid Mech.* **561** 329-58
[8] Marusic I and Heuer WDC 2007 Reynolds number invariance of the structure angle in wall-turbulence *Phys. Rev. Lett.* **99** 114504

Journal of Physics: Conference Series **506** (2014) 012010 doi:10.1088/1742-6596/506/1/012010

Turbulent pipe flow: Statistics, *Re*-dependence, structures and similarities with channel and boundary layer flows

George K El Khoury, Philipp Schlatter, Geert Brethouwer and Arne V Johansson

Linné FLOW Centre, KTH Mechanics, Royal Institute of Technology, SE-100 44, Stockholm, Sweden

E-mail: johansson@mech.kth.se

Abstract. Direct numerical simulation data of fully developed turbulent pipe flow are extensively compared with those of turbulent channel flow and zero-pressure-gradient boundary layer flow for Re_τ up to 1 000. In the near-wall region, a high degree of similarity is observed in the three flow cases in terms of one-point statistics, probability density functions of the wall-shear stress and pressure, spectra, Reynolds stress budgets and advection velocity of the turbulent structures. This supports the notion that the near-wall region is universal for pipe and channel flow. Probability density functions of the wall shear stress, streamwise turbulence intensities, one-dimensional spanwise/azimuthal spectra of the streamwise velocity and Reynolds-stress budgets are very similar near the wall in the three flow cases, suggesting that the three canonical wall-bounded flows share many features. In the wake region, the mean streamwise velocity and Reynolds stress budgets show some expected differences.

1. Introduction

The study of the flow of a viscous fluid along the surface of a solid body and its interaction with the wall, is among the most fundamental problems in fluid mechanics. In wall-bounded flows, there are three simple geometrical configurations which are referred to as canonical flows: the spatially evolving boundary layer, the channel and the pipe. The difference between these cases comes not only from the presence of curvature at the wall in pipe flow, but also from the spatial development of boundary layer flow and the mutual interaction between the internal viscous region and the external inviscid flow leading to intermittency effects in the latter case. Moreover, in the outer region the large turbulence scales are strongly confined in pipe flow as opposed to channel and boundary layer flow.

Some of the earliest research in the field of turbulent wall-bounded flows was purely theoretical, and a key point was the pioneering work of Prandtl who introduced concept of boundary layers in 1904. Another major breakthrough was the formulation of the law of the wall by von Kármán in the 1930s. According to the law of the wall, the wall region is divided into a viscous sublayer attaining a linear profile, a logarithmic-law region with a buffer layer in between, and finally an outer layer. Ever since the formulation of the preceding theories, there has been a debate about, on the one hand, the universality of turbulence and, on the other hand, the differences between the canonical wall-bounded flows. At the early stages, some of the controversies were about the effect of curvature in pipe flow on the near-wall region. Millikan (1938) derived the logarithmic law of the wall for pipe flow under the assumptions that the velocities close to the wall and in the core region are independent of the radius of curvature and viscosity, respectively. One of the first direct experimental comparisons between pipe and channel flows dates to Patel and Head (1969) who reported discrepancies in the mean velocity profile between the two cases and attributed them to the difference in geometrical configuration. Several years later, Huffman and Bradshaw (1972) examined the von Kármán constant by analysing data for pipe and channel flow, and stated that the transverse curvature in pipe flow may affect the viscous sublayer but not the rest of the inner layer.

It is hardly surprising that a clear understanding of the physical processes in wall-bounded flows could not be converted into a universal model for canonical flows for more than half a century. A point of dispute, which appeared from research on wall-bounded canonical flows, is the maximum of the streamwise/axial turbulence intensities. Channel flow and boundary layers show an increase in magnitude with increasing Re (Marusic et al., 2010). This is attributed to the rising influence of the outer layer as higher Reynolds numbers are approached. However, the "Superpipe" experimental investigations by Zagarola and Smits (1997) indicated that this peak holds a constant value with increasing Re. There remain many other controversies in this context as, for instance, the scaling laws and the presence of a logarithmic law or a power law for pipe, channel and boundary layer flows; see *e.g.* Zagarola and Smits (1998), McKeon et al. (2004) and George (2007). Kim (2012) reviewed the progress in pipe and channel-flow turbulence research over a fifty years period, and highlighted many of the still remaining questions. Other recent reviews of wall-bounded flows are given by Marusic et al. (2010); Smits et al. (2011); Jiménez (2012).

The first comparison between pipe and channel flows from a direct numerical simulation (DNS) study, came to light with the work of Eggels et al. (1994) who reported on low- and high-order statistics. At a friction Reynolds number of 180, Eggels et al. (1994) stated that the mean velocity profile in pipes does not follow the law of the wall, in contrast to channels. In addition, the skewness factor of the wall-normal velocity showed differences while other statistical quantities, including turbulence intensities and energy budgets, showed insignificant differences. Jiménez et al. (2010) conducted comparisons between turbulent boundary layers and channels at moderate Reynolds numbers to study the behaviour of the velocity and pressure fluctuations in the outer layer. On the experimental side, Monty et al. (2007) investigated the presence of large-scale structures in turbulent pipe and channel flows, and compared it with boundary layers. In this work, the spanwise two-point correlations showed that the large-scale structures are significantly wider in pipe and channel flows than in boundary layer flow. Another profound difference, reported in the same work, was about the rate of growth of these structures with increasing wall distance. It was noted that they grow at a slower rate with wall distance in boundary layer flow than in pipe and channel flows. In a subsequent attempt to clarify more in detail the extent of similarity between the three canonical flow cases, Monty et al. (2009) conducted streamwise velocity measurements on pipe, channel and boundary layer flows at the same Reynolds number and with the same measurement resolution. The key point in their findings was that the mean statistics, in terms of mean velocity profile and streamwise/axial turbulence intensity, are similar in the near-wall region in the three cases, and the one-dimensional premultiplied spectra yielded identical results between pipe and channel flows throughout the flow, but was still different from that in boundary layers even close to the wall. Sillero et al. (2013) compared one-point statistics of boundary layer DNSs at friction Reynolds numbers up to 2 000 with experimental and DNS data of pipe and channel flow. They found that the three normal Reynolds stresses are higher in boundary layers than in pipe and channel flow in the outer layer. These differences are independent of the Reynolds number, and were also observed by Lee and Sung (2013).

The present study aims, through highly accurate well-resolved DNS data, to concretely prove for the first time that pipe, channel and turbulent boundary layer flows are practically indistinguishable in the near-wall region. In particular, we focus our attention on the near wall region up to a wall-normal position of 100 wall units, highlighting the similarities and differences, if any, between these canonical wall-bounded flows using various statistical tools, including probability density functions, one- and two-dimensional spectral analysis and energy budgets, as well as advection velocities of the turbulent structures.

2. Data sets and numerical methods

For the pipe flow simulations in this study, we consider a straight circular pipe of radius $R = 1$ and length $25R$, with the pipe axis taken along the axial z direction. The incompressible Navier–Stokes equations are solved using the well-documented DNS solver `nek5000` (Fischer et al., 2008). `Nek5000` is a spectral element code in which the Navier–Stokes equations are solved using a Legendre polynomial based spectral element method (SEM), and are advanced in time with a semi-implicit scheme. The radial, azimuthal and axial grid spacing, measured in wall units, is set such that $\Delta r^+_{max} \le 5$ and $\Delta R\theta^+_{max} \le 5$ and $\Delta z^+_{max} \le 10$, respectively. A more detailed description, on the computational meshes and the usage of `nek5000` in the current pipe DNSs, can be found in El Khoury et al. (2013).

The other data sets that we consider are from well-resolved DNS of channel flow (del Álamo et al., 2004; Lenaers et al., 2012) and zero-pressure gradient boundary layer flow (Schlatter and Örlü, 2010). All these simulations were carried out with spectral codes employing Chebyshev polynomials in the wall-normal direction.

The friction Reynolds number for pipe flow is defined as $Re_\tau = u_\tau R/\nu$, where u_τ is the friction velocity and ν is the kinematic viscosity, whereas for channel and boundary flows layer Re_τ is based on the channel half-width and the 99% boundary layer thickness, respectively. From now on, r and y refer to the wall-normal direction, θ to the azimuthal or spanwise direction and z to the axial or streamwise direction.

3. Results

3.1. Axial/streamwise velocity statistics and visualisations

Figure 1 shows visualisations of the instantaneous axial or streamwise velocity field in a cross-stream plane of the pipe, channel and boundary layer DNSs at $Re_\tau \approx 1\,000$. We can observe the small-scale near-wall structures corresponding to the streamwise low- and high-speed streaks (Kline et al., 1967) as well as the large-scale turbulent structures in the outer regions. In particular, the visualisations illustrate the geometrical confinement of these large scales in the centre region of a pipe, while in channels and boundary layers there is no geometrical confinement in the spanwise direction. However, one should note that the periodic boundary conditions puts an upper limit on the size of the structures.

Figure 2(a) presents the mean axial or streamwise velocity profile obtained from the pipe, channel and boundary layer DNSs at $Re_\tau \approx 1\,000$. The mean velocity profiles in terms of wall units are very similar in the near-wall region and logarithmic-layer region but differ in the wake region. Here the boundary layer has the highest and the channel flow the lowest mean velocity. Monty et al. (2009) observed the same differences in the wake region in experiments of all three flow cases at a higher Reynolds number. The difference between pipe/channel flow and boundary layers is likely related to the intermittency caused by entrainment of irrotational fluid in the latter case (Jiménez et al., 2010), but there is no clear explanation for the difference between pipe and channel flow in the wake region.

Figure 2(b) shows the maximum of the root-mean-squared (rms) streamwise velocity fluctuations normalized by u_τ, $u^+_{z,rms}$, found in the buffer layer for pipe, channel and boundary layer flow. The estimated confidence intervals for our pipe flow DNS results and several other

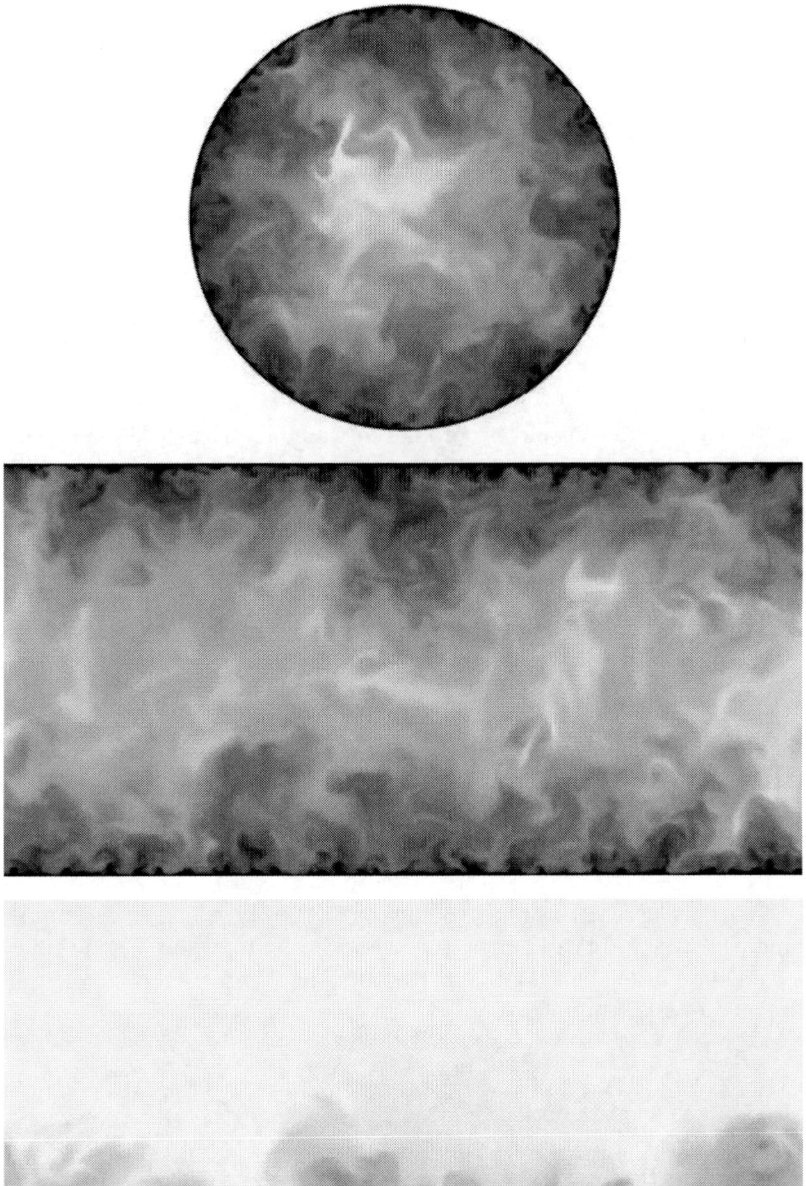

Figure 1. Visualisations of the instantaneous streamwise velocity field in a cross-stream plane in pipe (top figure), channel (middle figure) and boundary layer (bottom figure) at $Re_\tau \approx 1\,000$. The visualised spanwise extent of the channel flow plane is $3.85h$, where h is the channel half-width. The spanwise and wall-normal (vertical) extent of the boundary layer plane are 3.85δ and 2δ, respectively, where δ is the 99% boundary layer thickness. A lighter colour corresponds to a higher velocity.

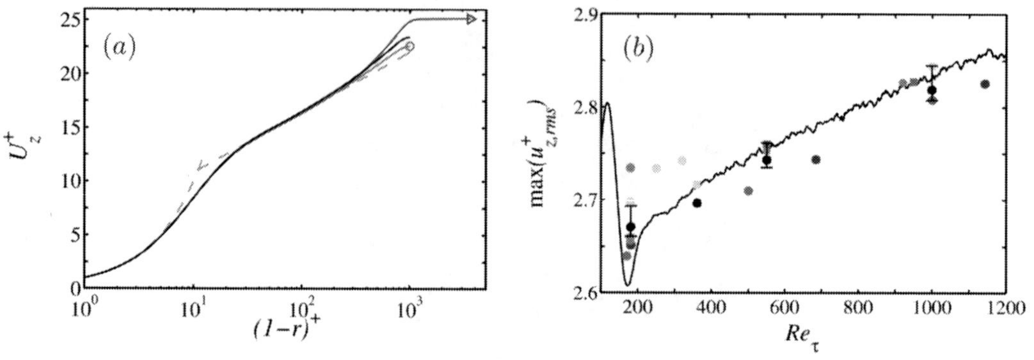

Figure 2. (a) Axial/streamwise mean velocity profiles in wall units in pipe (——), channel (——○) and boundary layer flow (——▷) at $Re_\tau = 1\,000$. Dashed lines indicate the linear and logarithm behaviours. (b) Maximum of $u_{z,rms}^+$ as function of Re_τ. Pipe flow: •, El Khoury et al. (2013); •, Chin et al. (2010), Klewicki et al. (2012); •, Boersma (2011); •, Wu and Moin (2008), Wu et al. (2012); •, Wagner et al. (2001); •, Eggels et al. (1994). Channel flow: ■, Lenaers et al. (2012); ■, del Álamo et al. (2004). Boundary layer flow: ——, Schlatter and Örlü (2010). The error bars indicate the statistical uncertainty in our pipe flow DNS results.

DNS data of pipe and channel flow are included in the figure as well. We can observe a similar increase of the maximum of $u_{z,rms}^+$ with Re_τ in all three cases in our DNS data, in agreement with the observations by Sillero et al. (2013). We also note that there is no statistically significant difference in the maximum $u_{z,rms}^+$ in the three flow cases at $Re_\tau = 550$ and $1\,000$ according to our DNS. However, the pipe flow DNS results of Chin et al. (2010); Klewicki et al. (2012); Wu and Moin (2008); Wu et al. (2012) appear to have a bit lower $u_{z,rms}^+$-maximum than channel flow and the boundary layer. The different pipe-flow DNS data show a significant scatter at low Re_τ, which could be caused by insufficient resolutions or statistical errors related to short sampling times in some of the simulations.

3.2. Wall region

We compare the canonical wall-bounded flows by looking at the wall region in more detail. The probability density function of the wall-shear stress PDF(τ_z^+) is shown in figure 3 for pipe, channel and boundary-layer flows for a wide range of Reynolds numbers. Such functions highlight the occurrences associated with instantaneous fields, and the importance of looking at the wall-shear stress stems from its direct connection to the drag of an immersed body. The characteristic behaviour of PDF(τ_z^+) is observed to be the same between the canonical flows. At approximately the same Reynolds number, the probability density functions of τ_z^+ collapse well, irrespective of the geometrical configuration, and indicate the presence of negative wall-shear stresses. This negative wall-shear stress is associated with negative axial/streamwise velocities, and is observed to occur more frequently and to increase in magnitude with increasing Reynolds number, as evidenced by the widening of the tail of PDF(τ_z^+). In this context, Lenaers et al. (2012) investigated in detail the presence of negative streamwise velocities in channel flow up to $Re_\tau = 1\,000$, and found that the widening of the negative tail of PDF(τ_z^+) with Re was solely due to the rise in the number of regions of backflow which descended from a strong oblique vortex outside the viscous sublayer. Very large positive values of τ_z^+ are also seen. These are likely caused by structures coming from outside the buffer layer that reach down into the viscous sublayer, see also figure 3 in Lenaers et al. (2012).

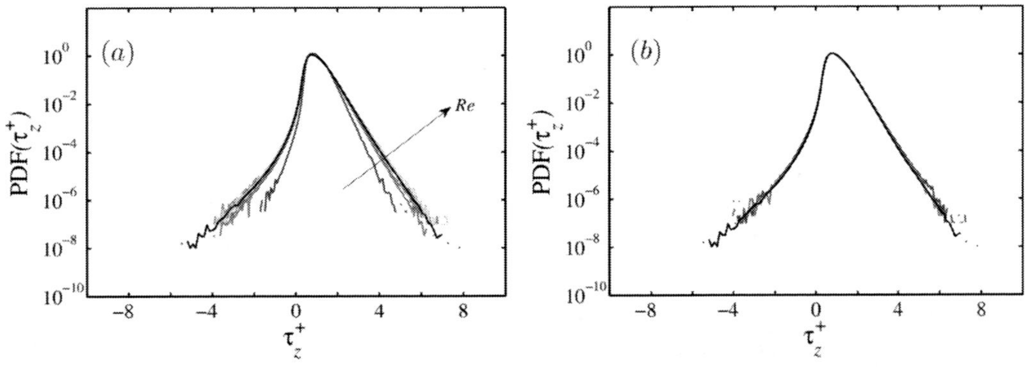

Figure 3. Probability density function (PDF) of wall-shear stress in wall units. Pipe by El Khoury et al. (2013); channel by Lenaers et al. (2012); boundary layer by Schlatter and Örlü (2010). (*a*) PDFs of pipe flows at $Re_\tau = 180$ (——), 550, 1 000 and of boundary layers at $Re_\tau = 400, 880, 1\,200$. The arrow indicates increasing Re_τ. (*b*) PDFs of pipe: ——, $Re_\tau = 1\,000$; channel: ——, $Re_\tau = 1\,000$; boundary layer: ——, $Re_\tau = 1\,200$.

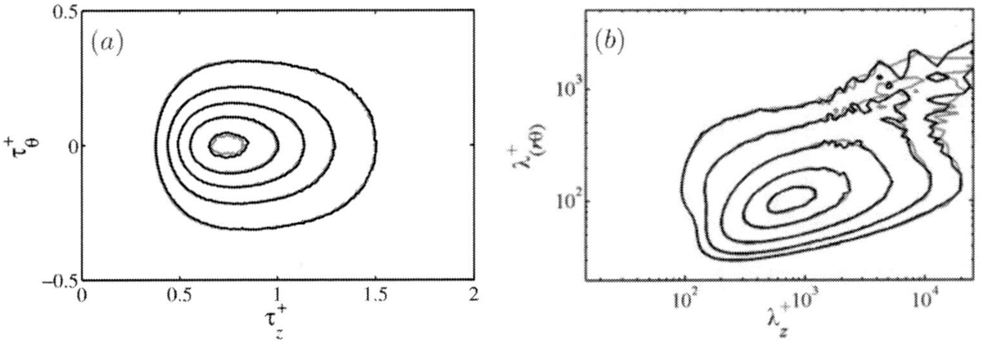

Figure 4. Pipe and channel flows at $Re_\tau = 1\,000$. (*a*) Joint probability density function of wall-shear stresses in the axial/streamwise (τ_z^+) and azimuthal/spanwise (τ_θ^+) directions. (*b*) 2*D*-premultiplied spectra of wall-shear stress $k_{(r\theta)}k_z\Phi_{(\tau_w\tau_w)}/(u_\tau^2/\nu)^2$. Pipe: —— (El Khoury et al., 2013); channel: —— (Lenaers et al., 2012).

The axial/streamwise wall-shear stress (τ_z^+) is generated by means of azimuthal vorticity impinging on the wall. There exist another wall-shear stress component linked with axial vorticity and that is the azimuthal wall-shear stress (τ_θ^+). The joint PDFs of these two wall-shear stresses, shown in figure 4(*a*) for pipe and channel flows at $Re_\tau = 1\,000$, are practically indistinguishable as well as the two-dimensional premultiplied spectra of τ_z for these two cases, see figure 4(*b*). Note that the imprint of large-scale turbulent structures on the wall shear stress in pipe and channel flow is visible in these spectra. A similar large-scale imprint near the wall was also observed in two-dimensional spectra of the streamwise velocity and enstrophy in channel flow at $Re_\tau = 2\,000$ by Hoyas and Jiménez (2008). The PDF of the pressure at the wall for pipe flow at $Re_\tau = 180$, 550 and 1 000 and channel flow at $Re_\tau = 1\,000$ are shown in figure 5. The intermittency of the pressure at the wall in pipe flow becomes stronger with increasing Re_τ, similar to the wall shear stress. Also in this case, the PDFs of pipe and channel flow are extremely similar.

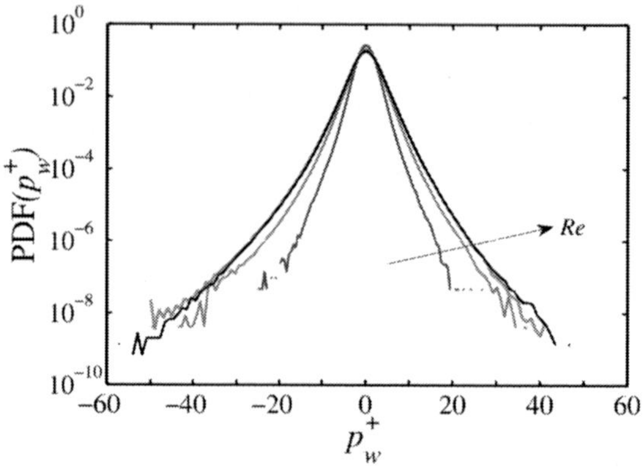

Figure 5. Probability density function of wall pressure in pipes (El Khoury et al., 2013) at $Re_\tau = 180$ (——), 550 (——) and 1 000 (——), and a channel (Lenaers et al., 2012) at $Re_\tau = 1\,000$ (——).

El Khoury et al. (2013) considered the development of the wall pressure fluctuations and wall-shear stresses with the Reynolds number for pipe, channel and boundary layer flows. In their results, there was a clear Reynolds-number dependence of $p^+_{w,rms}$, $\tau^+_{z,rms}$ and $\tau^+_{\theta,rms}$ for the three flow cases, being similar for pipe and channel flows, but lower than those in boundary layers. While the discrepancy in pressure fluctuations remained the same for boundary layer and pipe/channel flows as the Reynolds number increased, it was observed that the discrepancies in wall-shear stresses for the canonical flows tend to diminish with increasing Re. Therefore, it can be expected $\tau^+_{z,rms}$ and $\tau^+_{\theta,rms}$ would be the same for pipe, channel and boundary layer flows at sufficiently high Re. The higher pressure fluctuations in boundary layer flow, on the other hand, are linked to the intermittency of the potential and rotational flow regions for this configuration (Jiménez et al., 2010), and the difference to channels is thought not to disappear with Reynolds number.

3.3. Buffer layer

The buffer layer separates the viscous sublayer from the logarithmic-law region and sustains the presence of streaks responsible for the peak in axial/streamwise turbulence intensity and turbulence production. This layer extends between wall normal positions $y^+ = 5$ and $y^+ = 30$. In figure 6, the two-dimensional premultiplied spectra of the axial velocity and the co-spectra of Reynolds shear stress are shown for pipe and channel flows at $(1-r)^+ = 15$. There are no significant differences in the premultiplied spectra between the two flow cases. An inner peak is present in the two panels at axial/streamwise and azimuthal/spanwise wavelength of $\lambda_z^+ = 1\,000$ and $\lambda_{r\theta}^+ = 100$, respectively, which corresponds to the spacing of the above mentioned streaks (Kline et al., 1967). Meanwhile, an outer peak is observed at axial/streamwise and spanwise/azimuthal wavelengths of $10\,000^+$ and $1\,000^+$, respectively. This peak is the sign of the large-scale motions embedded in the logarithmic layer and is also visible in the two-dimensional premultiplied spectra of wall-shear stress in figure 4. El Khoury et al. (2013) reported that the inner peak of pressure fluctuations shows negligible differences between pipe and channel flows but is clearly lower than for the boundary layer. The outer peak is noticeably absent in the two-dimensional co-spectrum of the Reynolds shear stress (figure 6(b)) implying that the large-scale structures observed in figure 6(a) contribute little to momentum transfer and are in that sense inactive (Hoyas and Jiménez, 2008).

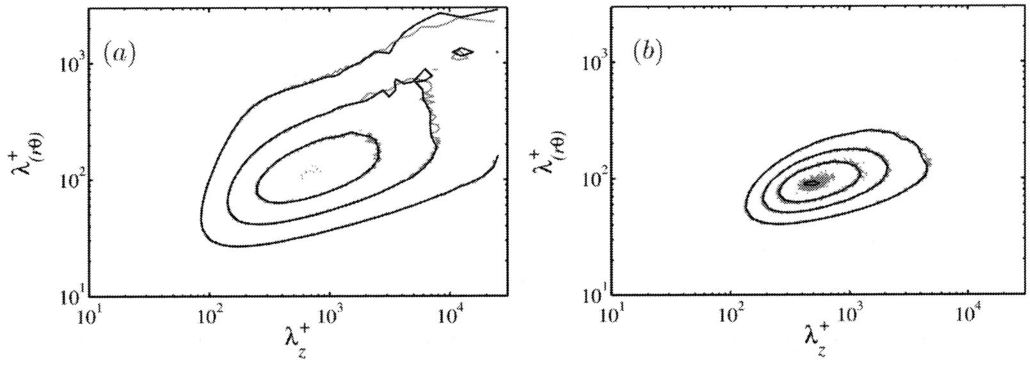

Figure 6. Pipe and channel flows in the buffer layer at $(1-r)^+ = 15$. (*a*) $2D$-premultiplied spectra of axial/streamwise velocity $k_{(r\theta)}k_z\Phi_{(u_z u_z)}/u_\tau^2$. (*b*) $2D$-premultiplied co-spectra $k_{(r\theta)}k_z\Phi_{(u_z u_r)}/u_\tau^2$. Pipe: —— (El Khoury et al., 2013); channel: —— (grey) (del Álamo et al., 2004).

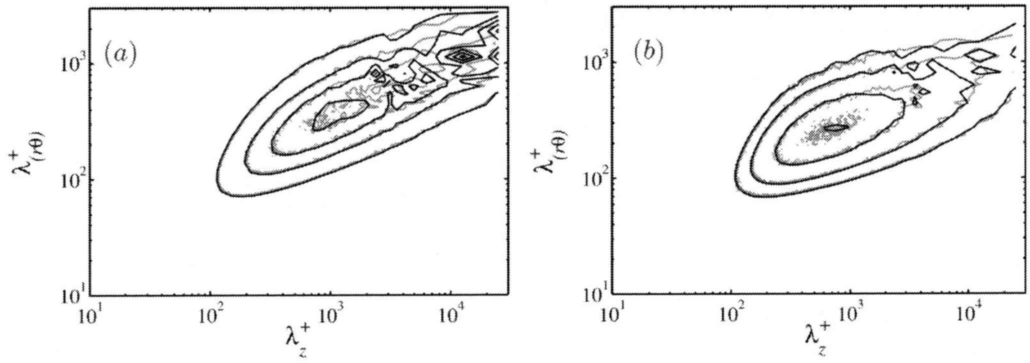

Figure 7. Pipe and channel flows in the logarithmic layer at $(1-r)^+ = 100$. (*a*) $2D$-premultiplied spectra of axial/streamwise velocity $k_{(r\theta)}k_z\Phi_{(u_z u_z)}/u_\tau^2$. (*b*) $2D$-premultiplied co-spectra $k_{(r\theta)}k_z\Phi_{(u_z u_r)}/u_\tau^2$. Pipe: —— (El Khoury et al., 2013); channel: —— (grey) (del Álamo et al., 2004).

3.4. Lower edge of the logarithmic-law region

In a similar fashion as in figure 6, the two-dimensional premultiplied spectra are presented in figure 7 at a wall-normal position of $(1-r)^+ = 100$. Here, it can be readily observed that the axial/streamwise location of the inner peak is not affected by the wall-normal displacement, whereas the spanwise wavelength has tripled to $\lambda^+_{r\theta} = 300$. This indicates that the width of the structures increases while moving away from the wall, and that the rate of increase is similar between pipe and channel flows. Another interesting feature, that is observed in the axial/streamwise velocity field is the presence of large-scale structures in both flows at the same axial and azimuthal wavelengths. Measured in wall units, the secondary peak is observed at $\lambda^+_z = 10\,000$ and $\lambda^+_{r\theta} = 1\,000$. This corresponds to a structure that is approximately 10 radii or half channel widths long in the streamwise direction and one radius/channel half-width wide in the spanwise one.

3.5. Overview of wall regions

The one-dimensional premultiplied spectrum of the axial velocity component is shown in figure 8(*a*) for pipe, channel and boundary layer flows as a function of azimuthal/spanwise wavelength. An interesting observation in this case is that the canonical wall-bounded flows agree

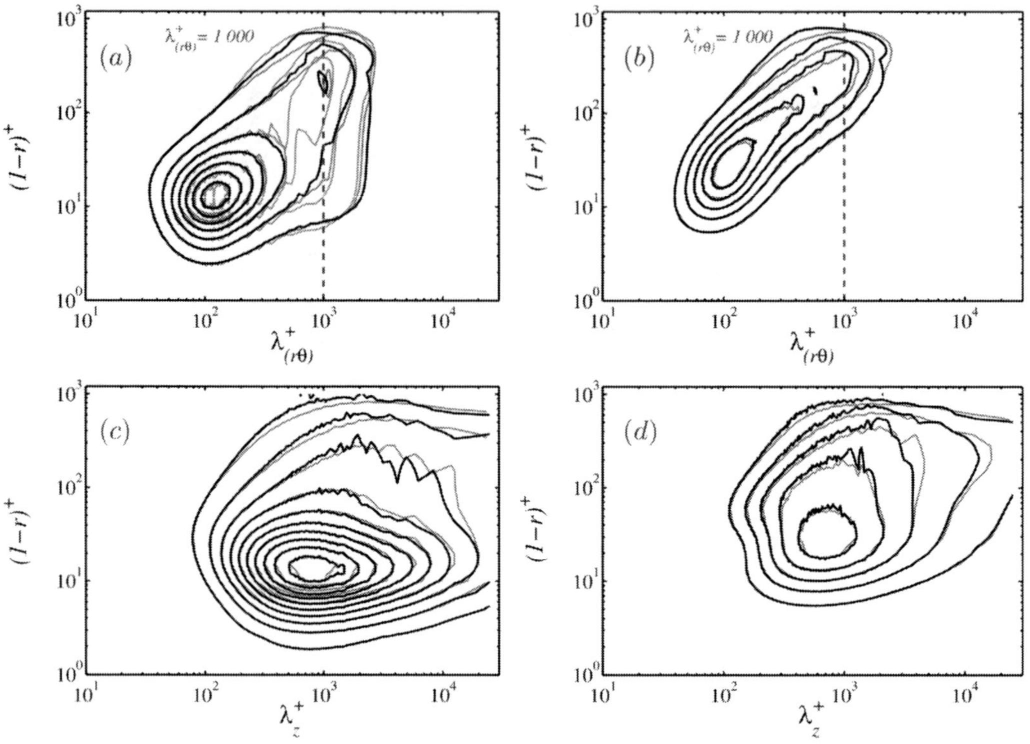

Figure 8. One-dimensional premultiplied spectra in pipe, channel and boundary layer flows. (a, b) As a function of azimuthal/spanwise wavelength $\lambda^+_{(r\theta)}$. (c, d) As a function of axial/streamwise wavelength λ^+_z. (a) Axial/streamwise velocity, $k_{(r\theta)}\Phi_{(u_z u_z)}/u^2_\tau$. (b) Co-spectra, $k_{(r\theta)}\Phi_{(u_z u_r)}/u^2_\tau$. (c) Axial/streamwise velocity, $k_z\Phi_{(u_z u_z)}/u^2_\tau$. (d) Co-spectra, $k_z\Phi_{(u_z u_r)}/u^2_\tau$. Pipe: —— (thick line), El Khoury et al. (2013), $Re_\tau = 1\,000$; channel: ——, del Álamo et al. (2004), $Re_\tau = 950$; boundary layer: —— *(grey)*, Schlatter and Örlü (2010), $Re_\tau = 1\,000$.

very well up to $(1 - r)^+ = 100$ for azimuthal wavelengths that are less than 300 viscous units. Just above the logarithmic layer and at a wall position of 200^+, a pronounced secondary peak, caused by large-scale structures, is observed for the three flow cases at an azimuthal/spanwise spacing of about one boundary-layer thickness or one pipe radius. For $Re_\tau = 1\,000$, this corresponds to a wavelength of 1 000 wall units. In contrast to channel and boundary layer flows, where the spanwise length is constant, the circumferential length of a pipe decreases as the pipe centre is approached. This circumferential length is $5R$ at the position of the secondary peak, $(1 - r)^+ = 200$, indicating the presence of five large-scale structures. Meanwhile, the co-spectrum for pipe and channel flows in figure 8(b) shows negligible differences for the two flow cases. While the first two panels display the energy as a function of azimuthal/spanwise wavelength, figures 8(c, d) show the energy content as a function of the axial/streamwise lengths. This helps us to look at the large-scale structures and quantify the differences associated in this direction. The axial/streamwise component give no clear discrepancies between pipe and channel flows, whereas the co-spectrum of the Reynolds shear stress in the outer layer indicates more momentum transfer at longer wavelengths for channel flow when compared to pipe flow.

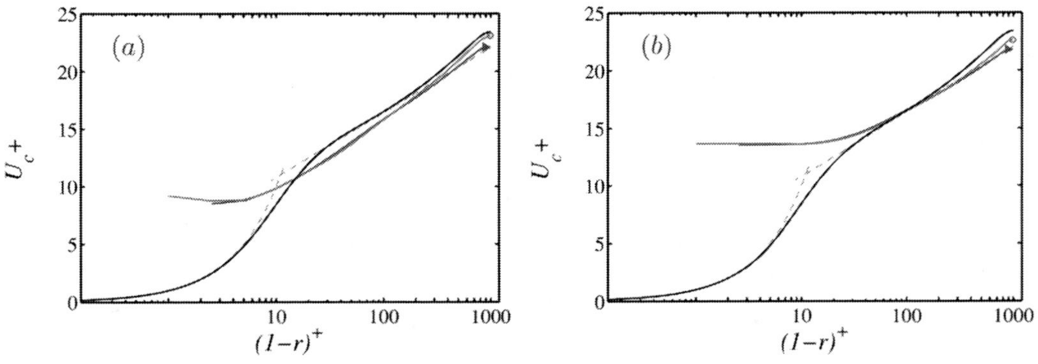

Figure 9. Advection velocity in pipe and channel flows at $Re_\tau = 1\,000$. (*a*) All Fourier modes. (*b*) Fourier modes with $\lambda_z > 2R$ (axial/streamwise) and $\lambda_{r\theta} > 0.4R$ (azimuthal/spanwise). Pipe: ——, mean velocity profile U_z^+; ——○, advection velocity U_c^+, (El Khoury et al., 2013). Channel: ——▷, advection velocity U_c^+, (Del Álamo and Jiménez, 2009). $U_z^+ = (1-r)^+$ and $U_z^+ = \kappa^{-1}\ln(1-r)^+ + B$ with $\kappa = 0.41$ and $B = 5.2$ are given as dashed lines.

3.6. Advection velocity

Several methods of different complexity have previously been used by researchers in order to calculate the advection velocity of the turbulent structures in channel and boundary layer flows. For the present pipe flow simulations, we calculate the advection velocity of the large-scale motions using the method introduced by Del Álamo and Jiménez (2009). In their work, Del Álamo and Jiménez (2009) defined the average velocity of each mode for the streamwise velocity component as

$$c_u(k_x, k_z, y) = -\frac{\langle \hat{u}\hat{u}^* \partial_t \psi_u \rangle}{k_x \langle \hat{u}\hat{u}^* \rangle} = -\frac{\operatorname{Im}\langle \hat{u}^* \partial_t \hat{u} \rangle}{k_x \langle \hat{u}\hat{u}^* \rangle}, \tag{1}$$

where $\hat{u}(k_x, k_z, y, t) = |\hat{u}(k_x, k_z, y, t)| \exp[i\psi_u(k_x, k_z, y, t)]$ with k_x, k_z being the streamwise and spanwise wavenumbers, respectively. Since the contribution to the streamwise velocity component, in a turbulent flow field, can be a combination of a range of wavenumbers, Del Álamo and Jiménez (2009) proposed a more natural definition for the advection velocity by taking into consideration a collection of modes Ω, and the advection velocity associated with Ω becomes

$$C_u(y) = \frac{\int_\Omega c_u(k_x, k_z, y)|\hat{u}(k_x, k_z, y)|^2 k_x^2 \mathrm{d}k_x \mathrm{d}k_z}{\int_\Omega |\hat{u}(k_x, k_z, y)|^2 k_x^2 \mathrm{d}k_x \mathrm{d}k_z}. \tag{2}$$

The same approach can be used for pipe flows where the streamwise and spanwise wavenumbers are replaced by the axial and azimuthal ones, respectively.

In figure 9(*a*) and (*b*) we show the axial advection velocity of the turbulent structures including all Fourier modes and only including the larger scales, respectively, and the mean velocity profile in pipe flow at $Re_\tau = 1\,000$ together with the streamwise advection velocity of the turbulent structures in channel flow at $Re_\tau = 950$ computed by Del Álamo and Jiménez (2009). In both cases the advection velocities are computed using equation (1) and (2). The advection velocity is more uniform, i.e. higher near the wall and a bit lower in the wake region, when the small turbulent scales are excluded. Figure 9 also shows that the advection velocities are very similar in pipe and channel flow in the near-wall region, while in the wake region they are somewhat higher in pipe flow. This difference can be explained by the higher mean velocity in the wake region in the latter case.

3.7. Reynolds-stress budget

The transport equation for the Reynolds stress tensor is given by

$$\frac{\mathrm{D}}{\mathrm{D}t}\langle u_i u_j\rangle = P_{ij} + \varepsilon_{ij} + \Pi_{ij} + D_{ij} + T_{ij}\ , \tag{3}$$

where the production, dissipation, pressure strain, divergence of pressure-velocity correlation, molecular diffusion and turbulent diffusion are defined as, respectively,

$$P_{ij} = -\langle u_i u_k\rangle \frac{\partial U_j}{\partial x_k} - \langle u_j u_k\rangle \frac{\partial U_i}{\partial x_k} \tag{4a}$$

$$\varepsilon_{ij} = -2\nu \left\langle \frac{\partial u_i}{\partial x_k}\frac{\partial u_j}{\partial x_k}\right\rangle \tag{4b}$$

$$\Pi_{ij} = \frac{1}{\rho}\left\langle p\frac{\partial u_i}{\partial x_j} + p\frac{\partial u_j}{\partial x_i}\right\rangle \tag{4c}$$

$$G_{ij} = -\frac{1}{\rho}\left(\frac{\partial}{\partial x_i}\langle p u_j\rangle + \frac{\partial}{\partial x_j}\langle p u_i\rangle\right) \tag{4d}$$

$$D_{ij} = \nu\frac{\partial^2}{\partial x_k^2}\langle u_i u_j\rangle \tag{4e}$$

$$T_{ij} = -\frac{\partial}{\partial x_k}\langle u_i u_j u_k\rangle\ . \tag{4f}$$

The budget and premultiplied budget terms for the axial/streamwise stress and the Reynolds shear stress are shown in figures 10 and 11 for pipe, channel and boundary layer flows at $Re_\tau = 1\,000$. Reynolds stress budgets for channel flow at $Re_\tau = 180$ and 2000 were presented by Mansour et al. (1988) and Hoyas and Jiménez (2008), respectively. It can be readily observed that the various budget terms (non-premultiplied) are almost indistinguishable for pipe, channel and boundary layer flows apart from the magnitude of the production and viscous dissipation of the Reynolds shear stress, which are slightly larger in the boundary layer than in pipe and channel flow in the buffer layer. The budget for the axial/streamwise stress is dominated by production and viscous dissipation with the expected large peak of positive production in the buffer layer. The negative diffusion terms in this region imply an extraction and transport of energy away from the buffer region, whereas the negative pressure-strain serves to distribute energy from the normal stress to the wall-normal and azimuthal/spanwise stresses. The most important terms in the Reynolds shear stress budget are production and pressure-strain (figure 11*a*). The production term, in this case, is given by $-\langle u_r u_r\rangle\,\mathrm{d}U_z/\mathrm{d}r$, $-\langle vv\rangle\,\mathrm{d}U/\mathrm{d}y$ and $-\langle uu\rangle\,\partial V/\partial x - \langle vv\rangle\,\partial U/\partial y$ for pipe, channel and boundary layer flows, respectively. The fact that the wall-normal Reynolds stress is identical between pipe and channel flow up to $(1-r)^+ = 200$ leads to the same production in these two flow cases, whereas the higher wall-normal stress $\langle vv\rangle$ in boundary layer flow is responsible for the higher production observed here. Note that $\langle uu\rangle\,\partial V/\partial x$ is negligible compared to $\langle vv\rangle\,\partial U/\partial y$ in a boundary layer.

The terms of the Reynolds-stress budget in the outer layer can be highlighted through premultiplication and the usage of logarithmic abscissa. The apparent area below the curves between two wall-normal positions directly corresponds to the integral of the various budget terms when plotted in this way. The premultiplication for channels and boundary layer flows only involves the inner-scale wall distance y^+ whereas for pipe flow, the premultiplication contains two terms which are related to the integration along a logarithmic abscissa, $(1-r)^+$, and to the circular geometry, r. The effect of premultiplication can be seen in the right panels of figures 10 and 11. In the wake region and for the streamwise/axial stress, the premultiplied production, viscous dissipation and pressure strain have the largest and smallest magnitude in

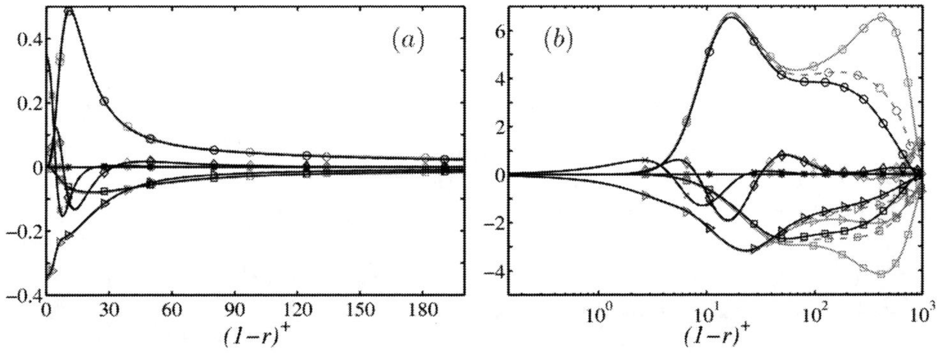

Figure 10. (*a*) Budget of axial/streamwise Reynolds stress $\langle u_z u_z \rangle$ or $\langle uu \rangle$ normalised by u_τ^4/ν for $Re_\tau = 1\,000$. (*b*) Premultiplied budget: by $r(1-r)^+$ in the pipe, and by y^+ in channels and boundary layers. Pipe: ——, present DNS; channel: – – – –, Lenaers et al. (2012); boundary layer: —— (*grey*), Schlatter and Örlü (2010). —○—, Production P_{ij}; —▷—, viscous dissipation ε_{ij}; —□—, pressure-strain, Π_{ij}; —∗—, divergence pressure-velocity correlation, G_{ij}; —×—, viscous diffusion D_{ij}; —◇—, turbulent diffusion T_{ij}.

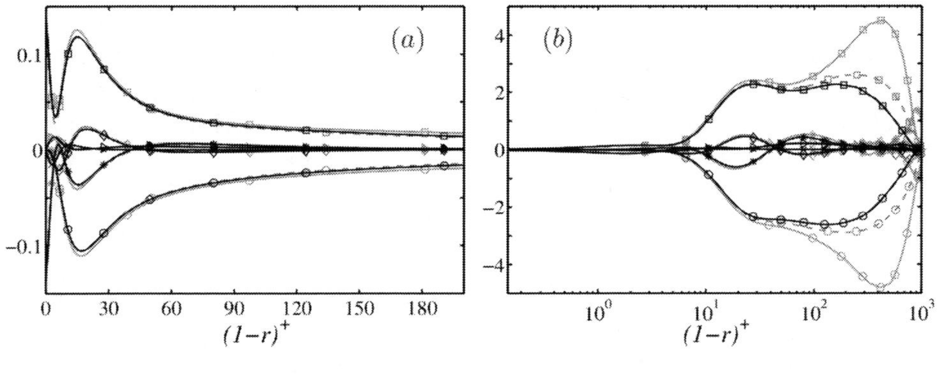

Figure 11. (*a*) Budget of Reynolds shear stress $-\langle u_z u_r \rangle$ or $-\langle uv \rangle$ normalised by u_τ^4/ν, for $Re_\tau = 1\,000$. (*b*) Premultiplied budget: by $r(1-r)^+$ in the pipe, and by y^+ in channels and boundary layers. Pipe: ——, present DNS; channel: – – – –, Lenaers et al. (2012); boundary layer: —— (*grey*), Schlatter and Örlü (2010). —○—, Production P_{ij}; —▷—, viscous dissipation ε_{ij}; —□—, pressure strain, Π_{ij}; —∗—, divergence pressure-velocity correlation, G_{ij}; —×—, viscous diffusion D_{ij}; —◇—, turbulent diffusion T_{ij}.

boundary layer and pipe flow, respectively. The pressure strain term in this case has a larger magnitude than the viscous dissipation. For the Reynolds shear stress, on the other hand, the most dominant terms are still production and pressure-strain, which tend to balance each other in the wake region and approach zero towards the centre (or towards the boundary-layer edge).

4. Concluding remarks

Single-point statistics, PDFs of the wall shear stress and pressure, spectra, advection velocities of the turbulent structures and Reynolds stress budgets obtained from our DNS data of turbulent pipe and channel flow up to $Re_\tau = 1\,000$ reveal a strong similarity up to at least $y^+ = 100$. The results strongly support the idea that the flow and dynamics, structure and even intermittency of the turbulence have a high degree of universality in the near wall region in these two flow cases. Results on the turbulence intensity and spanwise spectrum of the streamwise velocity, the PDF of the wall shear stress and the Reynolds stress budgets obtained from DNS of zero-pressure-gradient turbulent boundary layer at $Re_\tau = 1\,000$ are also very similar to the corresponding results for pipe and channel flow, showing that many features of the three canonical wall-bounded flows are alike. The observed similarity appears consistent with Monty et al. (2009) who reported for experiments a similar correspondence in the near-wall statistics of the three flow cases. Only the largest structures in pipe and channel flow differ from the ones in boundary layer flow in their experiments. Further DNS and experimental studies should indicate if the universality of the near-wall region also pertains to higher Reynolds numbers, but there is no obvious reason why this should not be the case.

In the wake region, our DNS data reveal differences in the mean streamwise velocity and the Reynolds stress budgets between the three flow cases. A comprehensive comparison of one-statistics of pipe, channel and boundary layer data by Sillero et al. (2013) showed that also the Reynolds stresses are different in the outer region. Some of the differences are related to the intermittency and entrainment of irrotational fluid in the outer part of a boundary layer and the absence of this intermittency in pipe and channel flow. However, further studies are needed to find the reasons for the differences observed in pipe and channel-flow statistics, i.e. the flows that are not affected by intermittency.

Acknowledgements

This research was funded in part by the Multiflow project of the ERC. We acknowledge PRACE for awarding us access, via the PIPETURB project, to resources *Hector* at the UK National Supercomputing Service, and *Louhi* at the IT Centre for Science in Finland. Computational resources at PDC were made available by the Swedish National Infrastructure for Computing (SNIC).The authors would like to thank Juan Sillero for insightful comments on the manuscript.

References

Boersma BJ 2011 Direct numerical simulation of turbulent pipe flow up to a Reynolds number of 61 000 *J. Phys.: Conf. Ser.* **318** 042045

Chin C, Ooi ASH, Marusic I and Blackburn HM 2010 The influence of pipe length on turbulence statistics computed from direct numerical simulation data *Phys. Fluids* **22** 115107

Del Álamo JC and Jiménez J 2009 Estimation of turbulent convection velocities and corrections to Taylor's approximation *J. Fluid Mech.* **640** 5–26

Del Álamo JC, Jiménez J, Zandonade P and Moser RD 2004 Scaling of the energy spectra of turbulent channels *J. Fluid Mech.* **500** 135–44

Eggels JGM, Unger F, Weiss MH, Westerweel J, Adrian RJ, Friedrich R and Nieuwstadt FTM 1994 Fully developed turbulent pipe flow: a comparison between direct numerical simulation and experiment *J. Fluid Mech.* **268** 175–209

El Khoury GK, Schlatter P, Noorani A, Fischer PF, Brethouwer G and Johansson AV 2013 Direct numerical simulation of turbulent pipe flow at moderately high Reynolds numbers *Flow Turbul. Combust.* **91** 475–95

Fischer PF, Lottes JW and Kerkemeier SG 2008 Nek5000 web page, http://nek5000.mcs.anl.gov

George WK 2007 Is there a universal log law for turbulent wall-bounded flows? *Phil. Trans. R. Soc. A* **365** 789–806

Hoyas S and Jiménez J 2008 Reynolds number effects on the Reynolds-stress budgets in turbulent channels *Phys. Fluids* **20** 101511

Huffman GD and Bradshaw P 1972 A note on von Kármán constant in low Reynolds number turbulent flows *J. Fluid Mech.* **53** 45–60

Jiménez J 2012 Cascades in wall-bounded turbulence *Annu. Rev. Fluid Mech.* **44** 27–45

Jiménez J, Hoyas S, Simens MP and Mizuno Y 2010 Turbulent boundary layers and channels at moderate Reynolds numbers *J. Fluid Mech.* **657** 335–360

Kim J 2012 Progress in pipe and channel flow turbulence, 1961–2011 *J. Turbul.* **13** N45

Klewicki J, Chin C, Blackburn HM, Ooi A and Marusic I 2012 Emergence of the four-layer dynamical regime in turbulent pipe flow *Phys. Fluids* **24** 045107

Kline SJ, Reynolds WC, Schraub FA and Runstadler PW 1967 The structure of turbulent boundary layers *J. Fluid Mech.* **30** 741–73

Lee JH and Sung HJ 2013 Comparison of very-large-scale motions of turbulent pipe and boundary layer simulations *Phys. Fluids* **25** 045103

Lenaers P, Li Q, Brethouwer G, Schlatter P and Örlü R 2012 Rare backflow and extreme wall-normal velocity fluctuations in near-wall turbulence *Phys. Fluids* **24** 035110

Mansour NN, Kim J and Moin P 1988 Reynolds-stress and dissipation-rate budgets in a turbulent channel flow *J. Fluid Mech.* **194** 15–44

Marusic I, McKeon BJ, Monkewitz PA, Nagib HM and Smits AJ 2010 Wall-bounded turbulent flows at high Reynolds numbers: Recent advances and key issues *Phys. Fluids* **22** 065103

McKeon BJ, Li J, Jiang W, Morrison JF and Smits AJ 2004 Further observations on the mean velocity distribution in fully developed pipe flow *J. Fluid Mech.* **501** 135–47

Millikan CB 1938 A critical discussion of turbulent flows in channels and circular tubes *Proc. 5th Int. Congr. Appl. Mech* eds. JPD Hartog and H Peters (Hoboken, NJ: Wiley) pp. 386–92

Monty JP, Hutchins N, Ng HCH, Marusic I and Chong MS 2009 A comparison of turbulent pipe, channel and boundary layer flows *J. Fluid Mech.* **632** 431–42

Monty JP, Stewart JA, Williams RC and Chong MS 2007 Large-scale features in turbulent pipe and channel flows *J. Fluid Mech.* **589** 147–56

Patel VC and Head MR 1969 Some observations on skin friction and velocity profiles in fully developed pipe and channel flows *J. Fluid Mech.* **38** 181–201

Schlatter P and Örlü R 2010 Assessment of direct numerical simulation data of turbulent boundary layers *J. Fluid Mech.* **659** 116–26

Sillero JA, Jiménez J and Moser RD 2013 One-point statistics for turbulent wall-bounded flows at Reynolds numbers up to $\delta^+ \approx 2000$ *Phys. Fluids* **25** 105102

Smits AJ, McKeon BJ and Marusic I 2011 High-Reynolds-number wall turbulence *Annu. Rev. Fluid Mech.* **43** 353–75

Wagner C, Hüttl TJ and Friedrich RJ 2001 Low-Reynolds-number effects derived from direct numerical simulations of turbulent pipe flow *Comp. Fluids* **30** 581–90

Wu X, Baltzer JR and Adrian RJ 2012 Direct numerical simulation of a $30R$ long turbulent pipe flow at $R^+ = 685$: Large- and very large-scale motions *J. Fluid Mech.* **698** 235–81

Wu X and Moin P 2008 A direct numerical simulation study on the mean velocity characteristics in turbulent pipe flow *J. Fluid Mech.* **608** 81–112

Zagarola MV and Smits AJ 1997 Scaling of the mean velocity profile for turbulent pipe flow *Phys. Rev. Lett.* **78** 239–42

Zagarola MV and Smits AJ 1998 Mean-flow scaling of turbulent pipe flow *J. Fluid Mech.* **373** 33–79

Scaling of pressure spectrum in turbulent boundary layers

Saurabh S Patwardhan and O N Ramesh
Department of Aerospace Engineering, Indian Institute of Science, Bangalore, India
E-mail: psaurabh@aero.iisc.ernet.in

Abstract. Scaling of pressure spectrum in zero-pressure-gradient turbulent boundary layers is discussed. Spatial DNS data of boundary layer at one time instant ($\mathrm{Re}_\theta = 4500$) are used for the analysis. It is observed that in the outer regions the pressure spectra tends towards the -7/3 law predicted by Kolmogorov's theory of small-scale turbulence. The slope in the pressure spectra varies from -1 close to the wall to a value close to -7/3 in the outer region. The streamwise velocity spectra also show a -5/3 trend in the outer region of the flow. The exercise carried out to study the amplitude modulation effect of the large scales on the smaller ones in the near-wall region reveals a strong modulation effect for the streamwise velocity, but not for the pressure fluctuations. The skewness of the pressure follows the same trend as the amplitude modulation coefficient, as is the case for the velocity. In the inner region, pressure spectra were seen to collapse better when normalized with the local Reynolds stress ($-\overline{u'v'}$) than when scaled with the local turbulent kinetic energy ($q^2 = \overline{u'^2} + \overline{v'^2} + \overline{w'^2}$)

1. Introduction

The pressure spectra are not studied and understood as extensively as the velocity spectra. The main reason for this is the difficulty involved in measuring pressure in laboratory experiments. However, pressure is an important variable and, in an incompressible flow, it is governed by a Poisson equation and it is non-local in nature. In recent years, there has been considerable interest in the study of spectra of static pressure. Just as in the case of velocity, it has been hypothesized that the similarity arguments of Kolmogorov [1] may also apply to pressure fluctuations. These arguments yield a $k^{-7/3}$ law for the longitudinal pressure spectrum in the wavenumber space. However, it is not entirely clear if the application of the local isotropy arguments to pressure spectrum is entirely justified [2] due to its non-local nature. This is because even regions far away from the point under observation make non-negligible contributions to the pressure fluctuations at the point of interest. Albertson *et. al.* [3] carried out static pressure measurements in an atmospheric boundary layer. They found that the large-scale motions exert strain on the inertial-range small-scale motions. This was expected to have an effect on the inertial-range pressure spectrum and hence to induce a deviation from the $k^{-7/3}$ law. To explain the lack of a -7/3 law and the straining of inertial range pressure fluctuations, they made a connection between the Townsend's "active-inactive motion" hypothesis and the straining of the pressure field. Townsend [4] hypothesized that the turbulent motion close to the wall consists of two parts, namely "active motion" and "inactive motion". Active motion is the one responsible for the Reynolds stress in the wall region, while the inactive part is the motion induced by large-scale-induced fluctuation generated close to the wall [5]. Hence, it is conceivable that significant interaction exists between the large- and small-scale wave numbers in the spectral space, and the local isotropy assumption may not be strictly valid very close to the wall in the case of boundary layers. Tsuji *et. al.* [6] attempted to verify the existence of a -7/3 law in a high-Reynolds-number laboratory experiment ($Re_\theta = 20000$). However, they observed a -1 power law in the low-wavenumber region, and a -7/3 law was not evident at any wavenumber at any location in the boundary layer. George *et. al.* [7] carried out experiments on turbulent jets to measure the static pressure fluctuations. They found a reasonable agreement between the experimental data and theoretical predictions at wavenumbers which correspond to -7/3 ranges. The existence of a -7/3 law in free shear flows and its non-existence in experiments in wall bounded flows warrant a further study of the spectra of relatively less understood variables like pressure. The use of direct numerical simulations of high-Reynolds-number turbulent boundary layers will be extremely useful in the evaluation of such laws, since the main difficulty in the experiments is the accurate measurement of static pressure inside the boundary layer [8]. In this work, we carry out an analysis of a well-resolved DNS data [9] of pressure fluctuations to gain insights into the spectral laws.

2. Details of direct numerical simulation

Although the work did not involve development of the computer algorithm, the details of numerical method for the simulation will be briefly discussed for the sake of completeness. The details of the algorithm can be found elsewhere [10, 11]. The code uses a fractional-step method [12] that solves for velocity using an estimated pressure gradient term, and then corrects it by enforcing the mass conservation condition with the help of a Poisson equation for the pressure. The time stepping involves three sub-steps of a Runge-Kutta third-order method with the wall-normal viscous terms treated in a semi-implicit way using a Crank-Nicholson scheme. The first- and second-order derivatives in streamwise (x) and wall-normal (y) directions are calculated using staggered three-point compact finite differences. A pseudo-spectral scheme is used to calculate derivatives in the homogeneous spanwise (z) direction, using the 2/3 rule to prevent aliasing. The Reynolds number (based on momentum thickness) of the simulated boundary layer range from 2800 to 6600. The number of grid points is $15360 \times 535 \times 4096$ in x, y and z directions respectively.

3. Results and Discussion

Recently, Mathis *et. al.* [13] carried out a study to investigate the relationship between large and small scales in high Reynolds number boundary layer flows. The inner and outer interaction was identified in terms of the imprint of the outer-layer (large-scale) flow on the inner-layer (small-scale) flow. The effect was quantified by calculating the correlation between the large-scale component of the flow close to the wall and the large-scale component of the Hilbert transform of the small-scale component. They observed that a strong correlation between the envelope of the small- and large-scale parts of the same flow close to the wall. This led them to the conclusion that very-large-scale motions from the outer layer tend to modulate the small-scale part of the flow close to the wall. They also found a strong resemblance between the wall-normal profile of amplitude modulation and the skewness of the velocity fluctuations [14].

As mentioned earlier, Albertson *et. al.* [3] found that the large-scale pressure fluctuations strain the small-scale fluctuations, which could cause pressure spectra to deviate from the one predicted by the isotropy arguments. This hypothesis is further supported by observations in jets, where, unlike in boundary layers, there are no eddy hierarchies, and pressure spectra reasonably agree with the -7/3 law [7]. In the following, we will try to assess the DNS data and look for the inner-outer interactions in pressure fluctuations.

Before looking at the results of amplitude modulation of pressure fluctuations, we present some basic data that will allow us to proceed with further analysis [9]. Figure 1 shows the

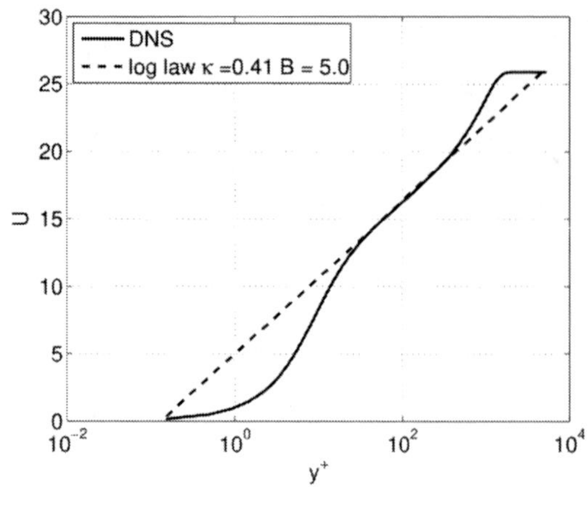

Figure 1. Mean velocity profile in zero-pressure-gradient turbulent boundary layer (Re_θ = 4500) [9].

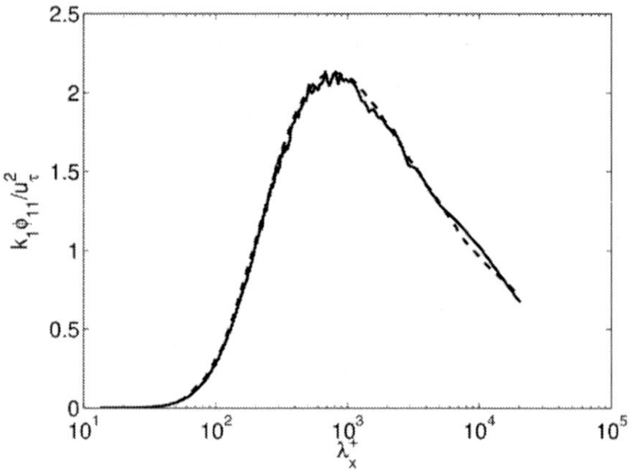

Figure 2. Premultiplied streamwise velocity spectra at $y_+ = 15$, from space series. Solid line: not time averaged. Dashed: time averaged from UPM data [9].

mean velocity profile of turbulent boundary layer flow. Also shown is the logarithmic-law profile with standard zero-pressure-gradient constants. The Reynolds number of the flow based on the momentum thickness is 4500. The spectra which will be presented in this paper are calculated from the spatial series at a given time, averaged in the homogeneous z direction. The streamwise extent of the space series is -7δ to 7δ. We expect the variation in boundary layer parameters, such as boundary layer thickness and skin friction coefficient, to be very small within that distance, allowing us to calculate the spectra assuming a locally parallel evolution of the boundary layer within a streamwise extent of 14δ. Also, we expect the correlations at such distances to be very small so that the flow does not suffer from memory effects. Note that the spectra are not time-averaged but the time averaging does not have a significant effect, as shown in figure 2. Apart from being a little wiggly in nature as expected, there is hardly any difference in shape and peak magnitude and hence it should not matter in the final conclusions of the study. Also, the amplitude modulation calculations are carried out on a space series and not on a time series as in original paper of Mathis *et. al.* [13], and the correlation is averaged in the z direction.

Figures 3 and 7 show pressure spectra at different wall-normal locations in Kolmogorov and outer scaling respectively. The spectra shown in all the figures here extend from $y_+ = 15$ to $y_+ = 1834$. Also shown in those figures are -1, -3/2 and -7/3 variations with wavenumber. Note that the pressure spectra obtained by Tsuji *et. al.* [6] showed k^{-1} to $k^{-3/2}$ like variation. It is interesting to note that the spectra very close to the wall neither behaves like a -1 nor a -3/2 power law, and even less like a -7/3 variation. As we move away from the wall, the -1 scaling region appears between wavenumbers $k_x\delta = 30$ to 60 in figure 7. Further away from wall, the slope continues to decrease and shows a -3/2 like behaviour. The decrease in slope continues further in the outer region and tends towards -7/3 far out in the boundary layer. To visualize this change in slope at different wall-normal locations, the premultiplied pressure spectra are plotted in figures 4-6 in Kolmogorov scaling, and in figures 8-10 in outer scaling. The pressure spectra in figures 4-6, and similarly in figures 8-10, are premultiplied with $(k\delta)^{7/3}$, $(k\delta)^{3/2}$ and $(k\delta)$ respectively. Premultiplication helps in a clear identification of the different slopes in the spectra by identifying the plateau region. In both these scalings, the region very close to the wall has a flat region in the spectra premultiplied with $k\delta$. As we progressively move out in the boundary layer, the plateau region is observed in the spectra premultiplied with $(k\delta)^{3/2}$. Far out in the boundary layer and close to the edge of the boundary layer the plateau is observed

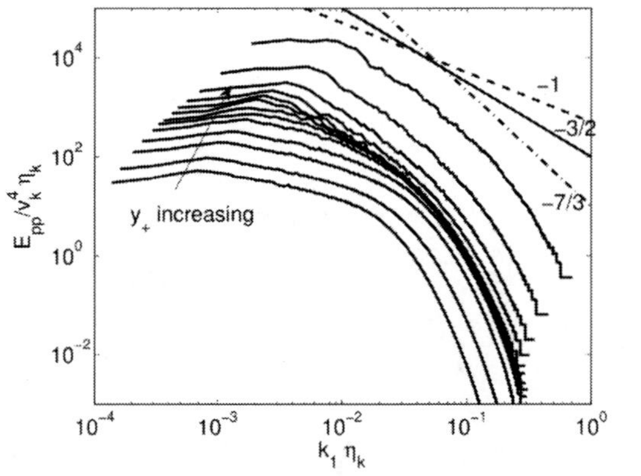

Figure 3. Pressure spectra at different wall-normal locations, normalized with Kolmogorov scale.

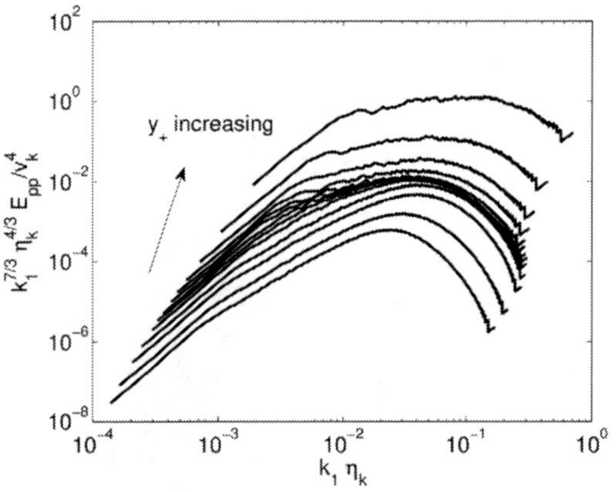

Figure 4. Premultiplied ($k^{7/3}$) pressure spectra at different wall-normal locations, normalized with Kolmogorov scale.

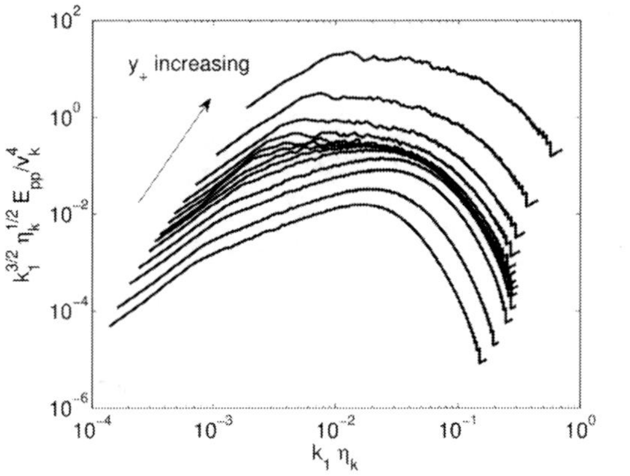

Figure 5. Premultiplied ($k^{3/2}$) pressure spectra at different wall-normal locations in normalized with Kolmogorov scale.

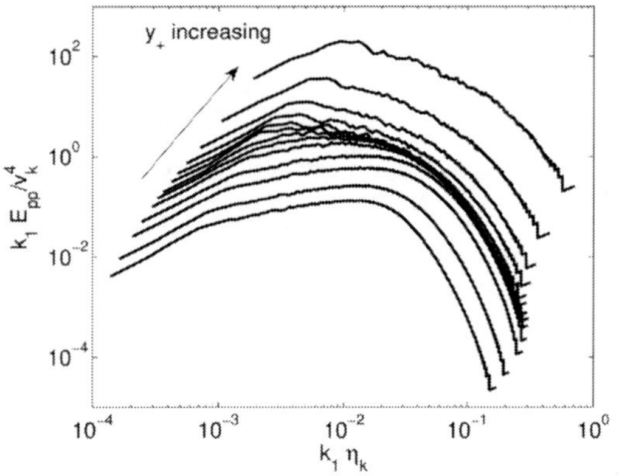

Figure 6. Premultiplied (k) pressure spectra at different wall-normal locations in normalized with Kolmogorov scale.

in the spectra premultiplied with $(k\delta)^{7/3}$.

The turbulent boundary layer has been traditionally divided into inner and outer regions. The inner region is largely influenced by the presence of the wall, while the outer region is largely free from the influence of the wall and to a certain degree acts like a free shear layer. In Townsend's [4] model, the inner region of the turbulent boundary layer consists of eddies that are attached to the wall while the outer layer is populated with eddies that are detached from the wall. However, recently it has been shown that the outer layer eddies or the superstructures show an imprint on the velocity fluctuations in the inner region [13]. Hence, the observation of a -7/3 law in the outer region, along with the observation of a -7/3 law in free shear layers [7], and the outer layer imprint on the inner region further suggests that inner-outer interaction might be playing a role in the disappearance of the -7/3 law in pressure spectrum.

From the above discussion, it is plausible that a significant amount of inner-outer interaction or straining of small scales by large-scale structures can cause deviations from the isotropic spectrum. Figure 11 shows spectra of streamwise velocity fluctuations at different wall-normal

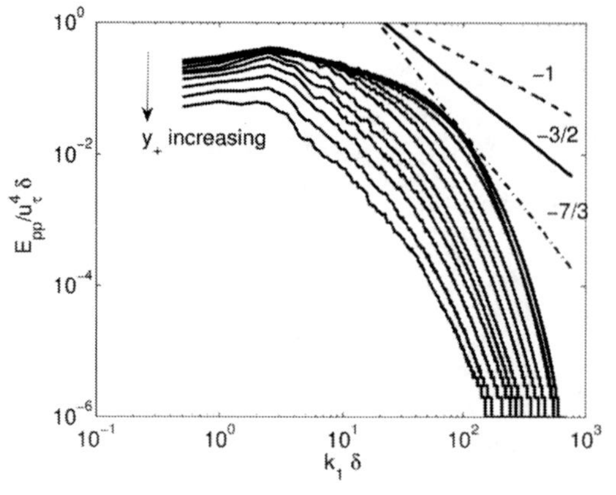

Figure 7. Pressure spectra at different wall-normal locations normalized with outer scales.

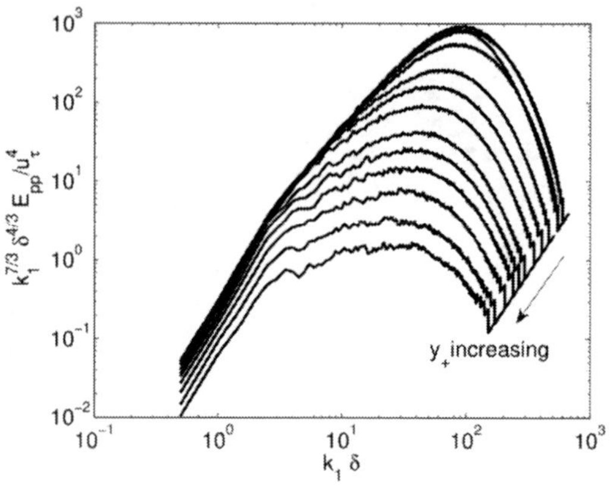

Figure 8. Premultiplied ($k^{7/3}$) pressure spectra at different wall-normal locations normalized with outer scales.

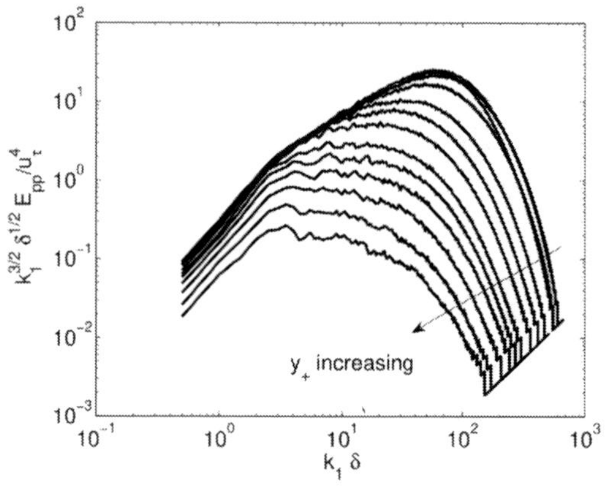

Figure 9. Premultiplied ($k^{3/2}$) pressure spectra at different wall-normal locations normalized with outer scales.

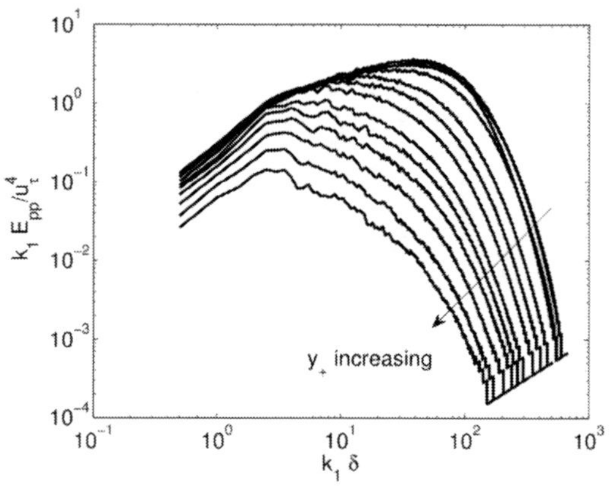

Figure 10. Premultiplied (k) pressure spectra at different wall-normal locations normalized with outer scales.

locations in the outer scaling. The celebrated -5/3 law of Kolmogorov is also shown in the figure as a dashed line. It is interesting to note that similar to the pressure, the velocity spectra does not show resemblance to -5/3 law close to the wall at $y_+ = 15$ to 96. Away from the wall, the slope decreases and it more looks like -5/3 in the outer layer, $y_+ = 846$. Note that the δ_+ value for this boundary layer is 1690. Similar to the pressure spectra, the premultiplied spectra for the streamwise velocity are plotted in figure 12 for clear visibility of the -5/3 slope in the outer region. The velocity and pressure spectra indicate a trend of moving near the isotropic prediction away from the wall, where the shear is less than in the near-wall region. To quantify the plausible inner-outer interaction in pressure, we carried out the amplitude modulation exercise on the pressure and velocity signals from the DNS data. The exercise makes use of space series and not time series as in the original procedure [13].

Here, we briefly describe the procedure for calculating the amplitude modulation which is essentially similar to the one used by Mathis *et. al.* [13] for time series. First we take the Fourier transform of the space series of the fluctuating quantity. The Fourier components are split into two parts: one representing the large-scale motion ($\lambda_x > \delta$) and the other representing small-

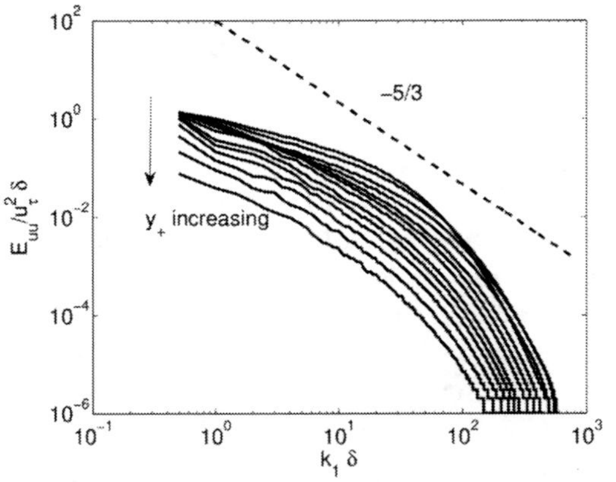

Figure 11. Streamwise velocity spectra at different wall-normal locations, normalized with outer scales.

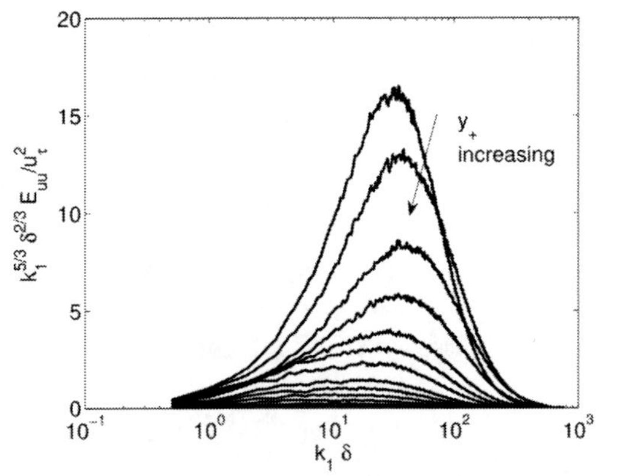

Figure 12. Premultiplied ($k^{5/3}$) streamwise velocity spectra at different wall-normal locations, normalized with outer scales.

scale motion ($\lambda_x < \delta$). The length scale δ is chosen as a threshold consistent with the choice of Mathis *et. al.* [13]. The original signal is then split into two parts by taking the inverse Fourier transform of the split Fourier components. Then, the envelope of the small-scale component is calculated by using a Hilbert transform technique. It is presumed that if there is an amplitude modulation effect of large scales on small scales then the large-scale envelope of the small scales should be well correlated with the large-scale part of the original signal. Hence, the envelope itself is split into small- and large-scale signals, and the correlation between the large-scale part of the envelope with the large-scale part of the original signal is calculated, which represents the amplitude modulation coefficient.

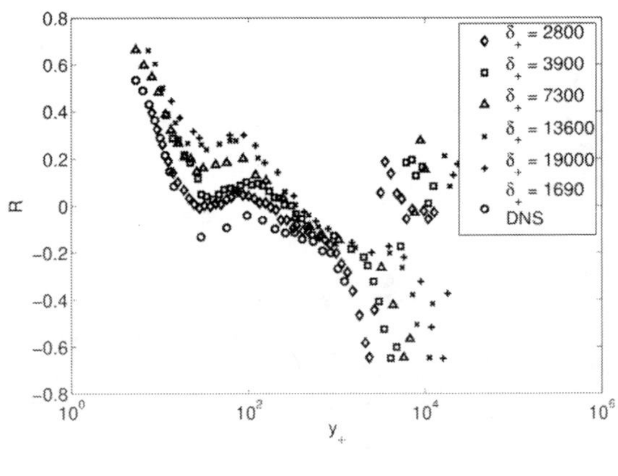

Figure 13. Amplitude modulation in the streamwise velocity signal.

Figure 13 shows the wall-normal variation of the amplitude modulation of the streamwise velocity. This plot represents the correlation between the large-scale part of the signal ($\lambda_x > \delta$) and the large-scale part of the Hilbert transform envelope of the small-scale part of the signal ($\lambda_x < \delta$) at a given wall-normal location. The circles are from DNS data at UPM [9], compared with the experimental data published in Mathis *et. al.* [13]. The experimental data have higher

Reynolds numbers than the DNS, the highest being an order of magnitude larger. The DNS data shows a correct trend both in the wall-normal profile and in the Reynolds-number effect. The velocity fluctuations close to the wall show a very high degree of amplitude modulation. The modulation effect decreases and becomes negative as we go away from the wall. The first minimum in the DNS data occurs below zero. The value of the first minimum is seen to increase with the Reynolds number, showing that the DNS data captures the right trend in regards to the variation of amplitude modulation. The result of the same exercise carried out on pressure fluctuations is shown in figure 14. Unlike the velocity, the pressure signals show no amplitude modulation close to the wall and the value of the correlation is zero. Surprisingly, the nature of the plot is similar to the velocity plot away from the wall, but the remarkable result is that there is no amplitude modulation close to the wall and that they are never positively correlated.

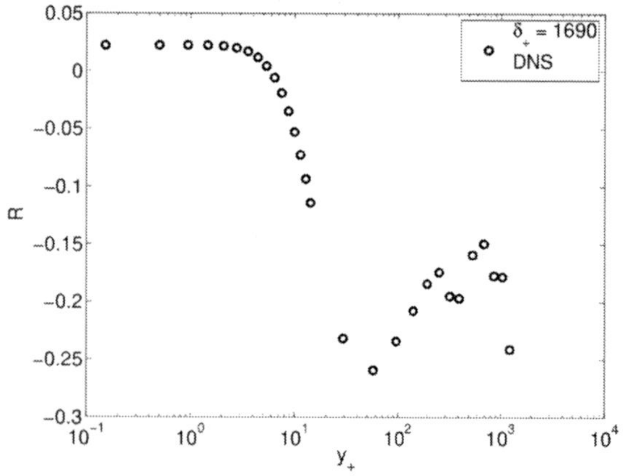

Figure 14. Amplitude modulation in the pressure signal.

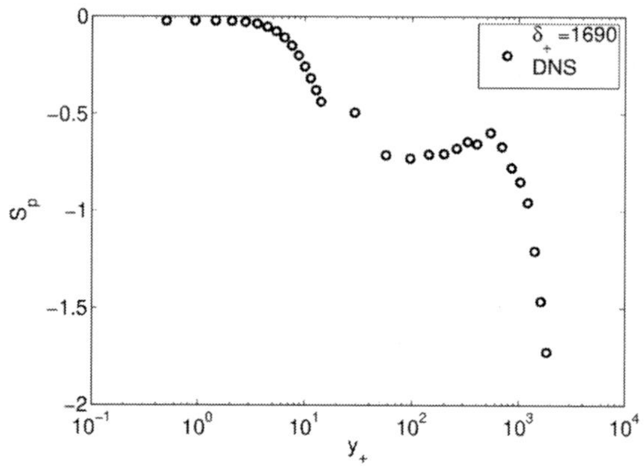

Figure 15. Variation of the skewness factor of pressure fluctuations.

Mathis *et. al.* [13] also observed a similarity between the amplitude modulation and the skewness factor for velocity. The same is evident in the case of pressure fluctuations, as seen in figure 15. The shape of the pressure skewness plot is similar to that of the amplitude modulation, which suggests that there is an inherent relationship between amplitude modulation and skewness for other variables as well.

3.1. Pressure spectra in other scalings

The result in the previous section that there is no modulating effect of the large-scale pressure fluctuations on the small-scale pressure fluctuations is surprising because the velocity shows a strong modulating effect. This raises the question as to how the outer region can influence the inner region in the absence of pressure modulation. It is possible that the modulation is mild (perhaps because of the moderate values of the Reynolds number) and hence velocity is more sensitive than pressure to large-scale influences. In order to understand further the inner-outer effect or "active–inactive" effect, we have plotted pressure spectra using normalizing variables other than u_τ. Figure 16 and 20 show pressure spectra normalized using $q^2 = \overline{u'^2} + \overline{v'^2} + \overline{w'^2}$ and $-\overline{u'v'}$ respectively. Two observations can be made from these plots. The spectra normalized by $\overline{u'v'}$ collapse well in the near wall region compared to the ones normalized by q^2. The region over which this good collapse is observed extends from $y_+ = 15$ to $y_+ = 329$. Also, away from the wall, spectra normalized with $\overline{u'v'}$ are closer to collapsing on a single line than the ones normalized by q^2. Again for better visualization of this, the compensated spectra are presented in figures 17-19 in q^2 scaling and in figures 21-22 in $\overline{u'v'}$ scaling. The original idea by

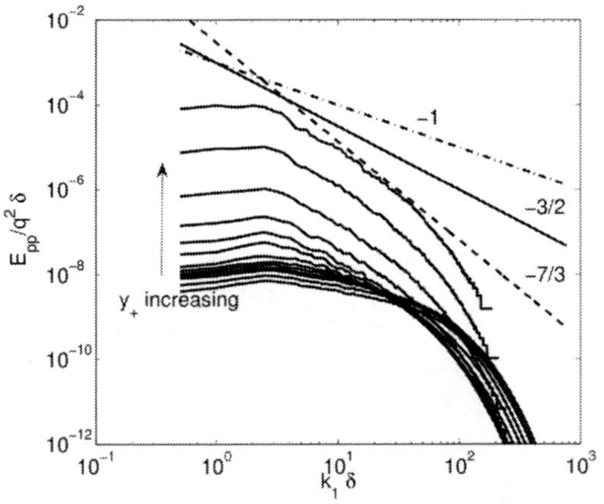

Figure 16. Pressure spectra normalized by q^2.

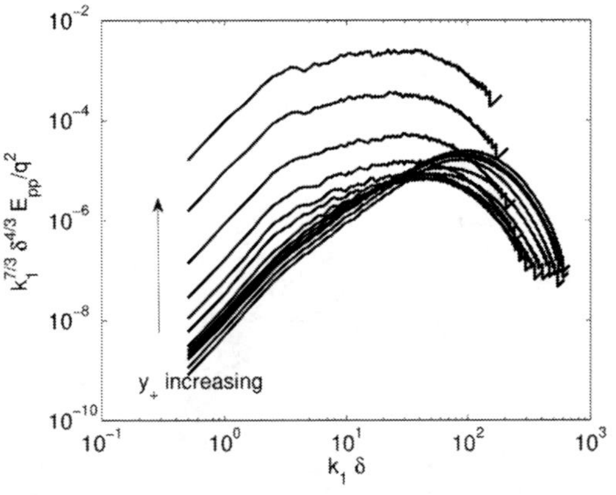

Figure 17. Premultiplied ($k^{7/3}$) pressure spectra normalized by q^2.

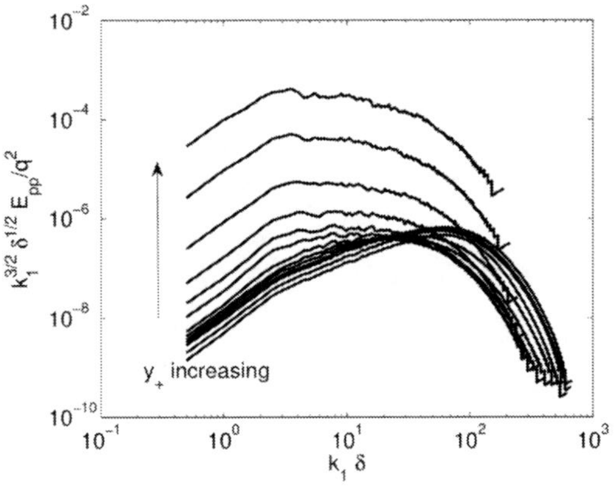

Figure 18. Premultiplied ($k^{3/2}$) pressure spectra normalized by q^2.

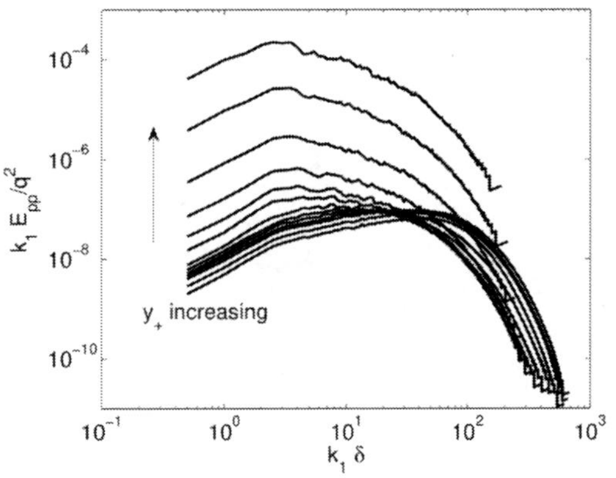

Figure 19. Premultiplied (k) pressure spectra normalized by q^2.

Townsend [4], that the inactive motion does not contribute to the shear stress but contributes to the kinetic energy, is appropriate in this context – a better scaling of pressure spectra with $\overline{u'v'}$ than q^2 is indicative of active motion. Such a better scaling of the pressure spectra with the Reynolds stress was previously observed in channel flows [15].

4. Conclusions

The main objective of this work was to analyze DNS data of pressure in a zero-pressure-gradient turbulent boundary layer in order to understand more about scaling of pressure spectra and the lack of a -7/3 law. Pressure spectra at different wall-normal locations in a boundary layer show that the slope decreases from -1 to lower values as we move away from the wall. Far out in the boundary layer at $y_+ = 1620$, the spectra is close to -7/3 line. The same trend is observed in the case of streamwise velocity spectra, which approach a -5/3 law in the outer region. This observation suggests that the shear near the wall is somehow responsible for the deviation from isotropic prediction. This also supports the possibility of inner-outer interaction, or active-inactive motion in Townsend's [4] words, which could cause large-scale straining of

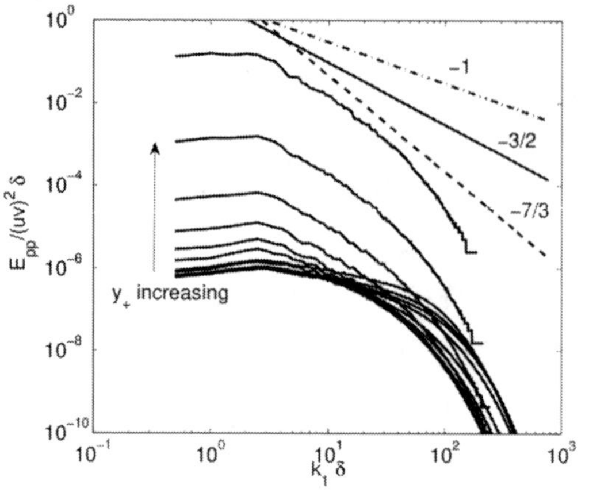

Figure 20. Pressure spectra normalized by $\overline{u'v'}$.

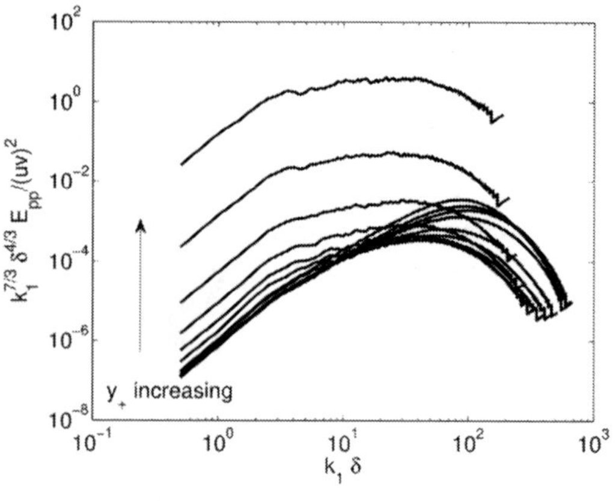

Figure 21. Premultiplied ($k^{7/3}$) pressure spectra normalized by $\overline{u'v'}$.

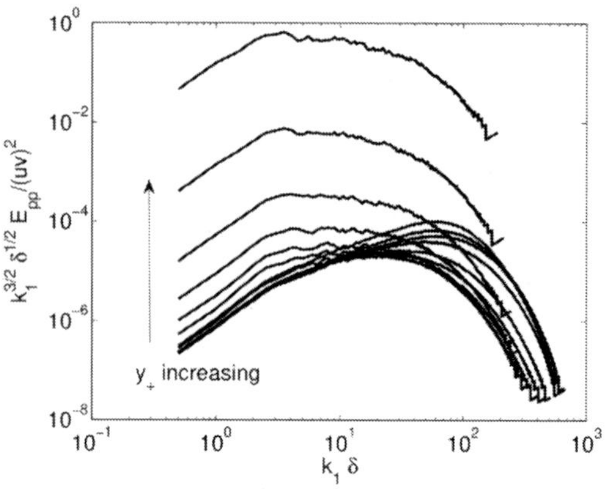

Figure 22. Premultiplied ($k^{3/2}$) pressure spectra normalized by $\overline{u'v'}$.

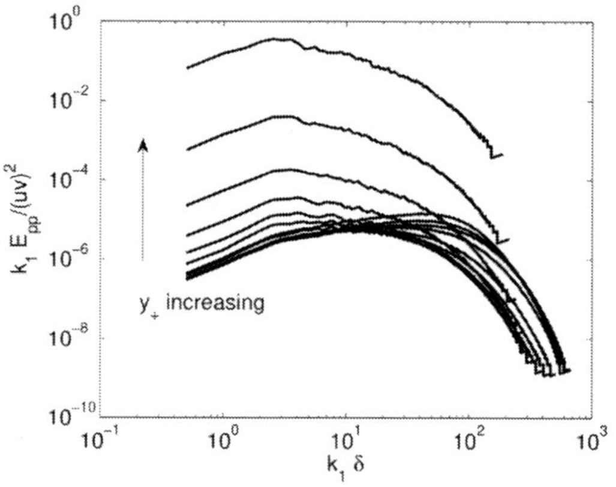

Figure 23. Premultiplied (k) pressure spectra normalized by $\overline{u'v'}$.

small scales in the near-wall region. To study this interaction, we followed the procedure of Mathis *et. al.* [13] to calculate the amplitude modulation effects on small scales by the large scales in the case of pressure fluctuations. However, to our surprise, we found that close to the wall there is no amplitude modulation of the pressure signal, unlike the streamwise velocity. The skewness of pressure also showed a shape similar to the amplitude modulation profile, as in the case of the velocity. Further, we investigated the pressure spectra in other scalings using as normalizing variables $\overline{u'v'}$ and q^2 (kinetic energy). The spectra scaled with $\overline{u'v'}$ show a better scaling compared to the ones scaled with q^2. This is consistent with the classical description by Townsend that inactive motion does not contribute to the Reynolds shear stress, but only to the kinetic energy.

Acknowledgments

This work was carried out as a part of MULTIFLOW workshop at Universidad Politécnica de Madrid and was funded in part by the Multiflow project of the European Research Council. We would also like to acknowledge Mr. Juan A. Sillero for helping us in accessing and processing the

DNS data. We would also like to acknowledge Professor Javier Jiménez and Professor Vassilis Theofilis for hospitality. We would like to thank Professor Paolo Orlandi for suggesting useful comments that have helped to improve the manuscript.

References

[1] Kolmogorov A N 1941 The local structure of turbulence in incompressible viscous fluid for very large Reynolds numbers *Dokl. Akad. Nauk SSSR* **30** [reprinted 1991 in *Proc. R. Soc. London, Ser. A* **434** 9–13]
[2] Monin A S and Yaglom A M 1975 *Statistical Fluid Mechanics* (Cambridge, MA:MIT Press), vol 2, p 874
[3] Albertson J D, Katul G G, Parlange M B and Eichinger W E 1998 Spectral scaling of static pressure fluctuations in the atmospheric surface layer: The interaction between large and small scales *Phys. Fluids* **10** 1725–32
[4] Townsend A A 1976 *The Structure of Turbulent Shear Flow* (Cambridge:Cambridge University Press)
[5] Bradshaw P 1967 Inactive motion and pressure fluctuations in turbulent boundary layers *J. Fluid Mech.* **30** 241–58
[6] Tsuji Y, Fransson J H M, Alfredsson P H and Johansson A V 2007 Pressure statistics and their scaling in high-Reynolds-number turbulent boundary layers *J. Fluid Mech.* **585** 1–40
[7] George W K, Beuther P D and Arndt R E A 1984 Pressure spectra in turbulent shear flows *J. Fluid Mech.* **148** 155–91
[8] Tsuji Y, Imayama S, Schlatter P, Alfredsson P H, Johansson A V, Marusic I, Hutchins N and Monty J 2012 Pressure fluctuation in high-Reynolds-number turbulent boundary layer: results from experiments and DNS *J. Turb.* **13** 50
[9] Sillero J A, Jiménez J and Moser R D 2013 One-point statistics for turbulent wall-bounded flows at Reynolds numbers up to $\delta^+ \approx 2000$ *Phys. Fluids* **25** 105102
[10] Simens M P, Jiménez J, Hoyas S and Mizuno Y 2009 A high resolution code for turbulent boundary layers *J. Comp. Phys.* **228** 4218–31
[11] Borrell G, Sillero J A and Jiménez J 2013 A code for direct numerical simulation of turbulent boundary layers at high Reynolds numbers in BG/P supercomputers *Comp. Fluids.* **80** 37–43
[12] Perot J B 1993 An analysis of the fractional step method *J. Comp. Phys.* **108** 51–58
[13] Mathis R, Hutchins N and Marusic I 2009 Large-scale amplitude modulation of the small-scale structures in turbulent boundary layers *J. Fluid Mech.* **628** 311–37
[14] Mathis R, Marusic I, Hutchins N and Sreenivasan K R 2011 The relationship between the velocity skewness and the amplitude modulation of the small scale by the large scale in turbulent boundary layers *Phys. Fluids* **23** 121702
[15] Jiménez J and Hoyas S 2008 Turbulent fluctuations above the buffer layer of wall-bounded flows *J. Fluid Mech.* **611** 215–36

Possible modification of the large-scale flow structures by vortical structural interactions

Manu V Goudar and Gerrit E Elsinga

Laboratory for Aero & Hydrodynamics, Delft University of Technology, 2628CA Delft, The Netherlands

E-mail: m.goudarvishwanathappa@tudelft.nl, g.e.elsinga@tudelft.nl

Abstract. We explore the mechanisms behind vortical structural interactions modifying large-scale structures in wall turbulence. The evidence for this in terms of vortex interactions, such as merging and intense vortex strengthening, is found in [4] in ideal flow conditions. Here, these interactions are studied experimentally and numerically in turbulent boundary layer and channel flows respectively. This is done by extracting statistical information from conditional averaging of different events based on the spanwise swirling strength. Experimental results showed vortex merger leading to vortex intensification. This was in good agreement with the results of [4]. However, numerical results did not show complete agreement with experimental results. This may be due to the difference in spatial resolution of experimental and numerical data. Furthermore, the peak Reynolds shear stress did reveal a relative increase in magnitude when two vortices merged in the numerical data.

1. Introduction

Understanding energy transfer and cascade processes in wall turbulence helps in developing better large-eddy-simulation models (LES). In LES models, large scales are explicitly computed and small scales are modelled. As the small scales contain only a small part of the turbulent kinetic energy, their effect on the computed large scales is considered limited, and they are modelled in simple ways based on scale invariance and on the universality of the small scales. Understanding the phenomenon behind the small-scale effects on the large scales may help in building better models. There has been some evidence of the small scale interactions with the large scales and of a reverse cascade in [1, 5, 11], but the mechanisms behind these cascade processes and interactions between scales is not completely understood.

Cascade processes and self-sustaining mechanisms in wall-bounded turbulence are likely dynamical, time-dependent processes. However, the study of the associated flow features has often concentrated on the description of the instantaneous structures and their spatial relationships. This can be attributed to the unavailability of time series of the three-dimensional flow-velocity field. Thanks to advances in both numerical simulations and experimental capabilities, such data are now becoming available, which allows us to consider the dynamical aspects of turbulent flow in a quantitative way.

Previous studies using the newly available time series have mainly dealt with the convection velocity and lifetimes of the various flow structures. They have also highlighted examples of vortex structures merging or splitting [2, 3, 8]. When presenting structural interactions, only the (small-scale) vortices were included in the discussion. But these interactions may also affect the larger scales responsible for most of the turbulent kinetic energy.

Cimarelli et al. [1] showed that the energy generated near the wall (buffer region) is transferred towards the outer region and the region very close to the wall. This suggests that the small scales in the near-wall region carry energy to, or interact with, the large scales in the outer regions. This would be an example of a so-called reverse cascade in which energy is transferred from small to large scales. The net average energy transfer is still from large to small scales, in the forward direction. Ishihara et al. [5] also observe a large transfer of kinetic energy down-scale (regular cascade) as well as a large up-scale transfer (reverse cascade) in homogeneous isotropic turbulence. A strong dynamic coupling between the scales was also noted in the simulations of [11].

If the energy is transferred from the small scales to the large ones, it implies that the small-scale interactions may modify the large scales. Evidence for this in an ideal flow was observed in [4]. They studied the dynamics of two vortex interactions similar to the approach followed in [12]. The initial velocity field whose dynamics were studied was a superposition of the perturbation velocity field corresponding to two vortices conditionally extracted from a fully developed turbulent channel flow and the mean turbulent velocity profile at the same Reynolds number. It was found that the two vortices (small scale structures) merge to form a stronger vortex and increase the strength of the low-speed streaks (large-scale structures). So the merging of two vortices and strengthening of low speed streaks can be considered as one of the interactions behind the reverse cascade. However, all the simulations in [4] were in an ideal flow environment at low friction Reynolds number. Observation of this interaction in real turbulent flows at high Reynolds numbers would further strengthen the hypothesis that the large scale structures can be modified by the vortical interactions.

The purpose of this paper is to study vortex interactions and their effect on the large scales in a high-Reynolds-number fully developed turbulent channel flow [9]. A new statistical tool is used to extract the information on vortical interaction. Section 2 describes the methodology. Results and conclusions are discussed in sections 3 and 4 respectively.

2. Methodology

To retrieve statistical information about vortex merging and intense vortex strengthening, a method based on conditional averaging is used. Here the conditional average is approximated by a correlation analysis based on a particular event. We denote by x, y, z the streamwise, spanwise and wall-normal directions, and by u, v, w the corresponding instantaneous velocity components, unless mentioned otherwise. The conditionally averaged fluctuating flow field $\tilde{u}_i' = (\tilde{u}', \tilde{v}', \tilde{w}')$ is estimated by

$$\tilde{u}_i'(\Delta x, \Delta y, z, \Delta t) = \frac{\langle u_i'(x+\Delta x, y+\Delta y, z, t+\Delta t)E(x, y, z_{ref}, t)\rangle}{\langle E(x, y, z_{ref}, t)^2\rangle} \tag{1}$$

where $u_i' = (u', v', w')$ are velocity fluctuations, z_{ref} is a chosen height in the boundary layer (or equivalently a selected wall-normal distance in the channel) and E is the particular event function. The brackets $\langle\cdot\rangle$ indicate averaging over the homogeneous flow directions (streamwise, x and spanwise, y) due to which the conditional average is only a function of the distances Δx and Δy between the reference and observation points.

Three different kinds of events are investigated in the present work. All of them are based on the two-dimensional spanwise swirling strength (λ_y). It highlights spanwise vortical elements which are correlated with the ejection events [4]. When λ_y increases, the ejection events are strengthened, and this strengthening is further correlated with intensification of the low-speed streaks or large-scale structures. The λ_y is the absolute value of the imaginary part of the eigenvalue of the matrix

$$J_{uw} = \begin{bmatrix} \frac{\partial u}{\partial x} & \frac{\partial u}{\partial z} \\ \frac{\partial w}{\partial x} & \frac{\partial w}{\partial z} \end{bmatrix}, \tag{2}$$

and is given a sign to indicate the direction of swirl using the local spanwise vorticity (ω_y). The signed spanwise swirling strength (or signed swirling strength) is given by

$$\lambda_y^* = \text{sign}(\omega_y)\lambda_y. \tag{3}$$

The first type of event is defined in such a way that the extracted velocity field represents the evolution of a single vortex with the same sign as the mean flow velocity. It is also referred to as a type I event, and is defined by

$$E(x, y, z_{ref}, t) = \begin{cases} \lambda_y^* & \text{if } \lambda_y^* > 0, \\ 0 & \text{otherwise.} \end{cases} \tag{4}$$

The second type of event (type II) is based on the Lagrangian time development of λ_y^*, i.e, the material derivative $D\lambda_y^*/Dt$, and is specified as

$$E(x, y, z_{ref}, t) = \begin{cases} \frac{D\lambda_y^*}{Dt} & \text{if } \frac{D\lambda_y^*}{Dt} > (a + b\lambda_y^*) \text{ and } \lambda_y^* > 0, \\ 0 & \text{otherwise.} \end{cases} \tag{5}$$

where a and b are constants with the same dimensions as $D\lambda_y^*/Dt$ and λ_y^*. The lines in the joint histogram (figure 1) of $D\lambda_y^*/Dt$, λ_y^* represent threshold levels for $D\lambda_y^*/Dt$ in the above event. Constant a is the offset from the origin and b is the slope of the line. This event is defined in such a way that at a given point the signed swirling strength is very low, but has a large material derivative, i.e. growth rate.

The conditional average based on the third type of event (type III) gives the evolution of two vortical elements separated by a streamwise ($2\Delta x_s$) and wall-normal ($2\Delta z_s$) distance. Type III events are estimated by the product of the signed swirling strength at both locations as shown below:

$$E(x, y, z_{ref}, t) = \begin{cases} \lambda_y^*(x - \Delta x_s, y, z_{ref} + \Delta z_s) \cdot \lambda_y^*(x + \Delta x_s, y, z_{ref} - \Delta z_s) \\ \qquad \text{if } \lambda_y^*(x - \Delta x_s, y, z_{ref} + \Delta z_s) > 0 \text{ and } \lambda_y^*(x + \Delta x_s, y, z_{ref} - \Delta z_s) > 0, \\ 0 \qquad \text{otherwise.} \end{cases} \tag{6}$$

Conditional averaging will next be carried out based on these events using experimental and numerical data of wall-bounded turbulent flows.

2.1. Numerical and Experimental data sets

The experimental dataset is a time resolved three-dimensional velocity field of a flat-plate zero-pressure-gradient turbulent boundary layer measured by tomographic-PIV [10]. The experiment was conducted in a water tunnel at TU Delft with free-stream velocity $U_c = 0.53$ m/s. Measurements were performed at height $0.1 < z/\delta < 0.3$, where the height of the boundary layer is δ. The streamwise location corresponds to a Reynolds number based on momentum thickness $Re_\theta = 2460$, or to a friction Reynolds number $Re_\tau = 800$. The spatial resolution was 55 wall units and the temporal resolution corresponds to the time taken to travel 10 wall units

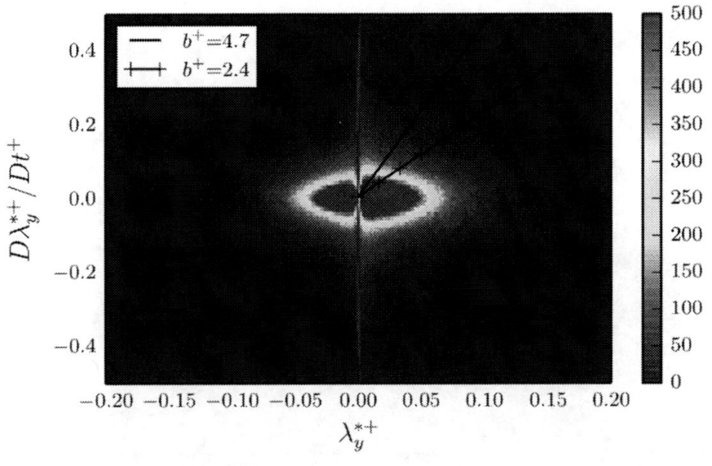

Figure 1. Joint probability density function of $D\lambda_y^*/Dt$, λ_y^* based on numerical data. Each line in the figure represents the limit of an event given by equation (5). Value of constant a^+ is fixed to 0.005.

in the convective direction, $\Delta t^+ = 0.47$ ('+' quantities represent wall units). The volume of each data set was $1.5\delta \times 1.5\delta \times 0.2\delta$ in streamwise, spanwise and wall-normal direction. The total number of data sets used to compute conditional average in terms of time series was equivalent to time taken to travel $100\delta/U_c$. More details on the experimental setup can be found in [10].

The numerical data are from a direct numerical simulation of fully developed turbulent channel flow at the friction Reynolds number $Re_\tau = 950$, based on the channel half-height (h) [9]. The domain size normalized by h is $[2\pi, \pi, 2]$ in streamwise, spanwise and wall-normal directions, respectively. A total of 41 samples of the above volume separated by a time difference $\Delta t^+ = 8$ are used to calculate conditional averages. All the variables throughout the paper are normalized by inner wall units except for the reference wall normal height (z_{ref}) which is 0.2δ in the case of the experiment data and $0.2h$ in numerical data.

2.2. Vortex Identification

Vortex identification is based on the local swirling strength suggested by [12]. It is defined as the imaginary part ($\tilde{\lambda}_{ci}$) of a complex eigenvalue of the velocity gradient tensor ($\partial \tilde{u}_i'/\partial x_j$). If all the eigenvalues are real, the local swirling strength is zero. Vortices are visualized by plotting the iso-surfaces of $\tilde{\lambda}_{ci}^2$ as shown in figure 5.

Vortices are also identified by the Q-criterion [6], as shown in figure 2,which is the second invariant of velocity gradient tensor ($\partial \tilde{u}_i'/\partial x_j$). Vortices visualized by both methods look very similar.

3. Results

In this section, statistical information extracted from experimental and numerical data is presented. This information is obtained by conditionally averaging to specific events, as mentioned in section 2. First, the analysis of experimental data is considered, and then the numerical results for different events are discussed. Mainly the observations connected with vortex merging and swirling strength intensification will be examined.

3.1. Experimental results

Velocity fields conditioned to type I events were extracted in time from the experimental data at $z_{ref} = 0.2\delta$. They represent a single vortex similar to the one shown in figure 5 for the numerical data. The vortex in figure 5 is visualized by iso-contours of the maximum swirling strength, and it is broadly similar to the vortices extracted by linear stochastic estimation in

Journal of Physics: Conference Series **506** (2014) 012012 doi:10.1088/1742-6596/506/1/012012

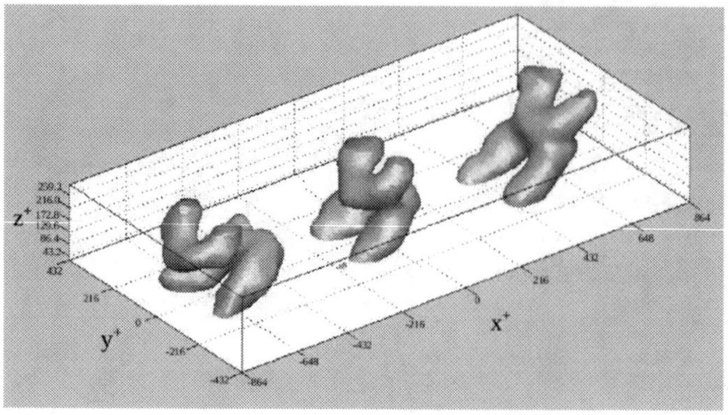

Figure 2. Green contours represent vortices visualized by Q-criterion. Figure shows three stages of the evolution and merging of two vortices at time $\Delta t^+ = -33.9$, -18.8 and 0 extracted from experimental data. Type II event is used to extract these vortices. $z_{ref} = 0.2\delta$, $a^+ = 0.0005$ and $b^+ = 0.21$.

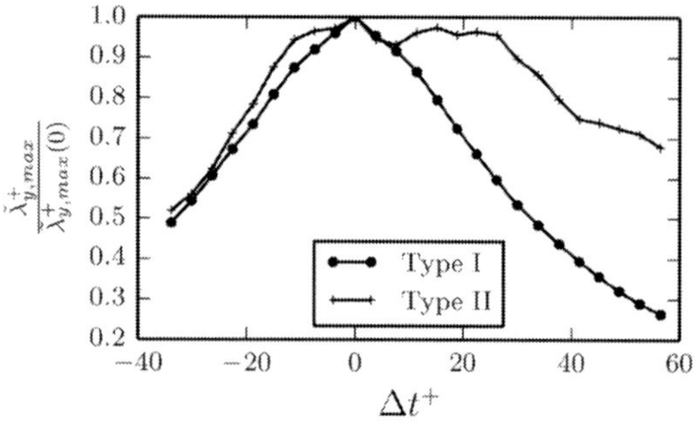

Figure 3. Temporal development of the maximum spanwise swirling strength in experimental data. Merging vortex (type II event) case shows an amplification in peak swirling strength after merging compared to single vortex (type I) case.

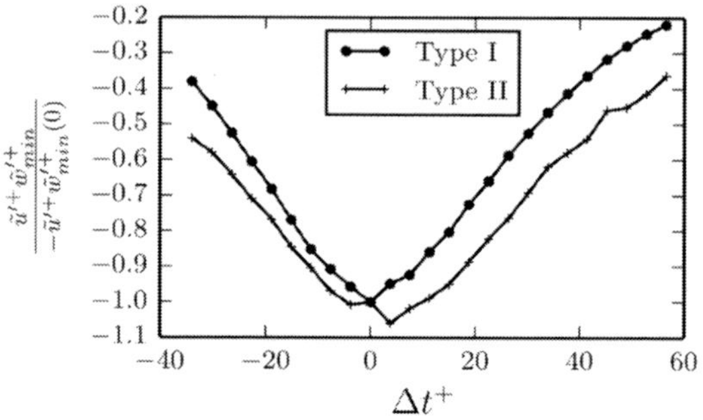

Figure 4. Temporal development of the peak negative Reynolds shear stress $\tilde{u}'^+\tilde{w}'^+$ in experimental data. Merging vortex (type II event) case shows slower decrease in the peak $\tilde{u}'^+\tilde{w}'^+$ after merging compared to single vortex (type I event) case.

[12] and [4]. The maximum spanwise swirling strength of the conditional average at time Δt is normalized by its value at $\Delta t^+ = 0$ and is plotted in figure 3. It can be observed that the swirling strength increases until $\Delta t^+ = 0$, i.e, the time at which the event was specified, and then decreases steadily with increasing time. This decreasing strength corresponds to a loss of correlation with increasing time shifts, which can be attributed to variations in convection velocities of the structures considered in the averaging [3]. Similarly, figure 4 shows that the peak negative Reynolds shear stress, normalized by its value at $\Delta t^+ = 0$, increases until $\Delta t^+ = 0$ and then decreases.

When conditional averaging is conditioned to a type II event, the extracted velocity field represents the merging of two vortices. The evolution of two vortices in time is shown in figure 2, where $z_{ref} = 0.2\delta$ and the conditioning constants are $a^+ = 0.0005$, $b^+ = 0.21$. Those values of a^+ and b^+ in the experimental joint probability density function of $D\lambda^*_y/Dt$, λ^*_y isolate extreme events similar to the lines representing the limit of an event in figure 1. The upstream vortex is located farther from the wall than the downstream one, and travels faster due to higher mean velocity at that position. This decreases the streamwise separation between the vortices, leading to merging. The evolution of the normalized maximum spanwise swirling strength ($\tilde{\lambda}^+_{y,max}$) and the peak Reynolds shear stress $\tilde{u}'^+\tilde{w}'^+_{min}$, conditioned on a type II event, are plotted in figure 3 and 4. It can be observed that both $\tilde{\lambda}^+_{y,max}$ and $\tilde{u}'^+\tilde{w}'^+_{min}$ decrease more slowly after time

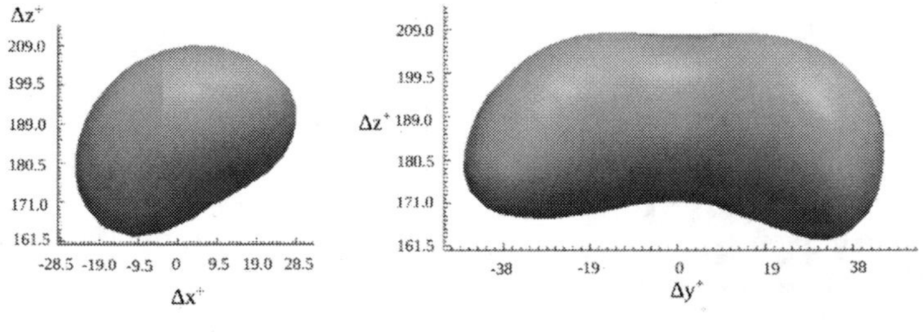

Figure 5. Green contour represents 5% of maximum swirling strength of a vortex extracted using type I event at $z_{ref} = 0.2h$ and $\Delta t^+ = 0$ from numerical data. Left figure is side view (xz plane) and right is front view (yz plane).

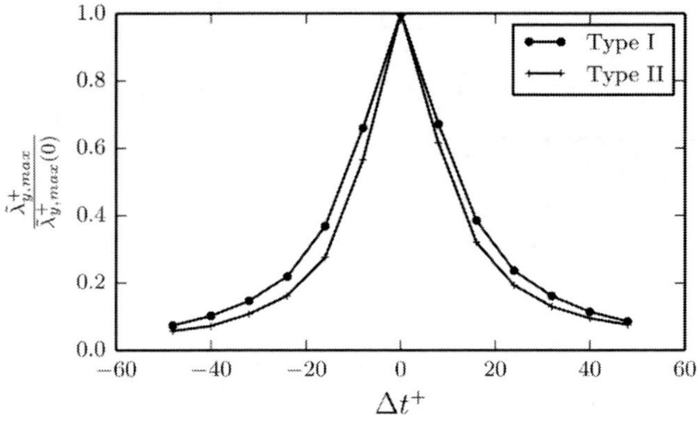

Figure 6. Temporal development of the normalized maximum spanwise swirling strength ($\tilde{\lambda}^+_{y,max}$) in numerical data. Type I and Type II refer to the kind of events used in conditional averaging.

$\Delta t^+ = 0$ than the single-vortex case. Figure 2 at the same time ($\Delta t^+ = 0$) shows merging of two vortices. Based on these observations, it can be inferred that the spanwise swirling strength and Reynolds shear stress for merging case strengthen or decrease slower than the single-vortex case, due to the merging. This is in agreement with the ideal simulation results in [4].

The same statistical procedure is used to extract structures from the numerical data, and the results are discussed in the following section.

3.2. Numerical results

The velocity field is estimated by conditional averaging based on type I events, and is visualized in figure 5 by iso-contours of the swirling strength at $\Delta t^+ = 0$. The single vortex looks similar to the experimental results, and to the conditionally averaged vortices in [12]. The evolution of the single vortex in terms of the maximum spanwise swirling strength ($\tilde{\lambda}^+_{y,max}$), which is the maximum value of the spanwise swirling strength in the whole domain at time t, is shown in figure 6 normalized by its value at $\Delta t^+ = 0$. The trend of increasing and decreasing normalized maximum spanwise swirling strength is similar to the observations found in experiments. The value of normalized maximum $\tilde{\lambda}^+_y$ decreases due to the loss of correlation with time.

The extracted velocity based on a rapidly increasing swirl event, i.e. a type II event at time $\Delta t^+ = 0$, can be seen in figure 7. It is visualized by iso-contours of the swirling strength. The type II event was specified at $z_{ref} = 0.2h$ with constants $a^+ = 0.005$ and $b^+ = 2.4$. Different values of z_{ref}, a^+ and b^+ were also used to compute the conditional averages, and are listed in table 1. For all the values in table 1 and at all the time steps, the extracted velocity field only represented a single vortex, which is completely different from the merging-vortex scenario in experiments (figure 2). Figure 6 shows the evolution of the vortex in terms of its maximum spanwise swirling strength ($\tilde{\lambda}^+_{y,max}$) normalized by its value at time $\Delta t^+ = 0$. The normalized $\tilde{\lambda}^+_{y,max}$ increases and decreases faster than the extracted flow field based on the type I event.

Journal of Physics: Conference Series **506** (2014) 012012 doi:10.1088/1742-6596/506/1/012012

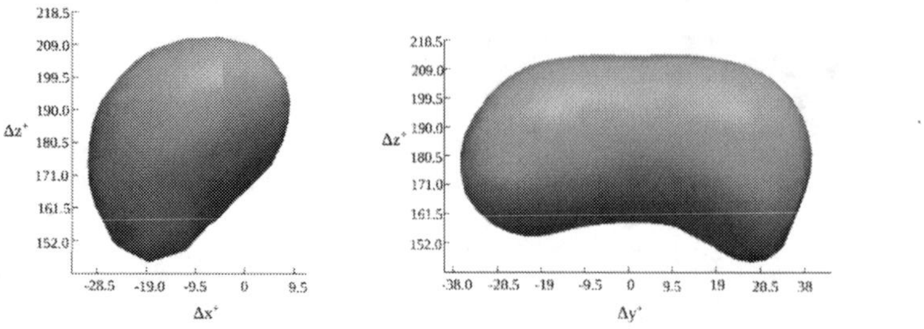

Figure 7. Green contour represents 5% of maximum swirling strength of a vortex extracted using type II event at $z_{ref} = 0.2h$ and $\Delta t^+ = 0$ from numerical data. Left figure is side view (xz plane) and right is front view (yz plane).

Table 1. Overview of all evaluations based on numerical data for type II event.

z_{ref}	a^+	b^+
$0.1h$, $0.2h$, $0.4h$	0.005	2.4
$0.2h$, $0.4h$	0.005	4.7
$0.2h$	0.027	4.7
$0.2h$	0.005	1.17

These results are in complete disagreement with the experimental results, perhaps due to the difference in the spatial and temporal resolutions. The spatial resolution is 55 wall units in all directions in the experimental data, compared to 7.6 and 3.8 wall units in the streamwise and spanwise direction in the numerical data. Moreover, due to an unfortunate error at the time of selecting the dataset, the temporal resolution in the numerical data corresponds to the motion of a fluid particle by approximately 140 wall units at that height, compared to 10 wall units in experimental data. The higher spatial resolution in numerical data results in vortices being sharper and smaller compared to the experiments. As the result of the smaller vortex size (approx 25 wall units) in numerical data, the fluid particle will move out of the vortex, or the material derivative de-correlates quicker in one time step. Similarly, due to coarser spatial resolution in experiments, the vortex size is larger and material derivative de-correlates slower. Hence, a different way of extracting two vortices was considered, to study whether merging leads to swirling strength intensification. One of the trivial ways was to define a type III event, which forces two initial vortices.

Velocity was extracted by conditional averaging based on a type III event at $z_{ref} = 0.2\delta$, with initial streamwise spacings $\Delta x_s^+ = 60$ and 120 and wall normal spacing $\Delta z_s^+ = 28$. The effect of streamwise spacing was also studied, along with merging and vortex intensification. Figures 8 and 9 show the evolution of the extracted velocity field visualized by iso-contours corresponding to 5% of the maximum swirling strength at time $\Delta t^+ = 0$, 8, 16 and 32. The two initial vortices were observed to merge for both the initial streamwise separation $\Delta x_s^+ = 60$ and 120 cases. The cause for merging was the higher velocity of the upstream vortex due to its higher wall-normal distance compared to the downstream vortex. This is similar to the idealized simulations of [4].

To study vortex intensification with the conditional eddies, the time evolution of maximum spanwise swirling strength ($\tilde{\lambda}^+_{y,max}$) and minimum of Reynolds shear stress ($\tilde{u}'^+\tilde{w}'^+_{min}$), normalized by their values at $\Delta t^+ = 0$, are plotted in figure 10 and 11 respectively. In figure 10, it can be observed that the normalized $\tilde{\lambda}^+_{y,max}$ for streamwise spacing $\Delta x_s^+ = 120$ decreases a little faster than for the spacing $\Delta x_s^+ = 60$. This is due to slower de-correlation with increasing time steps for $\Delta x_s^+ = 60$ compared to $\Delta x_s^+ = 120$ case. The drop in correlation is dominated by variations

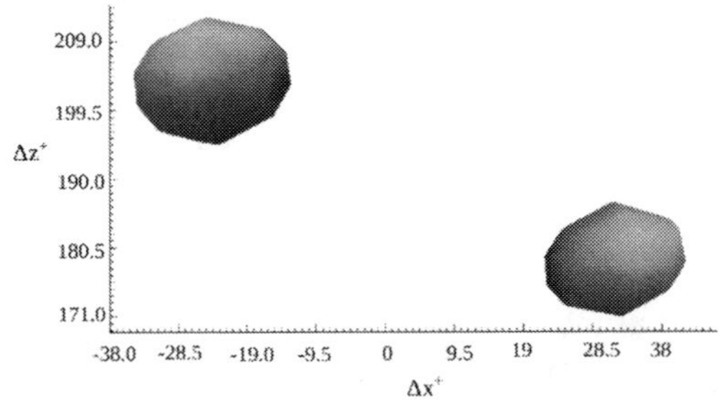

Figure 8. Green contour represents 5% of maximum swirling strength of vortices extracted using type III event at $z_{ref} = 0.2h$, $\Delta x_s^+ = 60$, $\Delta z_s^+ = 28$ and $\Delta t^+ = 0$ from numerical data.

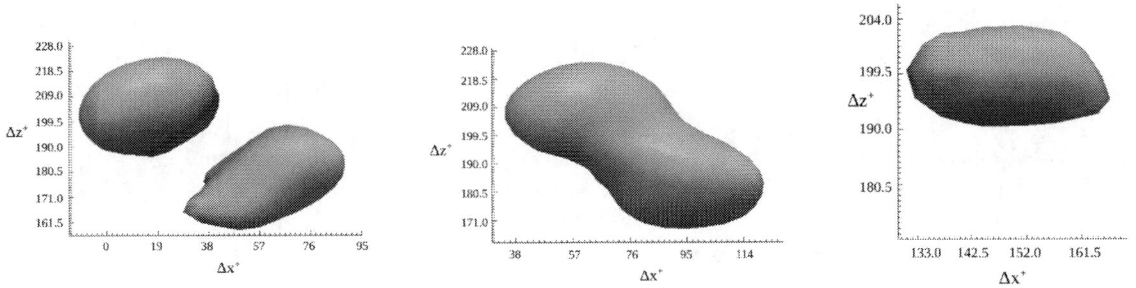

Figure 9. Green iso-contours represent vortices visualized by 5% of maximum swirling strength. Figure shows three stages of the evolution and merging of the two initial vortices shown in figure 8 at time $\Delta t^+ = 8$, 16 and 32 from left to right.

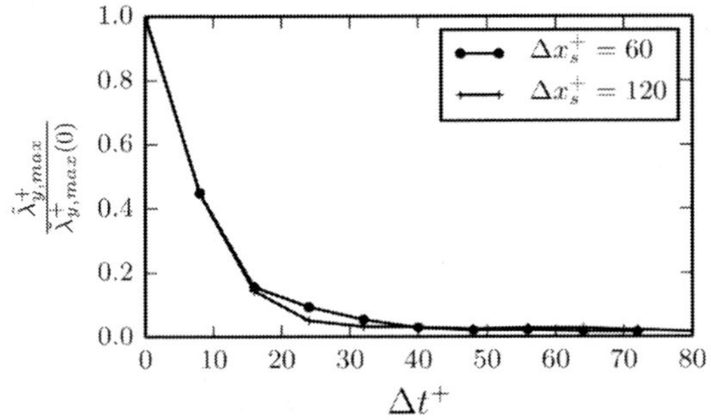

Figure 10. Temporal development of the normalized maximum spanwise swirling strength ($\tilde{\lambda}^+_{y,max}$) in the numerical data. Extracted velocity field is based on type III event. $\Delta x_s^+ = 60$, 120 represent initial streamwise separation between two vortices.

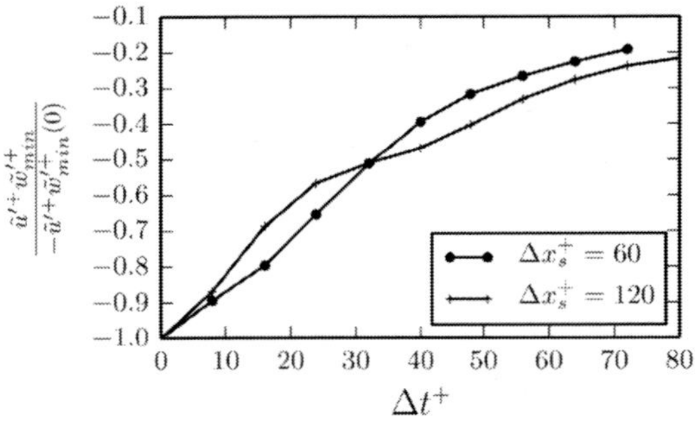

Figure 11. Temporal development of the minimum Reynolds shear stress ($\tilde{u}'^+\tilde{w}'^+_{min}$) in the numerical data. Extracted velocity field is based on type III event. $\Delta x_s^+ = 60$, 120 represent initial streamwise separation between two vortices.

in the convective velocities [3]. Therefore it is very difficult to interpret in terms of absolute or relative vortex strength.

Figure 11 shows a faster decrease in normalized $\tilde{u}'^+\tilde{w}'^+_{min}$ for $\Delta x_s^+ = 120$ before $\Delta t^+ = 32$ and slower decrease after $\Delta t^+ = 32$ compared to $\Delta x_s^+ = 60$. Merging occurs between $16 < \Delta t^+ < 32$ for $\Delta x_s^+ = 60$ case and between $24 < \Delta t^+ < 40$ for $\Delta x_s^+ = 120$. From these observations it can be inferred that merging does have a small effect on the normalized $\tilde{u}'^+\tilde{w}'^+_{min}$, but it is not sufficient to prove any vortex intensification or strengthening.

4. Conclusions

Small-scale vortical interactions were examined using a conditional averaging approach. The interactions were studied in terms of merging, leading to vortex intensification and possible changes in Reynolds shear stress $\tilde{u}'\tilde{w}'$, which is a large-scale property. The dynamics of vortex merging and vortical interactions were extracted from experimental and numerical turbulent flow data by conditional averaging based on events defined by the signed spanwise swirling strength (λ_y^*) and its material derivative ($D\lambda_y^*/Dt$). Statistics obtained by conditioning on the type I event represented the evolution of a single vortex for both experimental and numerical data. Both cases showed a qualitative agreement with the results obtained in [4] [12].

The velocity field extracted by conditional averaging using type II event showed different results for experimental and numerical data. Velocity fields extracted from experimental data showed the merging of two vortices. It also showed an increase in the vortex strength relative to the single vortex case, due to merging. This is in qualitative and quantitative agreement with the ideal simulation results by [4]. Conditional averages obtained from numerical data only showed a single vortex at all times, in contrast to merging-vortex scenario in the experimental data. Different values for z_{ref} and the constants a and b were tested for numerical data. None of the extracted velocity fields showed a merging-case scenario. This may be due to the coarser spatial resolution in experiments, that do not resolve scales below 55 wall units, resulting in larger vortex sizes and slower de-correlation of $D\lambda_y^*/Dt$, or to the coarser temporal resolution of the numerical data, which do not resolve times shorter than 150 wall units. A re-analysis using more-closely spaced numerical data is currently under way.

To test merging and vortex amplification scenario in the numerical data, the type III event was used to extract the flow field. Even though merger between two vortices was observed, there was no clear vortex amplification or even increase in swirling strength. The peak Reynolds shear stress ($\tilde{u}'\tilde{w}'$) however did reveal a relative increase in magnitude near the point of vortex merging.

To conclude, statistics based on the numerical database showed weak signs of vortical interactions modifying the large scales. More elaborate approaches are needed to get more detailed insights into the dynamic interactions between vortices and the large scales.

Acknowledgments

The authors thank Adrian Lozano–Durán for helping us with numerical data and stimulating discussions, along with the participants at the Multiflow summer program. We wish to acknowledge the financial assistance of the Multiflow program at Madrid of the European Research Council.

References

[1] Cimarelli A, de Angelis E and Casciola CM 2013 *J. Fluid Mech.* Paths of energy in turbulent channel flows **715** 436–51

[2] Elsinga GE and Marusic I 2010 Evolution and lifetimes of flow topology in a turbulent boundary layer *Phys. Fluids* **22** 015102

[3] Elsinga GE, Poelma C, Schroder A, Geisler R, Scarano F, and Westerweel J 2012 Tracking of vortices in a turbulent boundary layer *J. Fluid Mech.* **697** 273–95

[4] Goudar MV, Breugem WP and Elsinga GE 2013 Merging and auto-generation of vortices in wall-bounded flows *Proc. Turb. Shear Flow Phenomena (Poitiers)* **COH1D**

[5] Ishihara T, Kaneda Y and Hunt JCR 2013 Thin shear layers in high-Reynolds-number turbulence – DNS results *Flow Turb. Combust.* **91** 895–929

[6] Hunt JCR, Wray AA and Moin P 1988 Eddies, streams and convergence zones in turbulent flows *Proc. Summ. Prog. Center for Turbulence Research* (Stanford University) 193–208

[7] Lozano–Durán A, Flores O and Jiménez J 2012 The three-dimensional structure of momentum transfer in turbulent channels *J. Fluid Mech.* **694** 100–30

[8] Lozano–Durán A and Jiménez J 2012 Time-resolved evolution of coherent structures in turbulent channels *Bull. Am. Physical Soc.* **57** D20.1
[9] Lozano–Durán A and Jiménez J 2014 Effect of the computational domain on direct simulations of turbulent channels up to $Re_\tau = 4200$ *Phys. Fluids* **26** 011702
[10] Schröder A, Geisler R, Staack K, and Elsinga GE, Scarano F, Wieneke B, Henning A, Poelma C and Westerweel J 2011 Eulerian and Lagrangian views of a turbulent boundary layer flow using time-resolved PIV *Exp. Fluids* **50** 1071–91
[11] Yeung PK, Brasseur JG and Wang Q 1995 Dynamics of direct large-small scale coupling in coherently florced turbulence: concurrent physical- and Fourier-space views *J. Fluid Mech.* **283** 43–95
[12] Zhou J, Adrian RJ, Balachandar S and Kendall TM 1999 Mechanisms for generating coherent packets of hairpin vortices in channel flow *J. Fluid Mech.* **387** 353–96

Journal of Physics: Conference Series **506** (2014) 012013 doi:10.1088/1742-6596/506/1/012013

Characterization of the shear layer in a Mach 3 shock/turbulent boundary layer interaction

C Helm[1], M P Martin[1] and P Dupont[2]

[1] Department of Aerospace Engineering, University of Maryland, College Park, MD, 20742
[2] Institut Universitaire des Systèmes Thermiques et Industriels, UMR 7343 CNRS. Aix–Marseille Université, Marseille, France.

E-mail: pmartin@umiacs.umd.edu

Abstract. A detailed characterization of the shear layer in a direct numerical simulation of a Mach 3, separated shock/turbulent boundary layer interaction over a 24° compression ramp is presented. Similarity solutions are identified in the time-averaged shear layer and the growth of the shear layer is found to be approximately linear. In addition to the time averaged spatial organization of the shear layer, the characteristic frequencies and time scales of the large Kelvin–Helmholtz vortices are determined and are found to be consistent with the energized frequency content in the pre-multiplied power spectra of the wall pressure. The three-dimensional nature of the mixing layer vortices are identified using two-dimensional correlation contour plots and a technique for the identification of individual vortices. The large vortices shed from the separated region are found to lie in a plane parallel to the ramp surface, but angled at $\pm 45^o$ to the freestream direction.

1. Introduction

The shock/turbulent boundary layer interaction (STBLI) is an almost unavoidable phenomena in many compressible flow applications such as transonic airfoils, supersonic engine inlets, scramjet isolators, rocket nozzles and so forth. Strong STBLIs pose a significant problem in engineering applications because they are characterized by a low-frequency unsteadiness (two orders of magnitude lower than the turbulence motions of the incoming boundary layer) where the shock experiences large amplitude excursions from its mean position, causing potentially destructive fluctuations in temperature and pressure on the vehicle surface [1-4]. A thorough understanding of the dynamics of the STBLI will help in the development of more accurate predictive models for design applications and possible control solutions. Although the dynamics of separated STBLIs have been studied extensively [1,5-14], the overall understanding is still limited to only a general description of the statistical behaviour of the low-frequency cycle and statistical links between observed phenomena within the interaction. Models for the unsteadiness have been proposed in the past, however, an explicit explanation of the driving mechanism for the low-frequency shock motion has not yet been discovered and much work is still needed in this area of research.

In this paper the focus is on the behaviour and characterization of the shear layer that forms downstream of the shock and above the region of separation. Previous research has provided reason to believe that the shear layer plays a significant role in dictating the overall dynamics of the interaction. For example, Piponniau, Dussauge, Debieve, and Dupont [10] proposed a model of the low-frequency cycle by assuming that the depletion of the separation bubble is a result of

Table 1: Characteristics of the incoming boundary layer.

M	Re_θ	δ^+	$U_{\rm inf}$ ($\rm ms^{-1}$)	δ (mm)	δ^* (mm)	θ (mm)	H
2.91	2.9×10^3	3.4×10^2	610	7.1	2.58	0.47	5.49

fluid entrainment by the shear layer, and the authors were able to obtain a reasonable estimate of the low-frequency of their reflected shock interaction from experimental PIV data. Also, large vortical structures associated with the Kelvin-Helmholtz type instability are known to exist in both the reflected shock configuration (Dupont, Piponniau, Sidorenko, and Debieve [15] and Souverein *et al* [13]), and the compression ramp configuration (Priebe & Martin [14]). Agostini *et al* [16] used cross-correlations between the streamwise position of the shock and the pressure field in their large-eddy simulations of reflected shock interactions to provide further evidence that large, periodic vortices are generated in the shear layer and convect downstream. Priebe & Martin [14] observed a sawtooth-like motion of the reattachment point of the separation bubble in their direct numerical simulation of a Mach 3 compression ramp and, basing their conclusions on observations of similar behaviour in incompressible separation bubbles (Kiya & Sasaki [17]), associated this motion with the passage of these large vortical structures into the downstream flow.

To lend more insight into the behaviour of the shear layer and how it might contribute to the low-frequency unsteadiness of STBLI flows, a detailed characterization of the shear layer in the DNS of a Mach 3 STBLI over a 24° compression ramp is presented here. The paper is organized as follows. The DNS data set is introduced in section 2. In section 3, the time-averaged spatial description of the shear layer is presented, and in section 4 the timescales and spatial organization of the shear layer vortices are described. A summary of conclusions is given in section 5.

2. Flow and computational method

The analytical results presented in this paper were produced using data from a direct numerical simulation (DNS) of a shock-turbulent boundary layer interaction (STBLI) where the shock is produced by a 24° compression ramp. The incoming boundary layer is fully developed with a freestream Mach number of 2.9. This particular data set has been very well documented in the paper by Priebe & Martin [14] and the reader is referred to this paper for a complete description of the computational methods used and of the time-averaged and instantaneous organization of the flow, as well as a thorough analysis of frequency content and spectra. In summary, upstream of the shock the freestream velocity and temperature are $U_{\rm inf} = 610$m/s and $T_{\rm inf} = 107$K. The incoming boundary thickness is $\delta = 7.1$mm and the Reynolds number based on the momentum thickness is $Re_\theta = 2900$. A list of the important parameters of the incoming turbulent boundary layer are provided in Table 1. The interaction is fully separated in the time-averaged sense and the separation length defined as the distance along the wall between the mean separation point and the mean reattachment point is $L_{\rm sep} = 3.0\delta$. The computational grid is curvilinear with a resolution of $1024 \times 160 \times 128$ grid points in the streamwise, spanwise and wall-normal directions respectively. This corresponds to a physical domain size of approximately $(13.7 \times 2.0 \times 4.4)\delta$. The inflow conditions were prescribed from a separate auxiliary DNS of a spatially developing, flat-plate, turbulent boundary layer. The auxiliary DNS was computed on a Cartesian grid of size $410 \times 160 \times 112$ points with physical dimensions $(8.3 \times 2.0 \times 8.2)\delta$. The same computational

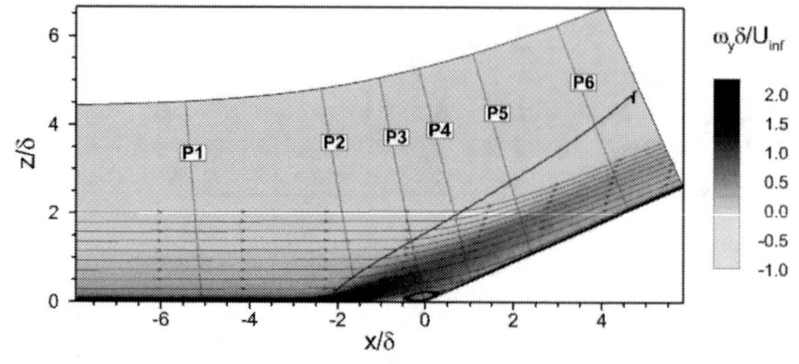

Figure 1: Compression ramp mean-flow organization.

method and boundary conditions as were used for the ramp simulation were used for the auxiliary DNS except that the inflow conditions were assigned using the recycling/rescaling method of Xu & Martin [18].

The three-dimensional space of the ramp domain is expressed in terms of the standard x, y, and z coordinate axes where x is in the streamwise direction with respect to the incoming flow, y is in the spanwise direction, and z is perpendicular to the wall surface ahead of the ramp. The velocity components in the three respective directions are indicated by u, v, and w.

The time- and spanwise-averaged organization of the ramp flow is shown in Figure 1. The average location of the shock is visualized in Figure 1 by the vertical location of maximum $|\nabla P|$ as a function of x. The boundary layer and the shear layer just downstream of the shock are visualized by the increased levels of the in-plane vorticity ω_{y}. Mean streamlines clearly demonstrate a region of recirculation at the ramp corner. In the DNS computational domain there are six yz grid planes from which time signals of the primitive flow variables are output at a high-frequency sample rate and the locations of these planes are indicated in Figure 1. The high frequency time signals are output at the sampling rate $f_{\mathrm{s}} = 200U_{\mathrm{inf}}/\delta$ for a duration of $766\delta/U_{\mathrm{inf}}$.

3. Time-averaged shear layer

The mean shear layer in the DNS data is visualized in Figure 2 by the contour plot of the averaged two-dimensional turbulent kinetic energy or TKE. In this section, an average quantity refers to an averaging in both time and the spanwise direction. The mean separation point is $x_{\mathrm{s}}/\delta = -2.1$ and the mean reattachment point is $x_{\mathrm{r}}/\delta = 0.9$. The white line marks the locus of points of maximum TKE found along the z-direction at each point in x. The location of maximum TKE can be considered as the approximate centre of the shear layer.

The qualitative organization of the mean shear layer can be described as follows: the shear layer appears to originate just downstream of the mean separation point, from where it grows in width until its growth is confined by the wall on the inclined side of the ramp. Its energy then smears out as it dissipates downstream of reattachment. The shear layer centre appears to form a straight line between a point near the wall just downstream of the shock foot to a point about 0.5δ above the mean reattachment. Downstream of reattachment, the line of maximum TKE runs essentially parallel to the ramp wall. This qualitative description is used to define a coordinate system for the developing portion of the shear layer. The axes of the shear layer coordinates are denoted by x' and z' where x' is oriented along the line of maximum TKE above the recirculating zone. The z'-axis is perpendicular to x' and pointed away from the wall. The origin of the mixing layer coordinates is the point at which the x'-axis intersects the wall upstream of the ramp. The distance of the shear layer coordinate's origin from the ramp corner is denoted by $x_{\mathrm{o1}} = -1.90\delta$. The shear layer coordinates are included in Figure 2 and they make an angle of 16.5^o to the computational coordinates. This angle is comparable but slightly

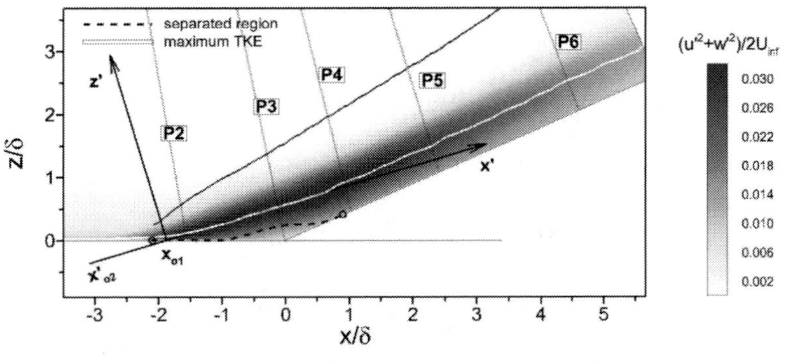

Figure 2: Spatial organization of the average shear layer. Contours of the time- and spanwise-averaged two-dimensional kinetic energy, $TKE = (u'^2 + w'^2)/2U^2_{\text{inf}}$. Time- and spanwise-averaged: shock location (solid black); approximate shear layer (white line); and dividing streamline for separated flow (dashed black).

higher than the shear layer angle of 11.5^o found by Dussauge, Piponniau, and Dupont [19] in the PIV data of reflected shock experiments at Mach 2.3 and $Re_\theta = 5100$ with flow deviation of 8.5^o and 9°. Delery & Marvin [1] also state that a typical experimental value of the shear layer's inclination to the wall is on the order of 11°.

If the shear layer does in fact grow linearly it should be possible to define a similarity variable, namely $\zeta = z'/(x' - x'_{o2})$, that will collapse the velocity profiles to a single similarity solution. The parameter x'_{o2} as indicated in Figure 2 is defined as the (imaginary) geometric origin of the shear layer and is found by varying the position of x'_{o2} along the x'-axis until what is deemed a good quality collapse of velocity profiles results. A value of $x'_{o2} = 1.41\delta$ was found to produce qualitatively the best collapse. In Figure 3 the profiles of the mean velocity and the two-dimensional Reynolds stresses $\overline{u'^2}$, and $\overline{u'w'}$ are plotted against the similarity variable ζ. Here the u' and w' indicate velocity fluctuations where the velocity components are in the direction of the x' and z' axes respectively. In Figure 3 the velocity terms are non-dimensionalized by $U_1 = 470$(m/s) which is the inviscid compression ramp solution for the freestream velocity behind a shock generated by a 24° ramp at $M = 2.9$. The velocity U_1 is therefore an estimate of

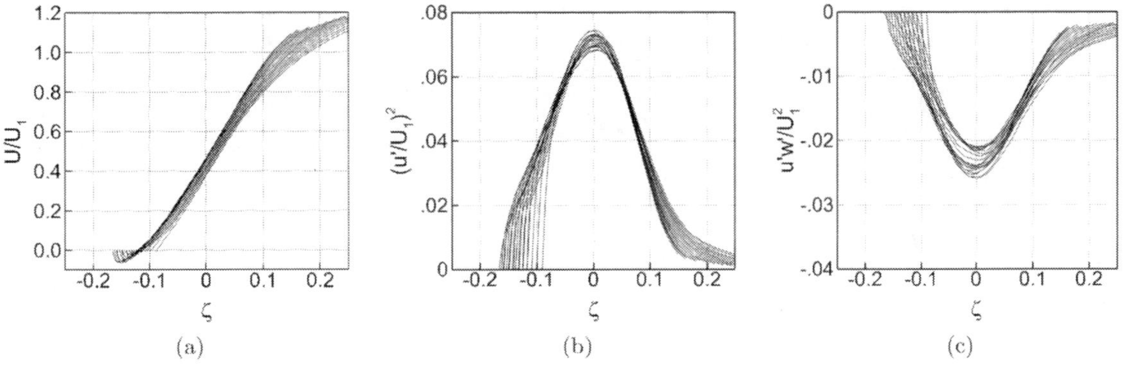

Figure 3: Collapse of profiles for (a) $\overline{U}$, (b) $\overline{u'^2}$, and (c) $\overline{u'w'}$ versus similarity variable ζ.

the freestream velocity on the shear layer's high-speed side. The profiles in Figure 3 are taken over the interval $-1.26\delta < x < 0.63\delta$. The streamwise and cross Reynolds stress profiles are well described as symmetrical about the mixing layer centre. In Figure 3, a spreading of the profiles on the low speed side is caused by the proximity of the wall surface. The profiles bear some resemblance to the profiles of a canonical mixing layer with self similarity as described by Townsend [20].

Dussauge *et al* [19] demonstrated with their experimental data of reflected STBLIs that is possible to estimate the spreading rate of the shear layer from the similarity profile of $\overline{u'^2}$. If it is assumed that the subsonic and supersonic mixing layers have the same $\overline{u'^2}$ profile shape, a comparison can be made between the widths of the two profiles and the compressible spreading rate can be determined from the incompressible spreading rate by the relation

$$\left(\frac{d\delta}{dx'}\right)_{\text{comp.}} = \frac{\zeta_{\text{comp.}}}{\zeta_{\text{incomp.}}}\left(\frac{d\delta}{dx'}\right)_{\text{incomp.}} \tag{1}$$

where $\zeta_{\text{incomp.}}$ and $\zeta_{\text{comp.}}$ are the widths of the two $\overline{u'^2}$ profiles at the same value of $\overline{u'^2}/max(\overline{u'^2})$. The spreading rate is denoted by $d\delta/dx$ where δ is the mixing layer thickness and is not to be confused with the boundary layer thickness from previous discussions. Using the subsonic mixing layer profiles of Mehta & Westphal [21], the linear spreading rate $d\delta/dx$ is estimated as 0.222. This result is comparable to the estimated spreading rates of 0.287 and 0.188 for 8° and 9.5^o reflected shocks respectively by Dussauge *et al* [19].

4. Large-scale vortices – time and space organization

The velocity profiles of the time-averaged shear layer shown in Figure 3 are characterized by a low-velocity side and a high-velocity side connected by a single inflection point. The single inflection point in the mean velocity is associated with an inviscid instability that produces the large Kelvin–Helmholtz vortices visualized in planar mixing layers (For example see the visualizations by Mehta & Westphal [21]). Large, spanwise-oriented vortices shedding downstream of the interaction have been identified in the reflected shock experiments of Dupont, Piponniau, Sidorenko, and Debieve [15] and Souverein *et al* [13]. These vortices have also been identified in incompressible separation bubbles, for example in the experiments of Hillier & Cherry (1981) and Kiya & Sasaki [17, 23] and in the discrete vortex simulations of Kiya, Sasaki, and Arie [24]. In addition, Priebe & Martin [14] identified strong, streamwise-oriented vortices in the same DNS data we use in this analysis. It is apparent that the large vortices occurring in these separated flows are originating from the shear layer and are being generated by a similar mechanism as in the planar mixing layer, that is by the inviscid instability in the mean profile. The characteristic time and length scales of the large-scale vortices are described in this section.

4.1. Vortex characteristic time scale

Information on the average time period of the large vortices can be estimated from cross-correlations in time between signals of streamwise mass flux $(\rho u)'$ and the wall pressure P'_{w} if the following thought experiment is considered. The vortices are assumed to appear in the shear layer with more or less a regular spacing between them, with one being followed by the next. Any one of these large vortices in the shear layer will produce a negative fluctuation in the pressure at the wall surface just beneath the vortex core. It might also be reasonably assumed that in between successive vortices there is on average a positive fluctuation in the wall pressure. If a velocity signal is taken from a stationary point in the low-speed side of the shear layer, a negative fluctuation in velocity will occur in the signal when a vortex lies just above the 'probe' location and conversely a positive fluctuation would occur in between successive vortices. With this reasoning, a correlation between a mass flux signal at a point below the average mixing

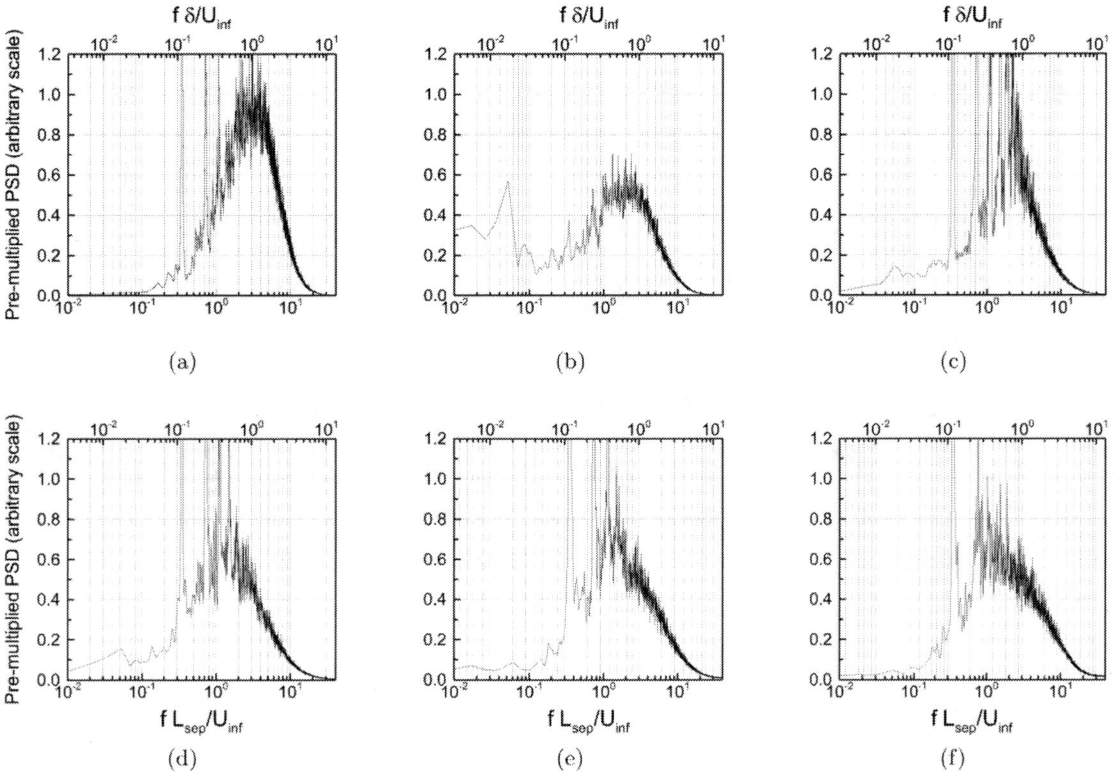

Figure 4: Pre-multiplied power spectrum plots for wall pressure signals at each of the grid planes indicated in Figure 1: (a)-(f) correspond to grid planes 1-6.

layer centre, and the signal of wall pressure just below that, should provide some information on the average periodic organization of the shear layer vortices. In fact, the auto-correlations of wall pressure by Hillier & Cherry [22] and the two-point correlations between velocity and wall pressure of Kiya & Sasaki [23] in incompressible separated shear layers show that these types of correlations produce a sinusoidal curve. The time period between successive vortices is the time period of the sinusoid. (See for example Figure 25 in Kiya & Sasaki [23].)

When performing two-point cross-correlations of the raw DNS time signals, both the smallest scales of the turbulence and the low-frequency motions of the shock tend to wash out the expected periodic shape of the mixing layer correlation curves. To retain only the signature of the large rollers, the data is first band-pass filtered before calculating the correlations. The band-pass filter is designed based on frequency information in the pre-multiplied power spectra of the wall-pressure signal. The pre-multiplied power spectral density (PSD) of the time signals of wall pressure at each of the high-frequency grid planes from Figure 1 are plotted in Figure 4. In Figure 4a, in the upstream boundary layer, a single broadband peak exists in the spectrum but, as the boundary layer passes through the interaction region (subsequent figures b-f), the centre of this peak undergoes a slight shift from a frequency $f \approx 5U_{\text{inf}}/L_{\text{sep}}$ to $f \approx 1U_{\text{inf}}/L_{\text{sep}}$. Just downstream of the shock at grid plane 2 shown in Figure 4b a second broadband energy peak appears at $f \approx 0.03U_{\text{inf}}/L_{\text{sep}}$. This second peak is associated with the low-frequency unsteadiness of the shock, and it is seen to disappear downstream of the separated region in Figures 4e and 4f. In addition to the broadband peaks, tall spikes are also apparent in the

Table 2: Summary of enhanced correlation results.

GridPlane	ΔT $(U_{\text{inf}}/L_{\text{sep}})$	f $(L_{\text{sep}}/U_{\text{inf}})$	z_{probe} (δ)	ζ_{probe}	N_e/N
2	0.547	1.828	0.089	-0.009	40 / 3072
3	0.691	1.447	0.196	-0.104	236 / 2720
4	0.849	1.178	0.571	-0.065	584 / 2720
5	0.832	1.202	1.189	—	740 / 2720
6	0.834	1.214	2.176	—	865 / 2720

spectra of Figure 4. As explained by Priebe & Martin [14], these are associated with a forcing of the turbulence by the recycling inflow method used in the simulation of the auxiliary DNS, which is in turn used as the inlet condition of the ramp simulation. As it turns out, if the first harmonic ($f \approx 0.35U_{\text{inf}}/L_{\text{sep}}$) of the rescaling frequency is filtered from the data, the subsequent harmonics do not effect in any significant way the results of the cross-correlation analysis. In order to remove from the data the effects of (1) the low-frequency cycle energy, (2) the first harmonic of the forcing introduced by the inflow condition, and (3) the finest scales of the turbulence, a finite impulse response (FIR) band-pass filter was designed with high- and low-frequency cutoff values of $0.4U_{\text{inf}}/L_{\text{sep}}$ and $10.0U_{\text{inf}}/L_{\text{sep}}$.

The band-pass filtering alone was not sufficient to bring out the desired sinusoidal shape in the cross-correlations and so a conditional averaging technique was used in addition to the filtering to produce 'enhanced' correlations. The enhanced correlations were computed as follows: the time signals of mass flux and wall pressure at each of the grid planes labelled in Figure 1 were first band-pass filtered, broken up into shorter time segments, and then cross-correlations were calculated between each pair of corresponding shortened $(\rho u)'$ and P'_{w} time signals. If the short correlation had the desired sinusoidal shape, it was retained for averaging, and if no such large-scale pattern was detected the time segments were discarded. This is similar to the method of enhanced correlations between mass flux and wall-shear stress used by Brown & Thomas [25] in their studies of the coherent structures in turbulent boundary layers.

Correlations were computed at grid planes 2-6 where streamwise mass flux signals were taken

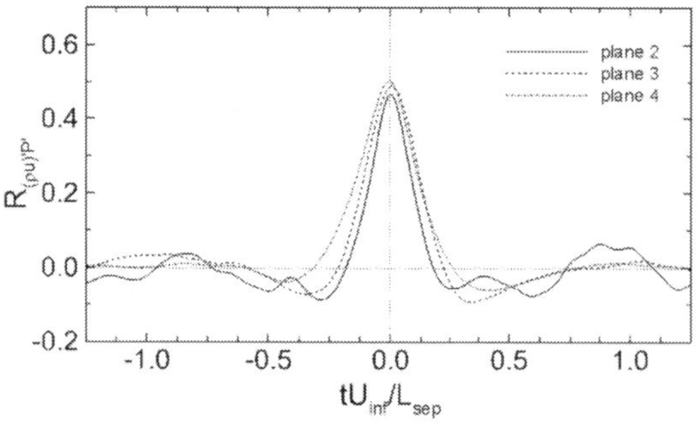

(a)

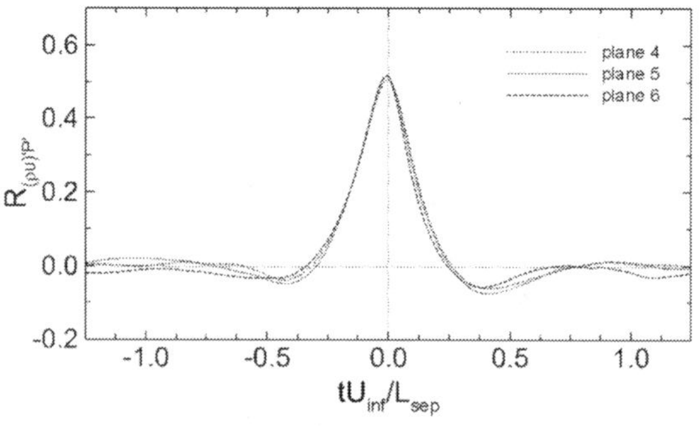

(b)

Figure 5: Enhanced correlations $R_{(\rho u)'P'}$ for (a) grid planes 2-4 and (b) 4-6.

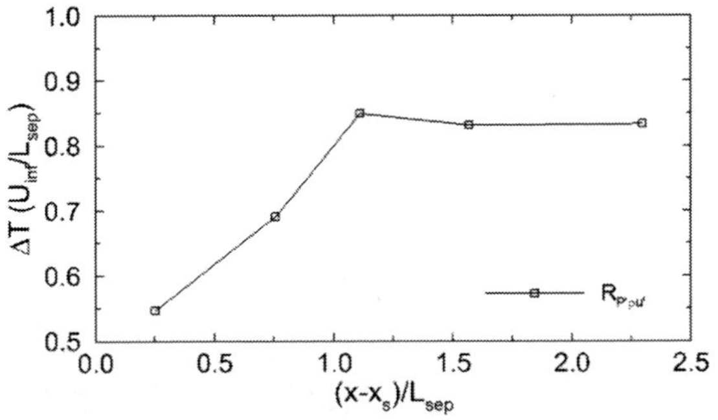

Figure 6: The time period ΔT between successive mixing-layer vortices versus downstream distance from the mean separation point x_s.

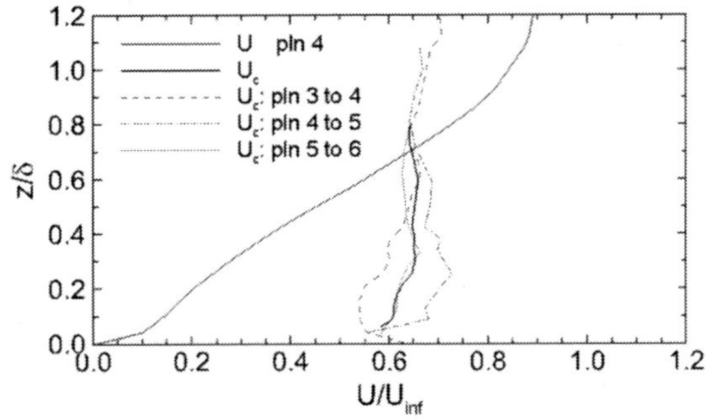

Figure 7: Average vortex convection velocity profile compared to the mean velocity profile at grid plane 4.

from 32 equally spaced points along the spanwise direction at a single wall-normal location. For grid planes 3-6 the time signals were broken up into $t = 5.8\delta/U_{\rm inf}$ long segments, approximately 2.5 times the period of the central frequency in the wall-pressure spectra. At grid plane 2 slightly shorter segments were used, as it was found that a time segment duration of $t = 5.0\delta/U_{\rm inf}$ produced stronger correlations at this plane. For a short correlation to be retained, two criteria had to be met. First, the cross-correlation curve at zero time lag must be greater than 0.4 (approximately twice the maximum of the non-enhanced correlations). Second, a minimum of less than -0.2 must occur at some point in the cross-correlation curve. The resulting enhanced correlations $R_{(\rho u)'p'}$ are plotted in Figure 5. The average time period ΔT between vortices is estimated as the distance between the two minima of the correlation curves, and are summarized in Table 2. For each grid plane, the timescale, the corresponding shedding frequency, the wall-normal location of the mass flux signals, and also the number of segments retained for the enhanced correlations over the total number of segments tested (N_e/N) are given. It is worth mentioning that no such periodic pattern was found in the two-point correlations in the incoming boundary later at grid plane 1.

Two important observations are made from this analysis. First, if the time scales in Table 2 are plotted versus $(x - x_s)/L_{\rm sep}$ (see Figure 6), it is apparent that the time scale grows almost linearly through the separated region and then levels off to approximately a constant value downstream of reattachment. The same trend in the development of the time scale was observed by Kiya & Sasaki [23]. Second, the frequencies reported in Table 2 are located at the approximate centre of the broad-band peak in the pre-multiplied PSD plots of the wall pressure at each grid plane. Kiya and Sasa [23] also showed that the broad-band peak in their wall-pressure power spectrum at reattachment was centred at the average vortex shedding frequency.

4.2. Vortex convection velocity

The frequency reported for grid plane 4 in Table 2 is interpreted as the shedding frequency of the large-scale vortices from the separated region of the interaction. It is assumed that these vortices propagate downstream of reattachment at approximately a constant convection velocity, as would be consistent with the levelling off of the vortex time scale observed in Figure 6. The convection velocity of the vortices can be estimated from two-point correlations calculated between mass-flux signals separated by some known distance in the streamwise direction. The average time it takes for an eddy to convect from the first signal location to the second is given by the offset of the peak in the cross-correlation curve. Thus it follows that the eddy convection velocity is this time lag divided by the distance between the two probes.

The two-point correlations are calculated between streamwise mass flux signals from adjacent high-frequency output planes. In this analysis, streamwise mass fluctuations $(\rho u)'$ now refer to fluctuations in the direction of the local mean streamline. Also, correlated signals between adjacent grid planes are chosen so that they lie on the same mean streamline. To determine an average convection velocity of the large vortices specifically, the time signals are filtered using a narrow band-pass FIR filter designed to retain frequencies between $f = 0.4U_{\mathrm{inf}}/L_{\mathrm{sep}}$ and $2.0U_{\mathrm{inf}}/L_{\mathrm{sep}}$, or roughly half to twice the average shedding frequency. Three separate estimates of convection velocity profiles were computed from correlations between grid planes 3 and 4, between grid planes 4 and 5, and between grid planes 5 and 6. The convection velocity profiles and the mean velocity profile at grid plane 4 are plotted together in Figure 7. A final value for the convection velocity is estimated by an average over the three resulting velocity profiles, or $U_{\mathrm{c}} = 0.65U_{\mathrm{inf}}$, where U_{c} is in the direction parallel to the ramp surface.

4.3. Vortex spatial organization

The convection velocity $0.65U_{\mathrm{inf}}$ multiplied by the vortex timescale of $0.85L_{\mathrm{sep}}/U_{\mathrm{inf}}$ at grid plane 4 yields an average spacing of $0.55L_{\mathrm{sep}}$ or 1.7δ between vortices downstream of reattachment. In their reflected shock experiments, Dupont *et al* [15] reported a spacing of approximately $0.4L_{\mathrm{sep}}$.

Additional qualitative information on the average spatial organization of the mixing layer vortices can also be observed from the two-dimensional (2D) correlations defined by

$$R_{(\rho \mathrm{u})'\mathrm{P}'} = \frac{\langle(\rho u)'(x+\Delta x, y+\Delta y, z)P'(x,y)\rangle}{\langle(\rho u)'_{\mathrm{rms}}P'_{\mathrm{rms}}\rangle}. \qquad (2)$$

The 2D correlations are calculated from the time signals at grid plane 4, where x is substituted with the quantity tU_{c}. The resulting correlation plots therefore describe vortex structures that are essentially frozen in space as they convect past grid plane 4. Before computing the

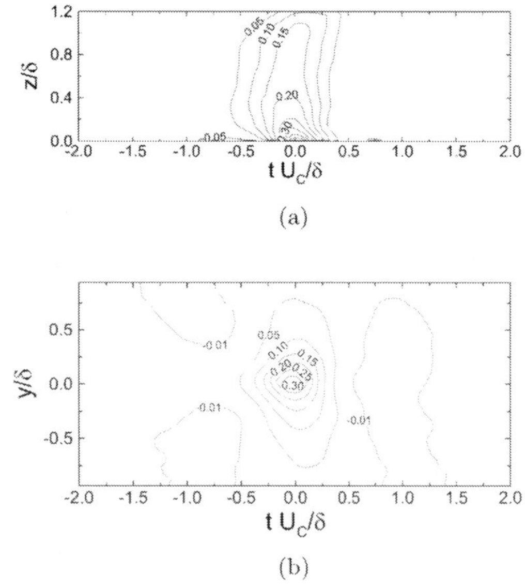

Figure 8: Contour plots of 2D correlations $R_{(\rho \mathrm{u})',\mathrm{P}'}$ (a) in the xz-plane at $y/\delta = 0$ and (b) in the xy-plane at $z/\delta = 0.1$.

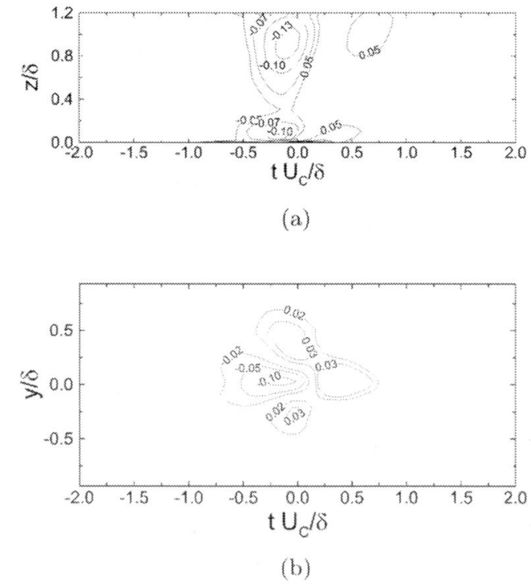

Figure 9: Contour plots of 2D correlations $R_{(\rho \mathrm{w})',\mathrm{P}'}$ (a) in the xz-plane at $y/\delta = 0$ and (b) in the xy-plane at $z/\delta = 0.1$.

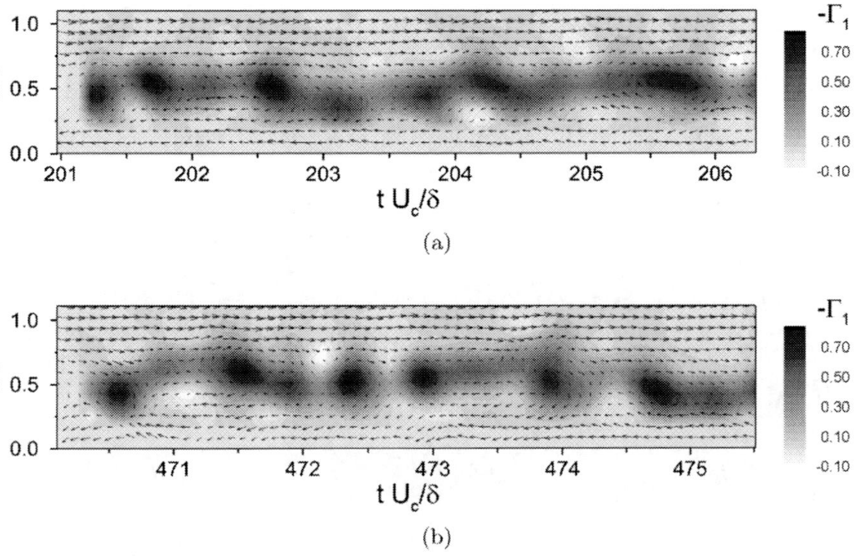

Figure 10: Contour plots of the vortex detector Γ_1 for two separate snapshots in time at grid plane 4. A value of $|\Gamma_1| > 0.63$ represents a vortex core.

correlations, the time signals of mass flux and wall pressure are band-pass filtered in time with a new filter that retains frequencies between $0.4U_{\text{inf}}/L_{\text{sep}}$ and $5.0U_{\text{inf}}/L_{\text{sep}}$. The same signature pattern of the large vortices is observed in the correlation plots, regardless of whether the high-frequency cutoff is set to $10.0U_{\text{inf}}/L_{\text{sep}}$ as in section 4.1, or to $5U_{\text{inf}}/L_{\text{sep}}$. However, the pattern is a bit more prominent when the high-frequency cutoff is set to $5.0U_{\text{inf}}/L_{\text{sep}}$. Contour plots of the 2D correlations between $(\rho u)'$ and P'_{w} are shown in Figure 8. In Figure 8a is plotted $R_{(\rho \text{u})',\text{P}'}$ in the xz-plane at $y = 0$ and in Figure 8b is plotted $R_{(\rho \text{u})',\text{P}'}$ in the xy-plane of $z = 0.1\delta$. If Taylor's hypothesis of frozen turbulence is assumed true for the large vortices, positive tU_{c}/δ in Figure 8 would correspond to structures downstream of plane 4, and negative tU_{c}/δ to structures upstream of plane 4. A large-scale structure is observed that extends in the wall-normal direction across the entire boundary layer. The 'footprint' for this structure is approximately 2δ in streamwise length. It is at once obvious that the large shear-layer vortices are three-dimensional in nature, since there appears to be an angle or kink in the structure pattern (see the xy-plane correlation plot shown in Figure 8b). This asymmetry is even more apparent in the 2D correlations between wall-normal mass flux and wall-pressure $R_{(\rho \text{w})'\text{P}'}$ that are plotted in Figure 9 for the same planes. In Figure 9b, an angle of about 45° to the x-axis is seen in the region of positive correlation.

The $R_{(\rho \text{w})'\text{P}'}$ correlations can be interpreted using the same arguments as those that were used in section 4.1. When a vortex exists just above a probe located in the low-speed side of the mixing layer, the wall-normal velocity will on average have a zero fluctuation. If the vortex is just ahead or behind the probe a negative or positive fluctuation respectively might be expected to occur. In addition, in between vortices, there is zero fluctuation in wall-normal velocity. As a result, the fluctuations in wall-normal velocity follow the same pattern as the fluctuations in wall pressure, only shifted ahead in time by a fourth of the time period of the vortices. Returning to Figure 9b, there is a positive lobe in the 2D correlations that is centred at approximately $tU_{\text{c}}/\delta = 0.4$, or about a fourth of the average vortex spacing of 1.7δ derived earlier, and a lobe of negative correlations just to the left of $tU_{\text{c}}/\delta = 0$. The results suggest that the vortices are angled in the xy-plane instead of spanning the width of the domain in a straight line.

An effort is made to explain the structure appearing in the 2D correlations of Figures 8 and 9 by visualizing individual (frozen) vortices in the time data of grid plane 4. A number of methods exist for the identification and visualization of vortical structures in turbulent flows. The goal here is to pick out specifically the large-scale vortices whose vortex cores are oriented in the spanwise direction in order to keep the pictures clear and not cluttered, for example by streamwise oriented vortices. Although it is a rather crude method for the identification of vortices and is generally used for post-processing coarser PIV vector field data (for example see Dupont *et al* [15]), the 2D vortex identification criteria Γ_1 of Graftieaux, Michard, and Grosjean [26] is used here primarily because of its simple and robust formulation. If P is a point in a 2D vector field, S is some area surrounding P, and M is a point inside S, the vortex detector Γ_1 is defined by

$$\Gamma_1(P) = \frac{1}{S}\int_{M\in S} \frac{(\mathbf{PM}\times\mathbf{U}_\mathrm{M})\cdot\mathbf{z}}{|\mathbf{PM}|\cdot|\mathbf{U}_\mathrm{M}|}dS = \frac{1}{S}\int_S \sin\theta_\mathrm{M} dS \tag{3}$$

where $\mathbf{PM}$ is the vector connecting points P and M, $\mathbf{U}_\mathrm{M}$ is the velocity vector at point M and θ_M is the angle between vectors $\mathbf{PM}$ and $\mathbf{U}_\mathrm{M}$. The parameter Γ_1 will be between -1 and 1 depending on the direction of rotation and it can be shown that a vortex exists when $|\Gamma_1| > 2/\pi$. For a rectangular interrogation area with N discrete points inside that area, equation 3 can be re-expressed as

$$\Gamma_1(P) = \frac{1}{S}\int_S \sin\theta_\mathrm{M} dS = \frac{1}{N}\sum_N \sin\theta_\mathrm{M} \tag{4}$$

The vector field is generated from time signals of u and w at grid plane 4 filtered using the same filter as for the 2D correlations, and then converted to space in the streamwise direction using again the variable $tU_\mathrm{c}/\delta \approx x/\delta$. The constant convection velocity $U_\mathrm{c} = 0.65U_\mathrm{inf}$ is subtracted from u and the velocity vectors are weighted by the local mean density ratio $\overline{\rho}(z)/\overline{\rho}_\mathrm{w}$ to account for compressibility effects across the shear layer in the form of a Morkovin-type density scaling [27]. Using a square interrogation area of size $0.15\delta^2$, the vortex detector Γ_1 is computed using equation (4) on the density-weighted, convecting velocity vector field in the xy-plane taken from the spanwise centre of grid plane 4. Two separate segments in time of the vector field and the corresponding contour plot of Γ_1 are plotted in Figure 10 to demonstrate typical results from the vortex detection method. The large vortices are visualized by the dark areas, which appear predominantly at the wall-normal distance of 0.5δ. This result is consistent with the location of the shear-layer centre at grid plane 4, shown in Figure 2. The individual vortices seem to be spaced apart by 0.5 to 2δ.

If Γ_1 is calculated at every xz-plane across the span of grid plane 4, a three-dimensional representation of the vortices can be constructed. The 3D isosurfaces of $\Gamma_1 = -2/\pi$ at the same time segments as in Figure 10 are plotted in Figures 11 and 12. From these images the individual vortices look to be about 0.5 to 1δ long, and almost all of the vortices are at an angle of approximately 45° to the flow. In fact, a number of these vortices are bent in the centre forming an arrowhead shape. This behaviour explains the structure angle seen in the 2D correlation plots of Figures 8b and 9b. From Figures 11 and 12 the direction of the angle is equally possible for forming arrows pointed upstream as for arrows pointed downstream. However, the 2D contour plots appear to be biased towards the upstream pointed vortices. The vortex cores also seem to be more or less confined to the xy-plane at $z/\delta \approx 0.5$.

5. Conclusions

In this paper a thorough characterization of the time-averaged organization of the shear layer in a Mach 3 STBLI over a 24° compression ramp is presented. This analysis consisted of defining

Journal of Physics: Conference Series **506** (2014) 012013 doi:10.1088/1742-6596/506/1/012013

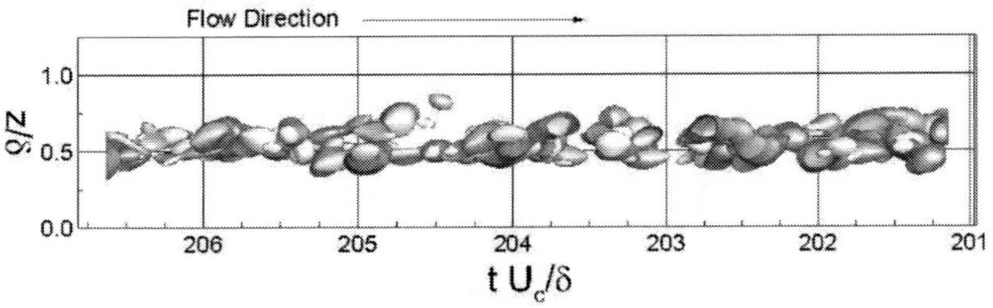

(a)

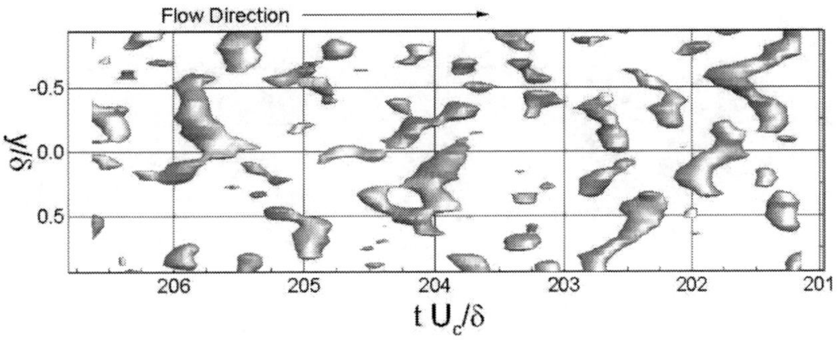

(b)

Figure 11: Vortex structure visualized the isosurface of $\Gamma_1 = -2/\pi$ at 'snapshot 1'.

the time- and spanwise-averaged position of the shear layer. Specifically, it was shown that the shear layer centre forms a straight line between the origin at the wall just downstream of the shock and a point just above the mean reattachment point. The angle between the shear layer and the wall is found to be 16.5^o. In addition, it was shown that, above the separated region, the shear layer grows approximately linearly up to the mean reattachment point. If a coordinate system is defined where the x-axis is pointed along the centre of the shear layer, a similarity variable can be defined. The similarity variable collapses the profiles of mean velocity and 2D Reynolds stresses to a single similarity profile. The linear spreading rate is determined from the profiles. This same analysis was first done by Dussauge, Piponniau, and Dupont [19] using their reflected shock PIV data and the current results are comparable to those from the reflected shock.

In addition to the time-averaged definition of the shear layer orientation, the frequency content of the shear layer is also analysed. Time and length scales of the large-scale vortices produced by the shear layer are estimated, as well as the average convection velocity of these vortices. It is shown that the time period between successive vortices increases almost linearly through the developing region of the shear layer and then becomes constant downstream of reattachment. It is apparent that the spreading of the shear layer is caused by the growth of the large-scale vortices that increase in size until they are confined by the surface of the ramp. The point of reattachment of the flow might therefore be dependent on the growth rate of the shear layer and also the angle the shear layer makes to the wall surface. In addition, the development of the average vortex timescale is consistent with the shift in frequencies that are seen in the pre-multiplied power spectra of the wall pressure.

Contour plots of correlations between mass flux and wall pressure resulted in a wall signature

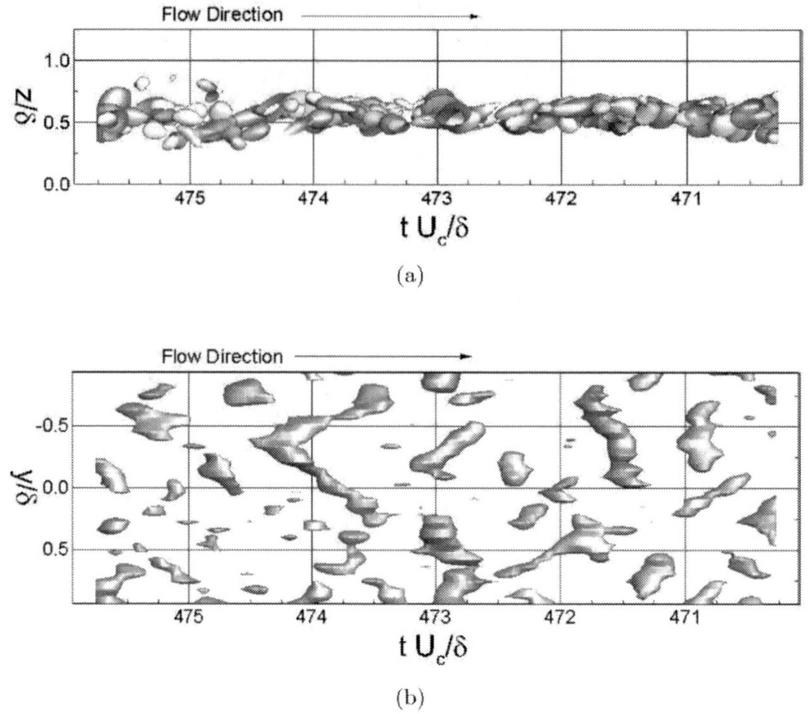

Figure 12: Vortex structure visualized the isosurface of $\Gamma_1 = -2/\pi$ at 'snapshot 2'.

that suggests the shear layer vortices are not two-dimensional in nature. The visualization of individual vortices using the vector detector of Graftieaux *et al* [26] provides proof that this is in fact the case. The vortex cores just downstream of reattachment are found to generally remain in a plane parallel to the wall surface, and are angled in either direction approximately 45° from the direction of the freestream. The shear layer shows resemblance to a canonical mixing layer in xyz. A distinct difference is that, in the STBLI case, the shear layer structures make the observed 45° angle with the flow, showing a crisscrossing pattern on streamwise-spanwise planes.

Acknowledgments

This work is supported by the Air Force Office of Scientific Research under grant AF/9550-10-1-0164, and in part by the Multiflow Program of the European Research Council at the Universidad Politécnica de Madrid (Spain).

References

[1] Delery J and Marvin J G 1986 Shock-Wave Boundary Layer interactions *AGARDograph* **280** (Paris: NATO)
[2] Dolling D S 2001 Fifty Years of Shock-Wave/Boundary-Layer Interaction Research: What Next? *AIAA J.* **39** 1517–31
[3] Smits A J and Dussauge J P 2006 *Turbulent Shear Layers in Supersonic Flow* (New York: Springer)
[4] Clemens N T and Narayanaswamy V 2009 Shock/Turbulent Boundary Layer Interactions: Review of Recent Work on Sources of Unsteadiness *AIAA Paper* 2009-3710
[5] Beresh S J, Clemens N T and Dolling D S 2002 Relationship Between Upstream Turbulent Boundary-Layer Velocity Fluctuations and Separation Shock Unsteadiness *AIAA J.* **40** 2412–22
[6] Dupont J, Haddad C and Debieve J-F 2006 Space and time organization in a shock-induced separated boundary layer *J. of Fluid Mech.* **559** 255–77

[7] Ganapathisubramani B, Clemens N T and Dolling 2007 Effects of upstream boundary layer on the unsteadiness of shock-induced separation *J. Fluid Mech.* **585** 369–94
[8] Robinet J Ch 2007 Bifurcations in shock-wave/laminar-boundary-layer interaction: global instability approach *J. Fluid Mech.* **579** 85–112
[9] Wu M and Martin M P 2008 Analysis of shock motion in shockwave and turbulent boundary layer interaction using direct numerical simulation data *J. Fluid Mech.* **594** 71–83
[10] Piponniau S, Dussauge J-P, Debieve J-F and Dupont P 2009 A simple model for low-frequency unsteadiness in shock-induced separation *J. Fluid Mech.* **629** 87–108
[11] Ganapathisubramani B, Clemens N T and Dolling 2009 Low-frequency dynamics of shock-induced separation in a compression ramp interaction *J. Fluid Mech.* **636** 397–425
[12] Humble R A, Scarano F, van Oudheusden B W 2009 Unsteady aspects of an incident shock wave/turbulent boundary layer interaction *J. Fluid Mech.* **635** 47–74
[13] Souverein L J, Dupont P, Debieve J-F, Dussauge J-P, van Oudheusden B and Scarano F 2010 Effect of interaction strength on unsteadiness in turbulent shock-wave-induced separations *AIAA J.* **48** 1480–93
[14] Priebe S and Martin MP 2012 Low-frequency unsteadiness in shock waveturbulent boundary layer interaction *J. Fluid Mech.* **699** 1–49
[15] Dupont P, Piponniau S, Sidorenko A and Debieve J-F 2008 Investigation by Particle image velocimetry measurements of oblique shock reflection with separation *AIAA J.* **46** 1365–70
[16] Agostini L, Larchevêque L, Dupont P, Debiève J-F and Dussauge J-P 2012 Zones of influence and shock motion in a shock/boundary-layer interaction *AIAA J.* **50** 1377–87
[17] Kiya M and Sasaki K 1985 Structure of large-scale vortices and unsteady reverse flow in the reattaching zone of a turbulent separation bubble *J. Fluid Mech.* **154** 463–91
[18] Xu S and Martin M P 2004 Assessment of inflow boundary conditions for compressible turbulent boundary layers *Phy. of Fluids* **16** 2623–39
[19] Dussauge J-P, Piponniau S and Dupont P 2010 *APS 63rd Annual DFD Meeting, Long Beach, CA*
[20] Townsend A A 1976 *Structure of Turbulent Shear Flow* (Cambridge: Cambridge Univ. Press)
[21] Mehta R D and Westphal R V 1984 Near-field turbulence properties of single- and two-stream plane mixing layers *AIAA Paper* 84-0426
[22] Hillier R and Cherry N J 1981 The effects of stream turbulence on separation bubbles *J. of Wind Eng. Ind. Aerodynamics* **8** 49–58
[23] Kiya M and Sasaki K 1983 Structure of a turbulent separation bubble *J. Fluid Mech.* **137** 83–113
[24] Kiya M, Sasaki K, and Arie M 1982 Discrete-vortex simulation of a turbulent separation bubble *J. Fluid Mech.* **120** 219–44
[25] Brown G L and Thomas A S W 1977 Large structure in a turbulent boundary layer *Phys. of Fluids* **20** S243–52
[26] Graftieaux L, Michard M and Grosjean N 2001 Combining PIV, POD and vortex identification algorithms for the study of unsteady turbulent swirling flows *Meas. Sci. Tech.* **12** 1422–9
[27] Morkovin M V 1962 *Mécanique de la Turbulence* (Paris: CNRS)

Influence of solid boundary conditions on the evolution of free and wall-bounded turbulent flows

Paolo D'Addio[1], **David Sassun**[1], **Oscar Flores**[2] **and Paolo Orlandi**[1]

[1] Dipartimento di Ingegneria Meccanica e Aerospaziale, Sapienza Università di Roma, Via Eudossiana 16, I-00184, Roma

[2] Dept. Bioingeniería e Ing. Aeroespacial. Universidad Carlos III de Madrid. Av. Universidad 30, Leganés. 28911 Madrid. Spain

E-mail: paolo.daddio@uniroma1.it, david.sassun@uniroma1.it, oflores@ing.uc3m.es, orlandi@dma.ing.uniroma1.it

Abstract. The effect of the boundary conditions on the organization of vortex clusters is analyzed in two separate cases: rough channel flow and grid-generated turbulence. The aim is to understand how far the fluid structures are affected by the presence of roughness and the geometry of the grid. The grid-turbulence cases show that the single- and multi-scale geometries generate a flow strongly dominated by the shear at the beginning. The shear is initially caused by the presence of the body and, for the multi-scale grids, subsequently by the large differences between the scales. Further downstream from the grid these shear-dominated structures break up and form more isotropic clusters, whose dimensions seem to depend little on the particular geometry of the grid. For fractal grids, clusters are formed right downstream of the grid, resulting in a flow with less inhomogeneities than for single- and multi-scale grids. Eight different rough surfaces have been analyzed. In the smooth channel, both attached and detached clusters have been found and, depending on the geometry, the roughness affects the attached structures. Roughnesses made of aligned obstacles with a large separation seem to reduce the number of these structures in the flow. When the roughness elements are closely packed, both for transverse and aligned obstacles, the attached clusters are not able to reach within the roughness elements, and they seem to be anchored to the plane of the crests.

1. Introduction

Many engineering applications require the manipulation of flow features like turbulence or mixing far from the solid boundaries where actuators are placed. Most of these cases fall into one of two categories: wall-bounded turbulent flows or turbulent wakes past an obstacle. The underlying question in both cases is the same: how far from the boundary condition (actuator on the wall or obstacle) are we able to modify a turbulent flow.

From the point of view of the turbulent wake past an obstacle, it is possible to draw an analogy between the boundary conditions and a Cauchy problem: the interaction of the object with the fluid particles sets their initial conditions, and they are then advected downstream, where they are free to evolve by themselves. One of the canonical cases in this category is grid turbulence, where a uniform flow is forced through a solid grid. In the region close to the grid the flow is highly anisotropic, being affected by the shape of the grid. Then, depending of the size of the vortices generated, an energy cascade process takes place: the larger eddies break-up, forming smaller scales. During this process the flow loses information about the initial condition and tends to isotropy. The resulting turbulent flow decays freely, in a similar way as in freely decaying turbulence simulations. In this case, the parameters of interest are the maximum level of isotropy and how quickly that condition is achieved in terms of distance to the grid. This problem was first studied by Taylor (1935) [8] and von Kármán (1937) [7]. Since then, several groups have analyzed the effect that the shape of the grid has on the rate of decay of the resulting turbulence: Lavoie et al.(2007) [9], Seoud and Vassilicos (2007) [10] and Ertunc et al. (2010) [11], among others. Specially interesting is the work by Seoud and Vassilicos (2007) [10], Hurst and Vassilicos (2007) [12] and Mazellier et al.(2010) [13]. Using multi-scale grids (grids with holes of different sizes) they were able to produce flows with unusual turbulence properties, which might lead to the design of new industrial mixers or combustion chambers.

From the point of view of the modification of a turbulent boundary layer using actuators at the wall, the problem is understanding the effect of the solid boundary conditions in wall-bounded turbulence. There has been extensive work on this problem, starting with the classical theory by Townsend (1976) [1]. Townsend postulated that, if Re is sufficiently high, the extent of the direct influence of the surface is limited to the near-wall region, while the rest of the flow only sees the slip velocity at the upper limit of the wall region. Most of the studies of turbulence over rough walls support this idea (see review by Jiménez, 2004 [2]): the difference between the mean velocity profiles of smooth- and rough-wall cases is the so called roughness function, a roughly constant velocity difference. This velocity difference has been linked to the root-mean-squared vertical velocity at the plane of the crests, a very important quantity that plays a major role in transition to turbulence (Orlandi (2011) [3]) and allows one to write a simplified expression for the velocity profile in the logarithmic layer (Orlandi and Leonardi (2006) [4]). Another very important aspect of the influence of the boundary condition is the forces that the wall imposes on the flow, namely the drag. Leonardi et al. (2007) [5] linked the drag components (friction and pressure) in each case with the behaviour of the velocity structures in the overlying flow (d-type or k-type roughness). This can be exploited also to reduce drag, as in riblets. Indeed, García-Mayoral and Jiménez (2011) [6] explained the mechanisms behind drag reduction and its disruption with the appearance of spanwise rollers close to the wall ($y^+ < 20$). It is then apparent that, in wall bounded turbulence, it is possible to modify the roughness function (which determines the mass flux) and the drag (which determines the pressure gradient) from the wall. However, the question of how far from the wall can we change turbulence remains open.

The macroscopic influence of the boundary conditions on the properties of the turbulent flows discussed above are due to the effects that both the roughness of the wall and the shape of the grids have on the velocity structures interacting with them. In this article, we study the statistical properties of these velocity structures, applying a coherent-structure tracking algorithm developed by del Álamo et al. (2006) [14] to our database of rough-walled channels and grid turbulence cases. This method allows to extract individual structures, and to acquire all the statistics necessary to understand the effect of the boundary conditions on the their formation and evolution, as well as to link the macroscopic effect of the boundary condition to these structures.

This clustering algorithm was originally applied by del Álamo et al. (2006) [14] to smooth-walled turbulent channels, and later by Flores et al. (2007) [15] to turbulent channels with disturbances at the boundary conditions at the walls (artificial roughness). To the authors knowledge, this is the first time that this algorithm is applied to turbulent channels with fully resolved roughness, and to grid turbulence cases.

This work is structured as follows: in section 2 we summarize the characteristics of the simulations and describe the numerical method applied. In section 3, the clustering technique is explained, and the results of the percolation analysis are presented. The joint probability density functions (p.d.f.s) of the invariant Q and of the clusters statistics are presented in section 4, along with three-dimensional representations of the structures. Finally, section 5 provides conclusions.

2. Numerical experiments

The non-dimensional incompressible Navier–Stokes equations, solved in this work, are

$$\frac{\partial u_i}{\partial t} + \frac{\partial u_i u_j}{\partial x_j} = -\frac{\partial p}{\partial x_i} + \frac{1}{Re}\frac{\partial^2 u_i}{\partial x_j^2} + \Pi\delta_{1i}, \quad \frac{\partial u_i}{\partial x_i} = 0,$$

where Π is the pressure gradient required to maintain a constant mass flow rate in the channel flow (Π is zero in the grid flow), u_i is the component of the velocity vector in the i direction, p is the kinematic pressure and Re is the Reynolds number. This set of equations has been discretized in an orthogonal coordinate system, with a staggered second-order finite-difference method. The integration in time uses a third-order low-storage Runge–Kutta algorithm, coupled with a second-order Crank–Nicolson scheme. To correct the non-divergence-free field, the fractional step procedure is used. Further details are found in Orlandi (2000) [16] and are not repeated here.

The interaction between the flow and the solid is reproduced with the immersed-boundary technique, which assumes that the velocity is zero inside the body, and applies a correction to the viscous term at the first grid point near the solid surface. This method is described in detail in Orlandi and Leonardi (2006) [4].

For the channel, the periodicity assumed in the streamwise and spanwise directions implies uniform grids in x and z. No-slip conditions at the walls and at the roughness requires a non-uniform grid in the wall-normal direction. In grid turbulence, the same resolution is used in the three directions. The solid grid is normal to the flow direction (x), and periodicity in y and z is imposed. Radiative conditions in x are applied at the outlet.

2.1. Simulation details for the grid-generated turbulence

In order to mimic wind tunnel experiments, the inlet flow is uniform with random fluctuations of small amplitude superimposed at $x = 0$. The large scales and hence the energy-containing eddies are generated by the solid grid, placed at $x_G = 2$. The shape of the grid dictates the velocity distribution at its exit, which has been checked to be independent of its location. The number of points and the dimensions of the domain are listed in table 1.

The solid grids considered are shown in figure 1 with the geometric parameters listed in table 1. For the regular grids G_{4-4-S} and G_{4-4-C}, and the multi-scale grids G_{12-4} and G_{32-4}, the mesh length of the grid is defined as $M = 2\pi/N_a$, where N_a indicates the number of holes in a row. The effective mesh length for the fractal grid is $M_{eff} = 4(L_2L_3)\sqrt{1-r}/P = 0.43$, where r is the solidity of the grid defined below and P is the fractal perimeter length. Further details can be found in Hurst and Vassilicos (2007) [12]. Table 1 also shows the solidity of the grid $r = V_{solid}/V_{tot}$, defined as the ratio between the solid and total volumes. All the grid turbulence simulations were performed at Reynolds number based on the domain length L_2, $Re_{L_2} = U_1L_2/\nu = 3000$, where U_1 is the uniform velocity imposed in the region upstream of the grid.

Figure 3(a) shows the decay of the turbulent kinetic energy with the distance x from the grid. It is possible to adjust this decay with a power law $\langle q\rangle = (X/M_L - X_0)^{-m}$ [12], where the value of the decay exponent m depends on the geometry. These exponents and the values of the turbulent kinetic energy at the origin $\langle q_0\rangle$ are also listed in table 1. The different origin for G_{fract} is a consequence of the different solidity of the grid r, smaller for the fractal grid. This was verified in preliminary simulations, where a uniform grid G_{4-4} with $r = 0.257$ yielded a value for $\langle q_0\rangle$ closer to that of G_{fract}. Physically, this simply states that the effect of the grid on the flow (i.e., the transfer of kinetic energy from the mean flow to turbulence) is smaller as the solidity decreases. The figure shows also that different geometries have different slopes of $\langle q_0\rangle$ at the exit from the grid. In particular, for the single scale grids, $d\langle q_0\rangle/dx$ is positive, as

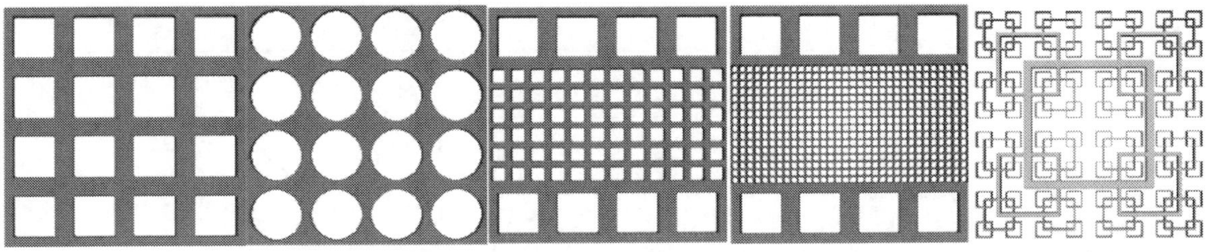

Figure 1. Geometry of the grids $a)G_{4-4-S}$, $b)G_{4-4-C}$, $c)G_{12-4}$, $d)G_{32-4}$, $e)G_{fract}$

Table 1. Parameters of the grids. M_L is the larger mesh; M_S is the smaller mesh; Re_{M_S} is the Reynolds number based on the small mesh; r is the solidity and m is the decay exponent of the kinetic energy. Re_{M_s} for G_{fract} is calculated with M_{eff} as reference length.

	$M_s : M_L$	r	Re_{M_s}	$L_1 \times L_2 \times L_3$	$n_1 \times n_2 \times n_3$	m	$\langle q_0 \rangle$
G_{4-4-S}	1 : 1	0.57	4710	$18\pi \times 2\pi \times 2\pi$	$1152 \times 192 \times 192$	2.39	2.18
G_{4-4-C}	1 : 1	0.57	4710	$18\pi \times 2\pi \times 2\pi$	$1152 \times 192 \times 192$	2.39	1.65
G_{12-4}	1 : 3	0.57	1560	$18\pi \times 2\pi \times 2\pi$	$1152 \times 192 \times 192$	1.11	2.33
G_{32-4}	1 : 8	0.57	588	$18\pi \times 2\pi \times 2\pi$	$1152 \times 192 \times 192$	1.09	2.10
G_{fract}	–	0.257	1290	$6\pi \times 2\pi \times 2\pi$	$768 \times 384 \times 384$	1	0.44

expected, because of the production term. For the multiscale and fractal grids $d\langle q_0 \rangle/dx < 0$ because the grid generates smaller scales closer to the grid that decay faster. Since the plot reports the average on a plane parallel to the grid, this effect is also taken into account in the decrease of total energy.

2.2. *Simulation details for the rough channel flow*

Regarding the channel simulations, $h = 1$ is the channel half-height, not including the roughness layer, and the dimensions of the computational domain are $L_1 = 8$ in the stream-wise direction and $L_3 = 4$ in the span-wise direction. The roughness is located only on the lower wall $(-1.2 < y < -1)$, while on the upper boundary the surface is smooth. All the simulations are performed at $Re_{bulk} = 2800$, corresponding to a smooth channel with $Re_\tau \approx 180$. The different values of Re_τ on the channel walls, along with other parameters of interest, are given in table 2. The different values of $Re_\tau|_R$ at the bottom wall are due to the different turbulent activity. The flow on the upper wall is not influenced by the lower wall, resulting in more similar values for $Re_\tau|_S$. At this Reynolds number, the roughness height is $k = 0.2$, and k^+ varies from 32.3 up to 60.1.

Several geometries have been reproduced (figure 2). The smooth channel (CHA) acts as both validation of the results and reference for the other cases. The rough walls include a three-dimensional geometry made of prisms (PRSH), transverse (SQWK1) or aligned square bars with two aspect ratios (SQBAR, SQBSM), and aligned triangular bars with three aspect ratios (TRBAR, TRBSM, TRBSSM).

An interesting aspect of these cases relates to the pressure gradient. As shown in the last column of table 2, the three-dimensional (3D) geometry causes a large drag increase, mainly due to the form drag. This term is also oscillatory, which is a sign of greater turbulent activity, and has been linked to strong vertical velocity events. The longitudinal square bars cause a moderate increase of the pressure gradient. One of the triangular-bars cases yields a drag reduction of

Table 2. Roughness parameters. Subscripts R and S are relative to the bottom rough wall and upper smooth wall respectively. $\overline{\Pi}$ is the pressure gradient, with the subscript i representing the geometry.

Geometry	k^+	$Re_\tau\vert_R$	$Re_\tau\vert_S$	$100\frac{\overline{\Pi}_i-\overline{\Pi}_{CHA}}{\overline{\Pi}_{CHA}}$
CHA	0.0	176.1	176.1	0.0
$PRSH$	60.1	295.9	189.5	85.4
$SQBAR$	39.7	199.8	177.2	9.9
$SQBSM$	41.1	204.4	178.1	13.7
$SQWK1$	40.8	203.7	177.3	12.5
$TRBAR$	37.6	190.2	174.3	−0.1
$TRBSM$	32.3	162.8	171.8	−13.7
$TRBSSM$	36.1	177.8	174.2	−3.9

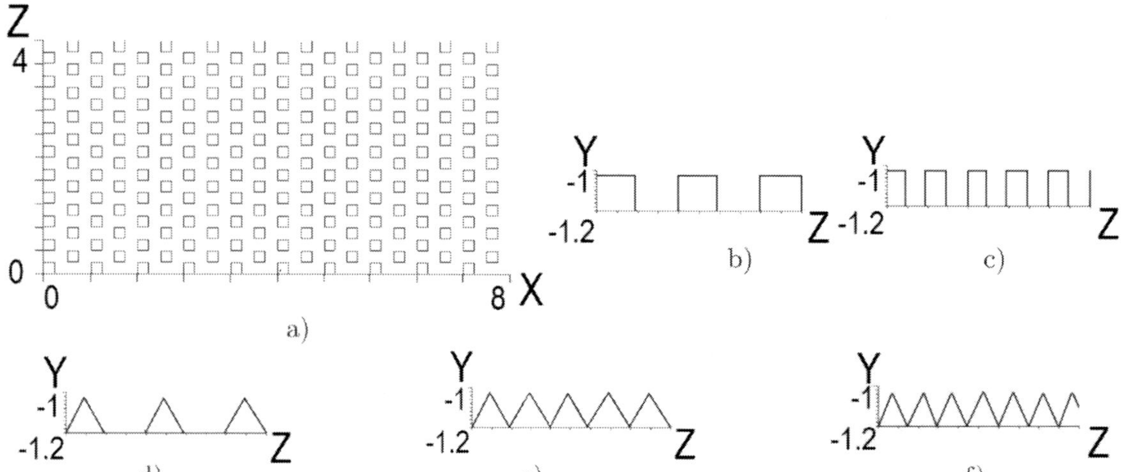

Figure 2. Roughness in the cases of *a*) PRSH, *b*) SQBAR, *c*) SQBSM, *d*) TRBAR, *e*) TRBSM, *f*) TRBSSM. From *b*) to *f*) only a portion of the domain is represented. The transverse square bars case (SQWK1) is the same as *b*), inverting the z and x axes.

nearly 14%, and the others are in line with the smooth channel case. In the transverse bars case, the form drag is low because $w/k = 1$ (w is the spacing between bars). Leonardi et al. [17] demonstrated that the form drag has a maximum for $w/k = 7$.

3. Clusters identification and classification

To identify and classify vortex clusters, the method described by del Álamo et al. (2006) [14] has been used. The same technique has been already adopted by Flores et al. (2006) [15] to investigate the vorticity organization in the outer layer of turbulent channels. This process is based on the vortex eduction method by Chong et al. (1990) [18]. According to those authors, a vortex core is a region where the velocity gradient tensor $\nabla\mathbf{u}$ is dominated by its rotational part. Expressed in terms of the second invariant of $\nabla\mathbf{u}$, this condition yields $Q > 0$. To identify regions in which $\omega_i\omega_i/2$ is significantly larger than $s_{ij}s_{ij}$, a threshold is used: $Q > Q^*$. This eduction method depends on the value of the threshold and, if the flow has an inhomogeneous direction, the dependence on the threshold becomes troublesome, complicating the comparison of data at different values of the coordinate. Nagaosa and Handler (2003) [19], applying the Q criterion to wall-turbulence, found that the p.d.f of the second invariant, normalized with its standard deviation, $Q/(\overline{Q'^2})^{1/2}$, is homogeneous everywhere except in the viscous sublayer.

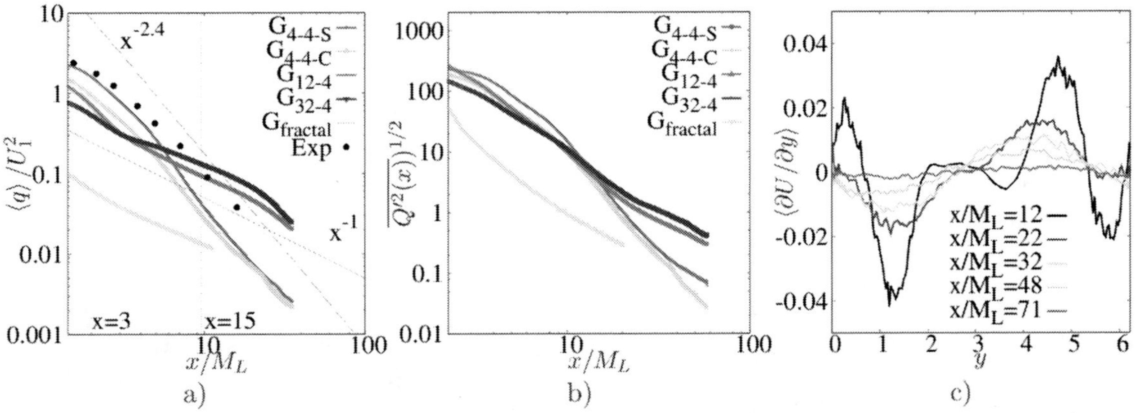

Figure 3. *a*) Dimensionless turbulent kinetic energy $\langle q\rangle/U_1^2$, as function of the non-homogeneous direction x/M_L. • is a perforated plate with circles ($r = 0.6$, $Re_M = 2200$) (Mazellier et al.(2010)) *b*) Standard deviation of the second invariant $(\overline{Q'^2(x)})^{1/2}$. *c*) Profiles of the shear $\left\langle \frac{\partial U}{\partial y} \right\rangle$ as function of the grid inhomogeneous direction y for the multi-scale grid G_{32-4}.

Starting from these results, del Álamo et al. (2006) [14], using D, the discriminant of $\nabla\mathbf{u}$, presented a procedure in which a point $\mathbf{x}$ belongs to a vortex if $D(\mathbf{x}) > \alpha(\overline{D'^2(y)})^{1/2}$, where α is the actual thresholding parameter and $(\overline{D'^2(y)})^{1/2}$ is the standard deviation of D. This procedure solves the problem of the inhomogeneity, and yields a vortex volume fraction which depends much less on the inhomogeneous direction than the one obtained using a uniform threshold.

In the present work, the clustering technique just described has been applied to grid generated turbulence and channel flows, where the inhomogeneous directions are respectively x (grid normal) and y (wall normal).

To identify the clusters, the second invariant of $\nabla\mathbf{u}$ has been used. A grid point belongs to a cluster if $Q(\mathbf{x}) > \alpha(\overline{Q'^2(\xi)})^{1/2}$, where ξ is the inhomogeneous direction. Averaging in the plane normal to the ξ axis, it is possible to obtain $(\overline{Q'^2(\xi)})^{1/2}$, shown in figure 3(b) for grid turbulence.

An appropriate value for the threshold α can be obtained through the percolation analysis (Moisy and Jiménez (2004) [20] and Stauffer (1985) [21]). Figures 4(a,c) represent the ratio between the volume of the largest cluster V_{max} and the overall volume V occupied by the clusters, as function of the thresholding parameter. When $\alpha > 1$, only a few small objects are identified. Decreasing α, new clusters appear while other coalesce, and the variation of V_{max}/V is the result of the trade-off between the two processes. For grid turbulence, if $\alpha \approx 0.85$, V_{max}/V remains roughly constant while the number of clusters that can be visualized is almost at its maximum (figures 4b,d). For the channel, the plot of the volume ratio is slightly more dispersed, while the peak in the number of clusters always occurs at $\alpha \approx 0.75$. Therefore, this value has been used in the subsequent analysis.

4. Clusters statistics

With the chosen value for the threshold, the identification algorithm has been applied to the domain. For grid turbulence, the interest is to detect clusters located immediately after the grid, and for this reason the analysis was performed on a domain ($2.25 < x < 18\pi$) which excludes the grid. In the rough channel, to investigate the interaction of the flow structures with the roughness elements, all the points, except those inside the solid body, are used for the clustering algorithm. In both cases, to compute the statistics, only clusters whose volume is larger than thirty grid points have been considered.

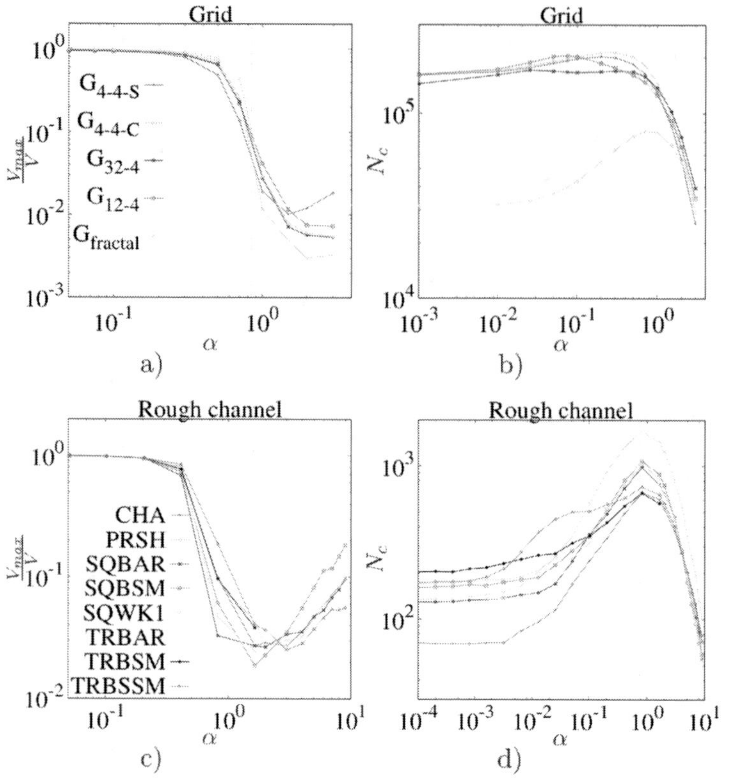

Figure 4. a) and c): Volume V_{max} of the largest cluster, normalized with the overall volume V occupied by all the clusters, as a function of the threshold α. b) and d): Number of clusters N_c as a function of the threshold α.

4.1. Grid turbulence

4.1.1. P.d.f.s of Q The clustering technique described in section 3 uses the standard deviation $(\overline{Q'^2})^{1/2}$ to normalize Q, so that the p.d.f. does not vary with the inhomogeneous direction. To verify this normalization for the grid turbulence cases, figure 5 shows the p.d.f.s of Q for all the geometries. The colours indicate the percentage of the data contained in the corresponding iso-probability contour. The good similarity in x confirms the suitability of this technique in the case of grid turbulence.

The figure also shows that the shape of the grid influences the p.d.f. of Q, which for the multi-scale (G_{12-4} and G_{32-4}) and single-scale grids (G_{4-4-S} and G_{4-4-C}) are skewed towards negative values. From the definition of $Q = \omega_i\omega_i/2 - s_{ij}s_{ij}$, this suggest that close to the grid the flow is dominated by the shear $s_{ij}s_{ij}$, consistent with the presence of a wake and shear layers downstream of the grid. In the case of single-scale grids, this shear is powered only by the wake of the grid, and its survival depends on the size of its holes. In the case of multi-scale grids, in addition to the small-scale shear there is a large-scale shear that is produced by the different velocities and lengthscales downstream of the two different meshes. This happens because these two flows have a similar behaviour to a jet interfacing with a different flow. This is the reason for the negative peak of Q that appears only in the p.d.f.s of G_{12-4} and G_{32-4}. Figure 3(c) shows the profiles of the shear $\langle \partial U/\partial y \rangle$ at different distances from the grid for G_{32-4}. Initially, two peaks are located at the two interfaces between the large- and small-scale flows, and these two homogeneous regions interact forming a mixing layer (Veeravalli and Warhaft (1989) [22]).

Note that, due to this shear-dominated region, few clusters (if any) will be identified close downstream of the grid. This is made explicit in figure 5 by the vertical black line, indicating the adopted threshold $Q/(\overline{Q'^2})^{1/2} = 0.85$. The only exception is the fractal grid, whose p.d.f. is roughly symmetric, suggesting that clusters will be found right downstream of the grid. This can be observed in figure 6, where a visualization of $Q/(\overline{Q'^2})^{1/2}$ is shown. While single-scale and

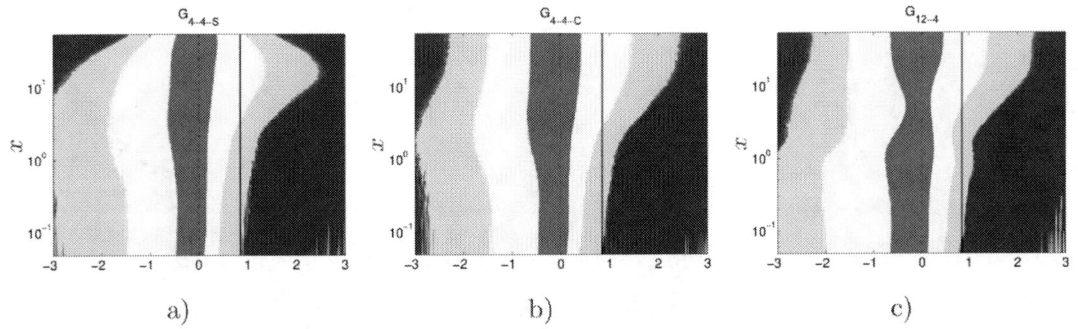

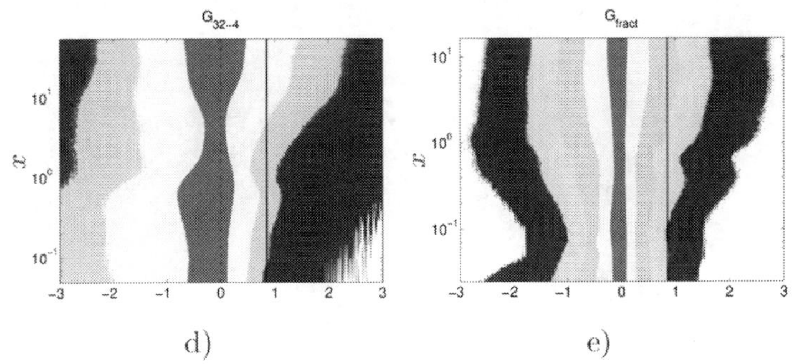

Figure 5. P.d.f.s of $Q/(\overline{Q'^2})^{1/2}$ as function of the distance from the grid x (log scale), for $a)G_{4-4-S}$, $b)G_{4-4-C}$, $c)G_{12-4}$, $d)G_{32-4}$, $e)G_{fract}$. The iso-probability coloured contours contain respectively: red 10% - yellow 30% - green 50% - cyan 70% - blue 95% of the data. The vertical black line is relative to the value adopted for the threshold $Q/(\overline{Q'^2})^{1/2} = 0.85$.

multi-scale grids only show vortices relatively far from the grid, the fractal grid shows evidence of vortices in close proximity to the grid.

4.1.2. Clusters distribution, orientation and dimensions The clustering algorithm was run for approximately forty fields (approximately one washout), to extract approximately $N_c \approx 10^4$ clusters with more than 30 data points. With this database, it was possible to obtain converged statistics of their geometry and distribution. The lengths of the box enclosing the cluster are L_x, L_y and L_z. Figure 7 shows joint p.d.f.s of the streamwise length of the structures (L_x) as function of the coordinate of its centre (x_c), evaluated as the average between the points of the cluster that are closest to and farthest from the grid. Note that the black dashed line ($L_x = 2x_c$)

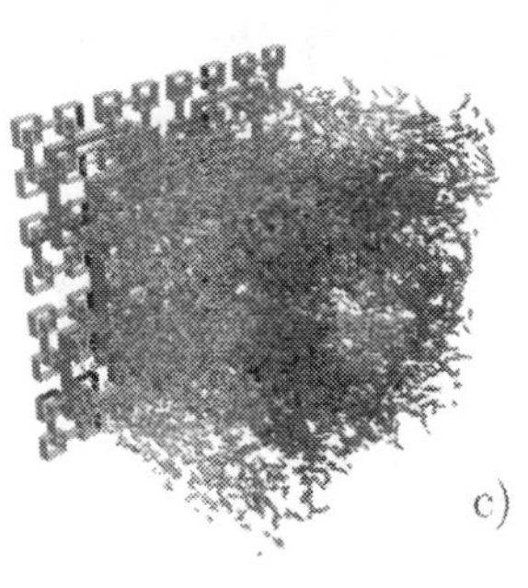

Figure 6. 3D contour plot of $Q/(\overline{Q'^2})^{1/2} = 2$, coloured by the distance from the grid for $a)G_{4-4-S}$ ($\frac{x_G}{M_L} < \frac{x}{M_L} < 17.2$), $b)G_{32-4}$ ($\frac{x_G}{M_L} < \frac{x}{M_L} < 17.2$), $c)G_{fract}$ ($\frac{x_G}{M_{eff}} < \frac{x}{M_{eff}} < 16.5$).

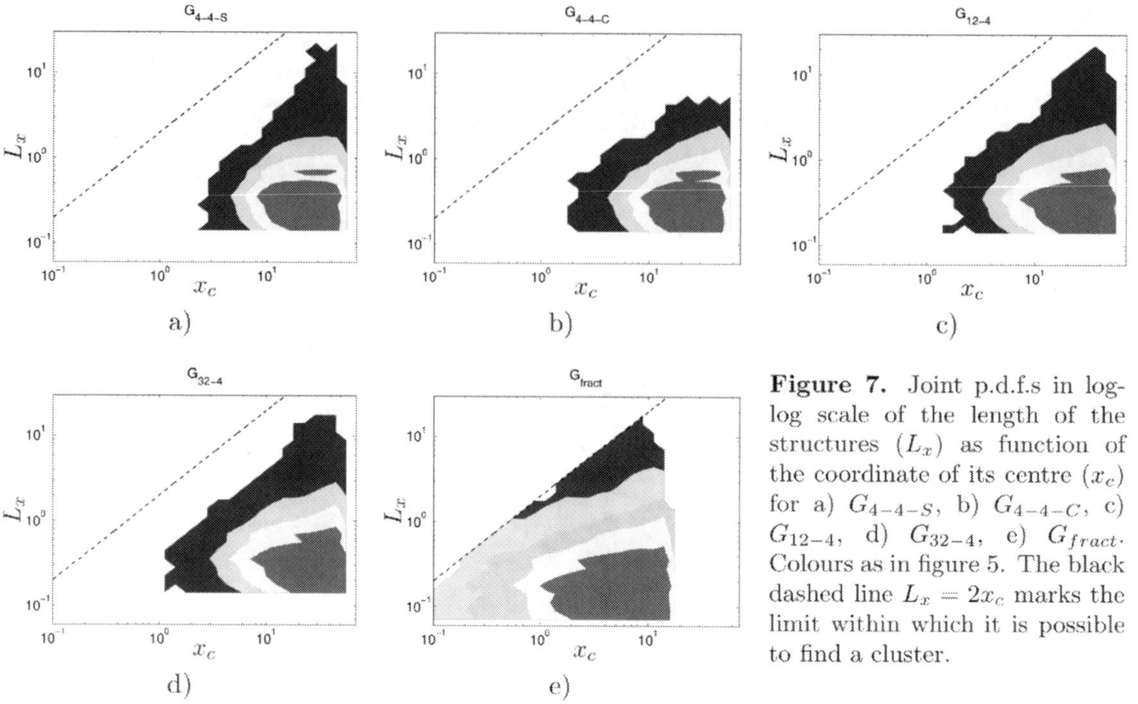

Figure 7. Joint p.d.f.s in log-log scale of the length of the structures (L_x) as function of the coordinate of its centre (x_c) for a) G_{4-4-S}, b) G_{4-4-C}, c) G_{12-4}, d) G_{32-4}, e) G_{fract}. Colours as in figure 5. The black dashed line $L_x = 2x_c$ marks the limit within which it is possible to find a cluster.

limits the plot, because the streamwise length of any cluster cannot be longer than twice the position of its centre.

Figure 7 confirms that there are no clusters attached to the solid grid for the single- and multi-scale grids. This supports that the geometry of the grid dictates whether or not the flow is initially dominated by the shear. For G_{fract} vortical structures are attached to the grid, while for single- and multi-scale grids the near region is shear-dominated and vortical structures (and clusters) are only formed downstream. It is interesting to note in figure 6 that the clusters are predominantly generated downstream of the larger holes, while the flow downstream of the finer mesh appears to be roughly irrotational.

A further interesting question is whether the size distribution of clusters is influenced by the geometry of the grid. Depending on the exponent n in $\Delta V/\eta^3 = (L_i/\eta)^n$, it is possible to infer the shape of these structures. If $n \approx 1$ the other two dimensions do not increase proportionally, meaning that the structure is elongated only in one direction (i.e. it is a tube). For $n \approx 2$ or 3 the structures have a sheet-like or cubical shape respectively and, if the exponent n is similar for all three L_i, there is no preferential direction of alignment. Figure 8 shows the joint p.d.f.s relative to the volume of the clusters in Kolmogorov units, as functions of one of the lengths of the cluster in the same units. For all cases, the joint p.d.f.s are well aligned with the power law $\Delta V/\eta^3 = (L_i/\eta)^{2.5}$. Interestingly, the same power law holds for the streamwise and lateral dimensions of the clusters, suggesting that the clusters are relatively isotropic objects, with circumscribing boxes that are approximately cubical. This observation can be made for all grids, suggesting that the influence of the grid geometry is small in the relative dimensions of the clusters.

It should be noted that figure 8 shows all the clusters, irrespective of their position downstream of the grid. In order to analyse the size distribution as a function of the distance to the grid, figure 9 shows the joint p.d.f. of volume and length of the clusters for three different regions downstream of the grid. These regions are the different stages of the decay of the turbulent kinetic energy $\langle q \rangle$ shown in figure 3(a), and can be identified by the changes in the

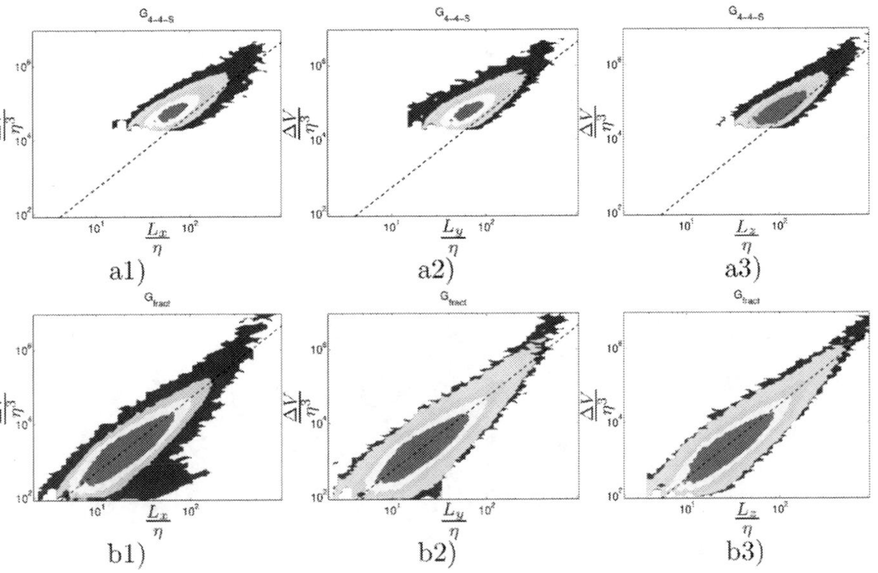

Figure 8. Joint p.d.f.s in log-log scale of the volume of cluster scaled with the Kolmogorov scale $\Delta V/\eta^3$, as function of the length: 1) L_x/η; 2) L_y/η; 3) L_z/η. For a) G_{4-4-S}; b) G_{fract} with $Q(\mathbf{x}) > \alpha(\overline{Q'^2(x)})^{1/2}$. Colours as in previous figures. The black dashed line indicates the power law $\Delta V/\eta^3 = (L_i/\eta)^{2.5}$.

decay exponent.

It is clear in the figure 9 that the shape of the grid has a strong impact on the development of the clusters downstream of the grid. Consistent with the discussion in the previous paragraphs, the fractal grid is the only geometry producing clusters attached to the grid. Interestingly, their aspect ratio changes little with the distance from the grid, showing clusters with the same dimensions in the first ($0 < x < 1$, figure 9e1) and second ($1 < x < 16.59$, figure 9e2) regions.

For the single-scale grids, there are basically no clusters in the first region, closer to the grid. Further downstream, the grid made of squares (G_{4-4-S}) shows only relatively large clusters, while the mesh made of circles (G_{4-4-C}) shows a broader range of scales. On the other hand, the multi-scale grids show a faster generation of clusters with a broader range of sizes.

Finally, it is interesting to note that both multiscale and fractal grids show a family of objects whose volume increase roughly like $\Delta V/\eta^3 = L_i/\eta$, suggesting tube-like clusters. These objects only appear in the most downstream regions, and represent a relatively small part of the population.

4.2. Geometry and distribution of the clusters for rough channel flows

Figure 10 shows the joint p.d.f. of the minimum and maximum wall distances of the points of the cluster. The diagonal black line indicates the limit within which it is possible to find a cluster, representing the simple condition $y_{max} > y_{min}$. The vertical black line corresponds to $y_{min} = -1$, i.e. the plane of the crests, which is also the position of the wall in the smooth case. The colours correspond to the logarithm of the p.d.f. (i.e. the number of clusters in each bin), and are on the same scale in all the images, while the black contour lines represent the points at constant probability that contain 30%, 50%, 70% and 95% of the data.

In the first image, relative to the smooth channel (CHA), two families of clusters appear: the wall-detached, lying on the $y_{max} = y_{min}$ line, and the wall-attached. Considering the bottom wall, this last family is made of clusters that start from the wall (small y_{min}) and can have any y_{max}. Among these the most interesting are those tall ones whose height is comparable to the

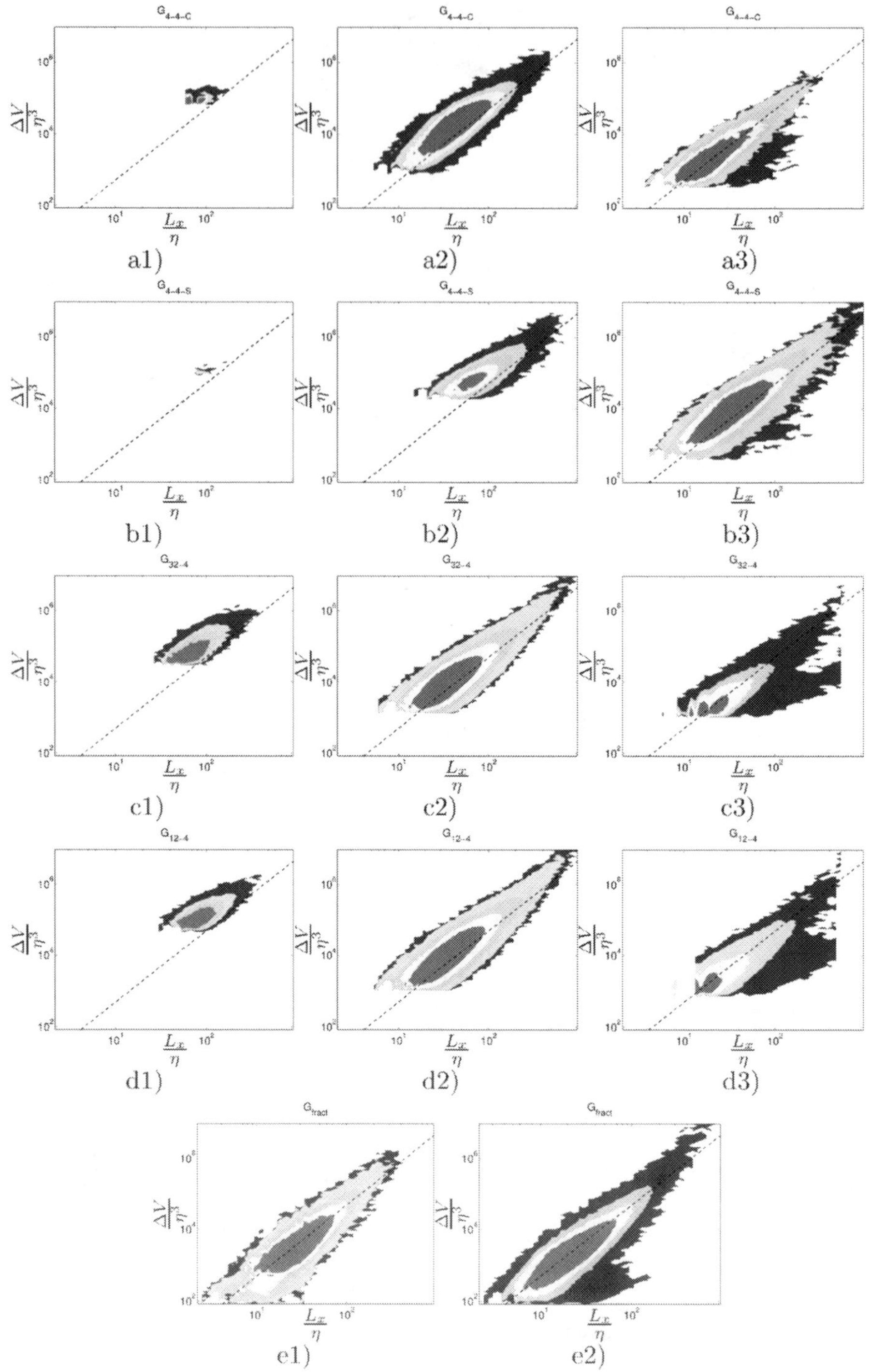

Figure 9. Joint p.d.f.s as in the previous figure as function of the length L_x/η. The p.d.f.s have been evaluated within three ranges for x. Single-scale: 1) $0 < x < 3$, 2) $3 < x < 15$, 3) $15 < x < 54.27$; Multi-scale: 1) $0 < x < 5$, 2) $5 < x < 30$, 13) $30 < x < 54.27$; Fractal: 1) $0 < x < 1$, 2) $1 < x < 16.59$.

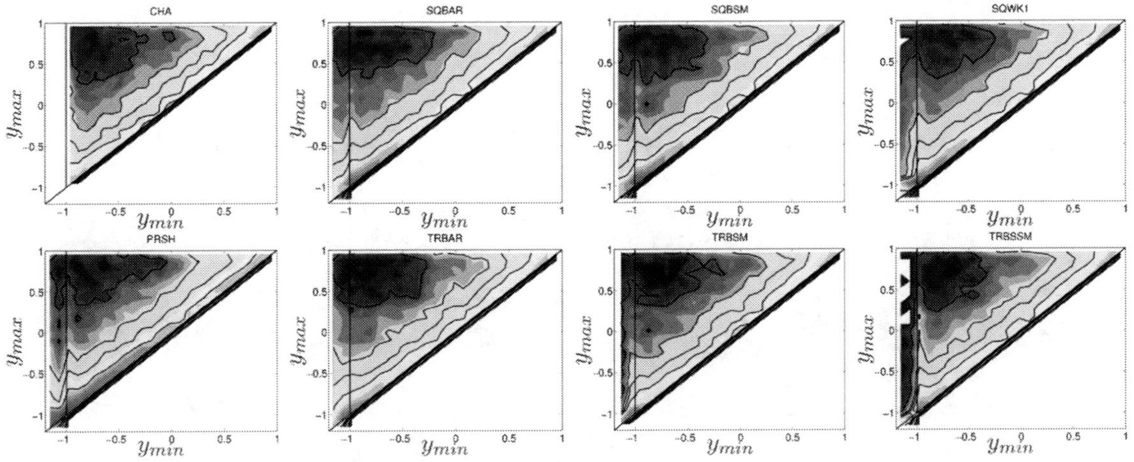

Figure 10. Joint p.d.f. of (y_{min}, y_{max}), for all the geometries. Colours represent the logarithm of the p.d.f. (i.e. the number of clusters). The black contour lines contain 30%, 50%, 70% and 95% of the data, the straight black lines are $y_{max} = y_{min}$ and $y_{min} = -1$

channel half-height ($y_{max} > -0.7$). All the attached clusters, both tall and short, are located near the y axis ($y_{min} \approx -1$): the smaller ones are in the lower-left corner and the tall ones are plotted above the former. For the top wall, this family of attached clusters is shown in the figures in the top-right corner ($y_{max} \approx 1$), and this group is not changed by the boundary condition on the lower wall: both the shape of the iso-probability lines and the number of clusters in this region are the same for all the geometries. The three-dimensional geometry (PRSH) shows a peak just above the plane of the crests. The greater turbulence intensity in this region causes a number of clusters so large that the colour in the plot is skewed towards the lower-left corner.

In the aligned-square-bars flows (SQBAR and SQBSM), the formation of clusters attached to the bottom wall is reduced with respect to the smooth channel. In fact, in the figures, the iso-probability lines assume a lower maximum value at $y_{min} \approx -1.2$ than in CHA. On the other hand, in these cases, the peak of the joint p.d.f. is shifted to $y_{min} \approx -1$, that is the plane of the crests. This means that the larger attached clusters are not able to reach into the grooves. In both cases, this decrease of clusters attached to the bottom wall occurs because, between the bars, the mean velocity is smaller. The figure relative to the transverse bars (SQWK1) shows a large number of clusters that are located between the bars ($y_{min} < -1$ and $y_{max} \approx -1$, in red), that are essentially isolated from the overlying flow. Besides these clusters, there are also attached clusters, even if they are lower in number when compared to the square-bars cases (30% and 50% lines).

Similar results are obtained for triangular bars. For the largest separation (TRBAR), the family of attached clusters seems also reduced with respect the smooth wall, and it is formed by structures with $y_{min} \approx -1.2$, attached to the groove between the bars. As the separation between bars decreases, the attached clusters move from the groove to the crests, and a family of clusters embedded in the grooves start to emerge. The triangular bars with an intermediate separation (TRBSM) show attached clusters both reaching into the groove and anchored in the crest of the triangles. For the triangular bars with the narrowest separation (TRBSSM), the attached clusters again seem as numerous as in the smooth case, but very few of them reach into the grooves.

The results found with the analysis of the joint p.d.f.s are qualitatively confirmed with the visualizations of $Q/(\overline{Q'^2})^{1/2}$ in figure 11. The threshold $\alpha = 0.75$ was chosen in the percolation analysis to maximize the number of structures, but using this value for the threshold would not

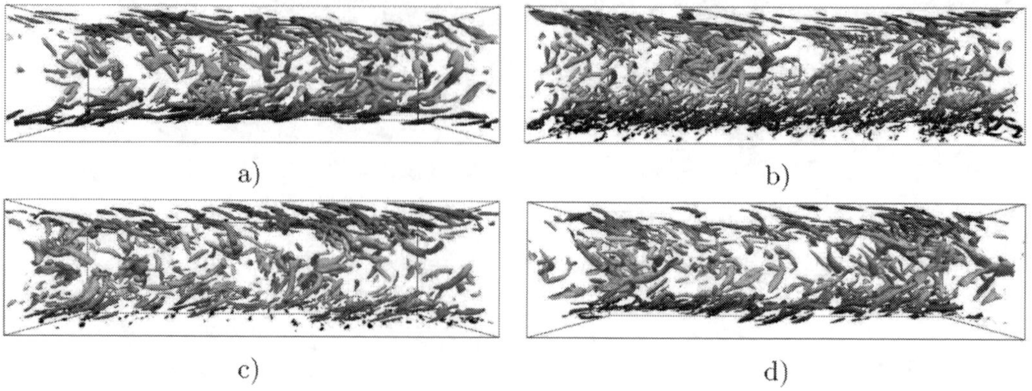

Figure 11. Three-dimensional contour visualizations of $Q/(\overline{Q'^2})^{1/2}$. Flow is left to right, and the wall-normal direction is vertical. The values for the isosurfaces are $Q/(\overline{Q'^2})^{1/2} = 2.5$ for a) CHA, b) PRSH, c) SQWK1 and d) TRBSSM.

have allowed to have a clear figure of the single structures. For this reason the value for the isosurface is $Q/(\overline{Q'^2})^{1/2} = 2.5$ for smooth (CHA), 3D (PRSH), transverse square bars (SQWK1) and triangular aligned (TRBSSM). In the smooth channel, the structures grow from the wall towards the centre of the channel, as expected. The structures are the same on both walls, and are very similar to those on the upper wall of the other cases. In the 3D geometry (PRSH), the clusters near the wall are much smaller with respect to CHA, showing that this region is characterized by a greater turbulence activity. The transverse bars (SQWK1) show structures similar to the smooth-wall case, that are generated on the plane of the crests. The image does not show the rollers inside the cavities because the value of the threshold is too large to allow their presence. The aligned-bars cases (only the triangular TRBSSM is reported) show that the clusters tend to reach the plane of the crests, but not the bottom wall.

The joint p.d.f.s of the height $y_c = (y_{max} - y_{min})/2$ versus the streamwise and spanwise lengths of the clusters, conditional on those that have $y^+_{min} < 20$, are shown in figure 12 and figure 13. It is important to stress the different origin of the outer and wall units. The outer unit y is traslated to have the origin on the bottom wall (located at $y = -1.2$), in order to use log-scale plots. The wall unit $y^+ = (y - y_w)Re_{\tau,R}$ has the origin on the plane of the crests to allow the comparison with the channel flow. In these plots, the black curve represents the iso-probability level that contains 98% of the data, and the straight lines represent the self-similar scaling found in [14] and [15]: $y_c = L_x/6 = L_z/3$.

In the smooth channel, the red region in figure 12 indicates that there are x-elongated structures and, while the spanwise length can grow up to the order of δ, for most of the clusters

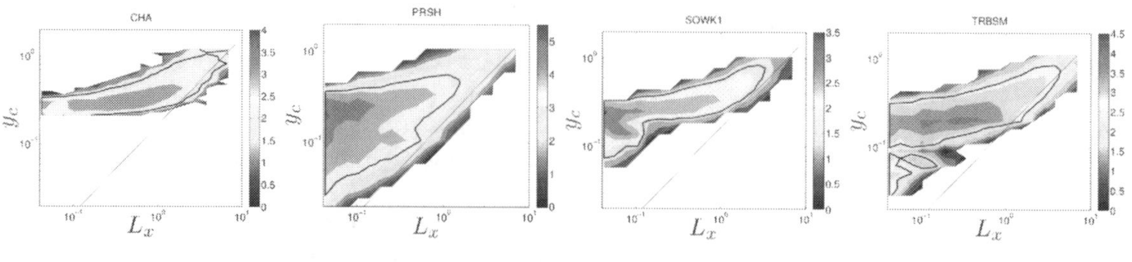

Figure 12. Joint p.d.f. of (y_c, L_x), conditioned to $y^+_{min} < 20$. The black curve contains 98% of the data, the straight line is $y_c = L_x/6$

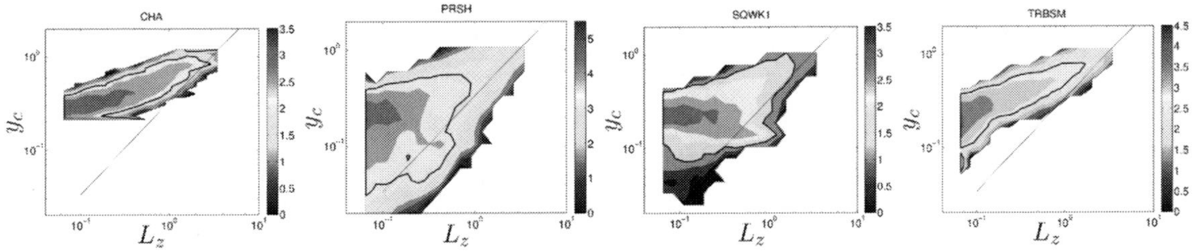

Figure 13. Joint p.d.f. of (y_c, L_z), conditioned to $y^+_{min} < 20$. The black curve contains 98% of the data, the straight line is $y_c = L_z/3$

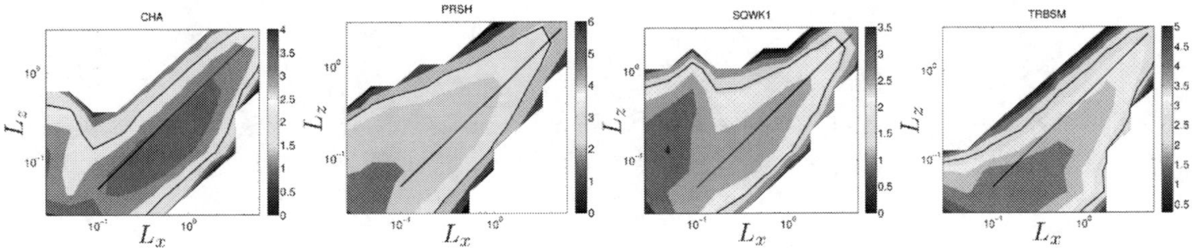

Figure 14. Joint p.d.f. of (L_x, L_z), conditional for $y^+_{min} < 20$. The black curve contains 98% of the p.d.f., the straight line is $L_z = L_x/2$

$L_z < 0.5 \approx 100\delta_\nu$, where δ_ν is the characteristic viscous length scale. Their height is lower than sixty plus units, suggesting that these are streak-like structures. It is not possible to find clusters with $y_c < 0.2$ because there is solid wall in that region. The prisms (PRSH) show a peak of small clusters lying just about the plane of the crests. The extension of the red contour near the origin is due to the formation of several structures within the roughness layer, which is an indication of an intense turbulent activity in this region. The aligned bars (only TRBSM is shown, for similarity of the results for the other cases) show longitudinal structures whose spanwise extent is very limited. It is possible to estimate from figure 12 that the footprint of the roughness is visible for approximately $h/10$ over the plane of the crests ($y^+ \approx 20$). In the roughness region, the structures are elongated in the direction of the bars. The opposite holds for the transverse bars, with spanwise structures in the roughness region (figure 13). In this geometry the structures above the plane of the crests have similar sizes in both directions. Direct visualizations have shown that the clusters come from the merging of the vortices inside the bars and those generated in the overlying region. In a few cases these clusters jump from a cavity to the next.

In figure 14, the joint p.d.f. of the length and width of the clusters is given. From the two self-similarity scalings found in the previous images, it follows that $L_z = L_x/2$, and this is indeed what we find in these plots. In the smooth channel, the intense red regions below the black line demonstrate that a large number of clusters have length about one order of magnitude greater than their width. The dimensions are in the range characteristic of streaks. A different family is identifiable in the figure, consisting of very short and wide structures that stay very close to the wall (see figures 12 and 13). In the 3D geometry (PRSH), L_x and L_z are of the same order of magnitude, and the clusters tend to be shorter than in the smooth wall case (CHA). In the transverse bars (SQWK1) we can identify relatively short and wide clusters, and figures 12 and 13 show that they correspond to clusters basically embedded in the grooves, or lying right on the plane of the crest and connecting one groove with the next. For the case of aligned bars (TRBSM), we find very long and narrow clusters, again mostly embedded within the grooves.

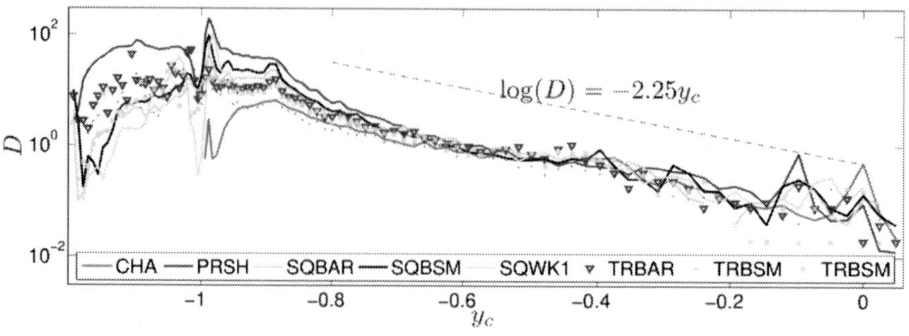

Figure 15. Density of clusters as a function of height, conditioned to $y^+_{min} < 20$

The density of the attached clusters as a function of height (figure 15) shows that the attached clusters are affected by the surface in the roughness sublayer, close to the roughness elements. However, the different cases collapse above $y_c^+ \approx 50$, suggesting that the attached clusters above the roughness sublayer are not influenced by the roughness elements, similar to the conclusions reported by Flores et al. (2006) [15] for channels with disturbances on the wall boundary conditions. The present result shows further evidence of the Townsend similarity hypothesis for turbulent flows with rough walls. The noise above $y_c \approx -0.2$ is due to the small number of attached clusters with heights of the order of the channel half-height.

5. Conclusions

The vortex clusters in rough-channel and grid-generated turbulence have been analyzed, defining them as regions of the flow satisfying a criterion based on the second invariant of the velocity gradient tensor. The inhomogeneity of the flows suggests to scale Q with its standard deviation, and the same percolation analysis proposed by [14] has been used to get the appropriate value of the threshold. This procedure has been applied to roughly 30 flow fields per geometry of DNS of grid-generated turbulence, and over 50 flow fields of DNS of turbulent channels with rough walls, to extract over 10^7 objects that have been used to produce p.d.f. of their sizes and spatial distribution.

The main conclusion of this study is that, for both flows, the effect of the boundary condition on the clusters seems to be limited to a region relatively close to the solid surfaces. For instance, in the grid turbulence case, the geometry of the grid is responsible for the flow right downstream of the grid to be dominated by shear (single- and multi-scale grids) or by vorticity (fractal grids). Of course, this affects the rate of decay of turbulent kinetic energy and the distance needed to develop an isotropic turbulent flow. But sufficiently far downstream of the grid, the distribution of clusters is very similar for all the grids considered here. These clusters seem to be isotropic, circumscribed by roughly cubical boxes. Their volume is roughly proportional to their lengths to the power of 2.5, suggesting relatively full objects.

The analysis of the multi-scale grids also shows that these grids produce a strong inhomogeneity between the regions with large and small holes. Relatively large clusters are observed downstream of the larger holes, while the clusters formed downstream of the smaller holes decay faster. This inhomogeneity in the flow results in a secondary source of shear, that is likely responsible for the higher turbulent kinetic energy levels of the multi-scale grids far downstream of the grid. On the other hand, the fractal grid produces a flow with little inhomogeneity that rapidly develops a wide range of clusters.

In terms of the rough channels, two families of clusters are found: wall-attached and wall-detached. The latter are little influenced by the details of the wall. The former have lengths and widths that are approximately proportional to their height, and they are somewhat affected by the details of the wall. For relatively open roughness, some attached clusters are found embedded within the grooves. When the roughness is aligned with the flow, they tend to form relatively long objects. When the roughness is transverse to the flow, they tend to form rollers. For very tightly spaced roughness, the distribution of attached clusters resembles that over smooth walls, but they are based on the crest of the roughness elements, not able to reach into the grooves of the roughness.

The differences between the distributions of clusters found over different roughness considered here mitigate when the tallest objects are considered. Indeed, the p.d.f of the density of attached clusters with a given height collapses for all rough surfaces once the clusters are larger than the roughness sublayer, in agreement with previous investigations [15]. This suggests, in agreement with the results of grid turbulence, that the influence of the boundary condition into the turbulent structures of the flow is limited to a region close to the boundary.

Acknowledgments
Funded by the Multiflow program of the European Research Council.

References
[1] Townsend A A 1976 *The Structure of Turbulent Shear Flows* (Cambridge University Press) 2nd Edition
[2] Jiménez J 2004 Turbulent flows over rough walls *Ann. Rev. Fluid Mech.* **36** 173–96
[3] Orlandi P 2011 DNS of transitional rough channels *J. Turbulence* **12** N29
[4] Orlandi P and Leonardi S 2006 DNS of turbulent channel flows with two- and three-dimensional roughness *J. Turbulence* **7** N53
[5] Leonardi S, Orlandi P and Antonia R A 2007 Properties of d- and k-type roughness in a turbulent channel flow *Phys. Fluids* **19** 125101
[6] García-Mayoral R and Jiménez J 2011 Hydrodynamic stability and breakdown of the viscous regime over riblets *J. Fluid Mech.* **678** 317–47
[7] von Karman T 1937 The fundamentals of the statistical theory of turbulence *J. Aer. Sci.* **4** 131–38
[8] Taylor G I 1935 Statistical theory of turbulence *Proc. R. Soc. Lond. A* **151** 421-54
[9] Lavoie P, Djenidi L and Antonia R A 2007 Effects of initial conditions in decaying turbulence generated by passive grids *J. Fluid Mech.* **585** 395–420
[10] Seoud R E and Vassilicos J C 2007 Dissipation and decay of fractal-generated turbulence *Phys. Fluids* **19** 105108
[11] Ertunc O, Ozyilmaz N, Lienhart H, Durst F and Beronov K 2010 Homogeneity of turbulence generated by static-grid structures *J. Fluid Mech.* **654** 473–500
[12] Hurst D and Vassilicos J C 2007 Scalings and decay of fractal-generated turbulence *Phys. Fluids* **19** 035103
[13] Mazellier N, Danaila L and Renou B 2010 Multi-scale turbulence injector: a new tool to generate intense homogeneous and isotropic turbulence for premixed combustion, ARXiv 1004.0074v1 [physics.flu-dyn]
[14] del Álamo J C, Jiménez J, Zandonade P and Moser R D 2006 Self-similar vortex clusters in the turbulent logarithmic region *J. Fluid Mech.* **561** 329–58
[15] Flores O, Jiménez J and del Álamo J C 2007 Vorticity organization in the outer layer of turbulent channels with disturbed walls *J. Fluid Mech.* **591** 145–54
[16] Orlandi P 2000 *Fluid Flow Phenomena: A Numerical Toolkit* (Kluwer Academic Publisher)
[17] Leonardi S, Orlandi P, Djenidi L and Antonia R A 2004 Structure of turbulent channel flow with square bars on one wall *Int. J. Heat Fluid Flow* **25** 384–92
[18] Chong M S, Perry A E and Cantwell B J 1990 A general classification of three dimensional flow fields *Phys. Fluids* A **2** 765–77
[19] Nagaosa R and Handler R A 2003 Statistical analysis of coherent vortices near a free surface in a fully developed turbulence *Phys. Fluids* **15** 375–94
[20] Moisy F and Jiménez J 2004 Geometry and clustering of intense structures in isotropic turbulence *J. Fluid Mech.* **513** 111–33
[21] Stauffer D 1985 *Introduction to percolation theory* (Taylor and Francis)
[22] Veeravalli S and Warhaft Z 1989 The shearless turbulence mixing layer *J. Fluid Mech.* **207** 191–229

AUTHOR INDEX

Atkinson, C 29
Borrell, G 29
Brethouwer, G 130
Cardesa, J I 115
Casciola, C M 61
Chong, M S 115
Cimarelli, A 61
Constantinou, N C 43
Cordier, L 71
D'Addio, P 178
De Angelis, E 61
Dupont, P 164
Eitel-Amor, G 100
El Khoury, G K 130
Elsinga, G E 154
Encinar, M P 16
Farrell, B F 43
Flores, O 100, 178
Frederiksen, J S 1
García-Mayoral, R 16
Goudar, M V 154
Gungor, A G 86
Hackl, J 29
Helm, C 164
Ioannou, P J 43
Jiménez, J 16, 43, 61, 71
Johansson, A V 130
Kitsios, V 1
Lozano-Durán, A 43, 71
Maciel, Y 86
Martin, M P 164
Monty, J P 115
Nikolaidis, M-A 43
Noack, B R 71
Orlandi, P 178
Patwardhan, S S 144
Ramesh, O N 144
Sassun, D 178
Schlatter, P 100, 130
Sillero, J A 1
Simens, M P 86
Soria, J 1, 29, 86, 115
Stegeman, P 29
Talamelli, A 61
Tissot, G 71

Institute of Physics
Dirac House, Temple Back
Bristol BS1 6BE UK

CURRAN ASSOCIATES INC
proceedings
.com